Metals and Micronutrients:
Uptake and Utilization by Plants

Annual Proceedings of the Phytochemical Society of Europe

ANNUAL PROCEEDINGS OF THE PHYTOCHEMICAL SOCIETY
OF EUROPE NUMBER 21

Metals and Micronutrients: Uptake and Utilization by Plants

Edited by

D. A. ROBB

*Department of Biochemistry, University of Strathclyde
Glasgow, UK*

and

W. S. PIERPOINT

*Department of Biochemistry,
Rothamsted Experimental Station, Harpenden,
Herts., UK*

1983

ACADEMIC PRESS

A Subsidiary of Harcourt Brace Jovanovich, Publishers

London   New York
Paris   San Diego   San Francisco   São Paulo
Sydney   Tokyo   Toronto

V

ACADEMIC PRESS INC. (LONDON) LTD.
24–28 Oval Road
London NW1

United States Edition published by
ACADEMIC PRESS INC.
111 Fifth Avenue
New York, New York 10003

British Library Cataloguing in Publication Data

Metals and micronutrients.—(Annual proceedings of
the Phytochemical Society of Europe, ISSN
0309–9393; no. 21)
1. Plants—Nutrition—Congresses
I. Robb, D. A.   II. Pierpoint, W. S.   III. Series
581.1′335   QK899

ISBN 0–12–589580–1

Text set in 10/12 pt Linotron 202 Times, printed and bound
in Great Britain at The Pitman Press, Bath

Contributors

I. AHMAD, *Department of Botany, University of Toronto, Toronto, Ontario, M55 1A1, Canada.*

J. W. ANDERSON, *Botany Department, La Trobe University, Bundoora, Victoria 3083, Australia.*

D. A. BAKER, *Department of Biological Sciences, Wye College, University of London, Ashford, Kent, TN25 5AH, UK.*

H. F. BIENFAIT, *Laboratory for Plant Physiology, Kruislaan 318, 1098 SM, Amsterdam, The Netherlands.*

M. J. CORMIER, *Department of Biochemistry, The University of Georgia, Athens, Georgia 30602, USA.*

I. R. COTTINGHAM, *Biochemistry Department, School of Biological Sciences, University of Sussex, Falmer, Brighton BN1 9QG, UK.*

T. DA SILVA, *Department of Biochemistry, The University of Georgia, Athens, Georgia 30602, USA.*

A. GILDON, *Rothamsted Experimental Station, Harpenden, Herts AL5 2JQ, UK.*

E. J. HEWITT, FRS, *University of Bristol, Long Ashton Research Station, Bristol BS18 9AF, UK.*

M. F. HIPKINS, *Department of Botany, Glasgow University, Glasgow G12 8QQ, UK.*

H. W, JARRETT, *Department of Biochemistry, The University of Georgia, Athens, Georgia 30602, USA.*

O. T. G. JONES, *Department of Biochemistry, University of Bristol, Medical School, Bristol, BS8 1TD, UK.*

R. T. LEONARD, *Department of Botany and Plant Sciences, University of California, Riverside, CA 92521, USA.*

B. C. LOUGHMAN, *Department of Agricultural Science, University of Oxford, OX1 3PF, UK.*

A. L. MOORE, *Biochemistry Department, School of Biological Sciences, University of Sussex, Falmer, Brighton BN1 9QG, UK.*

B. A. NOTTON, *University of Bristol, Long Ashton Research Station, Bristol, BS18 9AF, UK.*

A. J. PARR, *University of Oxford, Botany School, Downing Street, Cambridge, UK.*

P. J. PETERSON, *Department of Biological Sciences, Chelsea College, University of London, London SW10, UK.*

N. W. PIRIE, FRS, *42 Leyton Road, Harpenden, Herts., UK.*

A. R. SCARF, *Botany Department, La Trobe University, Bundoora, Victoria 3083, Australia.*

G. R. STEWART, *Department of Botany, Birkbeck College, University of London, London, WC1E 7HK, UK.*

P. B. TINKER, *Rothamsted Experimental Station, Harpenden, Herts AL5 2JQ, UK.*

F. VAN DER MARK, *Laboratory for plant Physiology, Kruislaan 318, 1098 SM, Amsterdam, The Netherlands.*

Preface

Metallurgy, the extraction and utilization of metals, is an important aspect of human civilizations; past epochs have been characterized by the use of bronze and iron, and future ones may be characterized by their use or misuse of uranium and plutonium. Metals useful to man may be essential or poisonous to life and the renewed interest in the interactions between metals and man has become an important facet of inorganic biochemistry. Perhaps the current vigour of inorganic biochemistry is provided by chemists—certainly the majority of treatises in the field seem to be written by them.

Often such books are neglectful of the mineral nutrition of plants and in view of the rapid progress in this area, it appeared appropriate to devote a symposium to the topic. Thus the 1981 Easter meeting of the Phytochemical Society of Europe brought together chemists, biochemists, physiologists, and agronomists to discuss aspects of phytometallurgy—how plants extract, accumulate, and use metals. Because boron and selenium are required in trace amounts, like many metals, and because, like molybdenum, they are available to the plant as anions, it seems logical to include these elements in the discussion also.

Thus, this book is based on the contributions of the invited speakers at the meeting, with the addition of an epilogue by N. W. Pirie. The authors prepared their manuscripts at the conclusion of the meeting and we have edited them so that they may appeal to senior undergraduates and postgraduates in general. Each article is written by an expert and summarizes well-established fields of research and also highlights more recent endeavours. The order of chapters is meant to emphasize stages in the sequence, uptake—incorporation—function, but the breadth of some articles makes their location rather arbitrary.

The meeting was judged a success both scientifically and socially. The latter was due largely to the careful arrangements made by Dr. Threlfall, Professor Friend, and their colleagues, and also, it must be remembered, to the community singing led by the President of our Society, Professor C. van Sumere. We are grateful to these people and also the staff of Academic Press for the assistance given in preparing this book.

D. A. R
W. S. P

Contents

Part 1

CHAPTER 1

Uptake of Cations and their Transport within the Plant

D. A. Baker

CHAPTER 2

Mycorrhizal Fungi and Ion Uptake

P. B. Tinker and A. Gildon

CHAPTER 3

Adaptation to Salinity in Angiosperm Halophytes

G. R. Stewart and I. Ahmad

CHAPTER 4

Adaptation to Toxic Metals

P. J. Peterson

CHAPTER 5

Potassium Transport and the Plasma Membrane-ATPase in Plants

R. T. Leonard

CHAPTER 6

Boron and Membrane Function in Plants

A. J. Parr and B. C. Loughman

Part 2

CHAPTER 7

Phytoferritin and its Role in Iron Metabolism

H. F. Bienfait and F. Van der Mark

CHAPTER 8

Ferrochelatase

O. T. G. Jones

Part 3

CHAPTER 9

Metals and Photosynthesis

M. F. Hipkins

CHAPTER 10

Characteristics of the Higher Plant Respiratory Chain

A. L. Moore and I. R. Cottingham

CHAPTER 11

Calmodulin Activation of NAD Kinase and its Role in the Metabolic Regulation of Plants

H. W. Jarrett, T. Da Silva, and M. J. Cormier

CHAPTER 12

Micronutrients and Nitrate Reductase

B. A. Notton

CHAPTER 13

Selenium and Plant Metabolism

J. W. Anderson and A. R. Scarf

CHAPTER 14

A Perspective of Mineral Nutrition: Essential and Functional Metals in Plants

E. J. Hewitt

Epilogue

PART 1

To Draw Nutrition

Processes in the absorption of metals and
micronutrients by plants, and the influences
of the environment

Fixed like a plant to his peculiar spot,
To draw nutrition, propagate and rot;

(Pope, Essay on Man)

CHAPTER 1

Uptake of Cations and their Transport within the Plant

D. A. BAKER

*Department of Biological Sciences, Wye College, University of London,
Ashford, Kent, England.*

I. Introduction

Green plants are the miners of the earth's crust, being the major selective accumulators of inorganic nutrients upon which other life forms are directly or indirectly dependent. The initial accumulation of ions by terrestrial plants is a function of the root system which exhibits an extensive ramification through the soil. For example, a maize seedling can produce more than 2000 roots with a combined length exceeding 50 m in six to eight weeks. The area of contact between the root and soil particles is often increased by the production of root hairs, which can exceed 1000 million in number (e.g. 10 000 to 15 000 million on a single rye plant) although some roots, such as those of onion, are completely devoid of root hairs.

The root is separated from the regions of photosynthesis, necessitating a transport of inorganic nutrients to those aerial parts. Although plants are autotrophic the roots are, in fact, normally heterotrophic, requiring a supply of assimilates from the photosynthetic shoot. This transport of organic assimilates also involves some inorganic ions which are thus circulated within the plant. The uptake and translocation of inorganic solutes within the whole plant is therefore an essential and major aspect of the process of plant nutrition.

In this paper the translocation of inorganic cations will be considered for the whole plant, although inevitably the nature of the balancing anion has an important bearing on this process and will be included where relevant. The processes involved are the initial acquisition of cations by the plant root, their transport from root to shoot through the xylem, the subsequent accumulation by the tissues of the shoot, and the redistribution of these cations within the plant through the phloem pathway. This topic, which has long been a primary subject for research in plant physiology, has been extensively reviewed in recent years (Weatherley, 1969; Pitman, 1975; Sutcliffe, 1976; Pitman and Cram, 1977).

II. The Entry and Translocation of Cations

A. AMOUNT OF IONS IN PLANTS

Plants will absorb to some extent any element presented to them in nutrient media but the ionic content of plants will vary, reflecting selectivity and species variation, the availability of nutrients in different rooting media, or the stage of development of the plant. The approximate range of various ions or elements within plant shoots is presented in Table I. However, the distribution of cations within the plant may be heterogenous; certain elements such as lead, nickel, copper, zinc, iron, manganese, chromium, and vanadium being preferentially retained by the root system, whereas calcium and potassium are transported in major part to the shoot system. The pattern of distribution of these elements may be considerably affected by their concentration in the external medium. Examples of how plant growth and nutrient content are influenced by external concentration has been provided by a series of studies by Asher and co-workers, two of which are presented here to illustrate general trends (Tables II and III). Increasing the external potassium concentration above 0·024 mM has little effect upon growth and relative potassium content of the shoots, although the level of potassium in the roots continues to rise. It is probable that the increased level of potassium in the root is stored within the vacuoles and is not being utilized for growth and metabolism. The effect of the external concentration of calcium is more complex (Table III). The levels of calcium are substantially greater in legumes than in grasses and a different pattern is observed between these

TABLE I

Typical concentrations of mineral elements in foliage of normal plants (from Hewitt and Smith, 1975)

Element	p.p.m. in dry matter	mM in cell sap*	Nutrient solution† (mM)
N	15 000–35 000	150–350	15
P	1500–3000	7–14	1
S	1000–3000	7·5–140	1·5
Ca	10 000–50 000	35–175	5
Mg	2500–10 000	15–60	1·5
K	15 000–50 000	55–180	5
Na	200–2000	1–12	1
Fe	50–300	0·15–0·75	0·1
Mn	25–250	0·06–0·6	0·01
Cu	5–15	0·01–0·03	0·001
Zn	15–75	0·03–0·15	0·002
Co	0·2–29	0·005–0·05	0·0002
B	15–100	0·2–1·3	0·05
Mo	0·5–5	0·004–0·075	0·0005
Cl	100–1000	0·4–4	0·1

* This value merely indicates in round numbers the total aqueous concentration possible without regard to insoluble, lipid, or other structural fractions.
† Typical nutrient solution concentrations.

TABLE II

Relative content of potassium and relative yield of 14 species (data from Asher and Ozanne, 1967)

K^+ in solution	Relative yield	Relative contents $(\mu mol(gFW)^{-1})$		Root:shoot
(μM)	(%)	Tops	Roots	K^+ ratio
1	7	22	12	0·55
8	64	110	42	0·38
24	86	144	66	0·46
95	91	162	83	0·51
1000	97	190	112	0·59

two groups. At low external concentrations of calcium the relative calcium content of roots and of shoots remained constant, while at higher concentrations those internal values rose proportionately. Magnesium is reported to behave in a similar manner to calcium (Pitman, 1975). The uptake of most other nutrient elements reflects the patterns observed with potassium and calcium.

TABLE III

Relative content of calcium in 16 legume species and 11 grass and cereal species. (data calculated by Pitman, 1975, from Loneragan and Snowball, 1969, and Asher and Ozanne, 1967)

Ca^{2+} in solution (μM)	Relative contents of Ca^{2+} (μmol (gFW)$^{-1}$)			
	Legumes		Graminae	
	Tops	Roots	Tops	Roots
0·3	4·5	1·1	1·7	0·9
0·8	4·4	1·4	1·5	0·8
2·5	4·7	1·6	1·7	0·7
10	12	1·7	3·3	0·9
100	39	2·2	10·5	1·5
1000	54	3·3	18	2·3

B. UPTAKE OF CATIONS BY THE ROOT

The uptake of cations by the plant root has been widely studied but little understood. This is particularly because the area has become blurred by the different concepts of carriers evolved from either the analysis of carrier kinetics of from the biophysical approach, i.e. pumps.

Almost all workers agree that ions cross the plasma membrane of a root cell at some point in their passage across the root. This movement across a membrane must be facilitated by some form of carrier mechanism to permit uptake to occur at measured rates. Evidence is accumulating that cation uptake is energized by an ATPase system which extrudes H$^+$ creating a gradient ($\Delta\bar{\mu}_{H^+}$) which may be utilized to drive uptake (see Leonard, this volume).

It has been observed by a large number of investigators that the uptake of certain monovalent cations by plant root tissue appears to obey a relationship analogous to the Michaelis-Menten equation used in analysing the kinetics of enzyme reactions

$$v = \frac{[S]\ V_{max}}{K_s + [S]} \tag{1}$$

where V_{max} is maximum rate of uptake when all carrier is saturated, and K_s a constant, characteristic of a particular ion crossing a specific membrane expressed in units of concentration. The reciprocal of equation (1) gives a linear relationship:

$$\frac{1}{v} = \frac{K_s}{V_{max}[S]} + \frac{1}{V_{max}} \tag{2}$$

Plots of $1/v$ versus $1/[S]$ often yield a curve comprising two or more straight lines from which different K_s and V_{max} values may be obtained. This has been interpreted as evidence for separate carriers within the membrane, each with a different affinity for a particular ion. Often two such carriers are postulated for a particular ionic species; one functional at low external concentrations (< 0.5 mM) with a low K_s and thus high affinity, termed system I, the other functional at higher concentrations (> 0.5 mM) with a higher K_s and lower affinity, termed system II. The system II uptake isotherm is frequently irregular and may be further subdivided (Nissen, 1974).

The location of these two uptake systems has been hotly debated, some investigators suggesting that system I is on the plasma membrane, system II on the tonoplast, and they are thus arranged in series (Laties, 1969). Others subscribe to the view that both mechanisms are located on the plasma membrane, arranged in parallel (Epstein, 1976).

Alternative models have been proposed to explain the dual isotherms without invoking the concept of two or more carrier sites. Only one uptake system may be involved, the dual mechanisms reflecting either allosteric structural changes in the membrane (Nissen, 1974), the membrane behaving as a semi-conductor (Thellier, 1973) or a direct contribution of the membrane potential (Gerson and Poole, 1971). Invagination of the plasma membrane on which ions are selectively bound has been considered and the dual isotherms interpreted in terms of the delivery of ions into different cytoplasmic compartments (Baker and Hall, 1973).

Some investigators have cast doubt on the usefulness of the application of Michaelis-Menten kinetics to the uptake of cations by plant tissues (see Wyn Jones, 1975). Ion absorption isotherms, when they are the result of active processes, do not always follow the Michaelis-Menten formalism while passive ion movements, particularly if carrier-mediated, may sometimes manifest saturation kinetics. It is, therefore, necessary to approach results obtained from this "carrier" type of analysis with a degree of caution.

C. RADIAL TRANSPORT ACROSS THE ROOT

In the uptake of cations by the young roots of plants the cortical cell walls form a hydrated free space continuum between the external medium and the cortical cell membranes (the apoplast), the inner limit of this free space being the endodermis. The plasma membrane of the cortical cells may thus be bathed in the external solution and a large surface is available for selective ion uptake into the cortical cytoplasm.

This concept of the even distribution of the uptake process over all the cortical plasma membrane has been questioned by some authors (see van Iren and Boers-van der Sluijs, 1980). Particularly when external concentrations are low the uptake of ions takes place only in the outer cell layers and the pathway for radial ion transport towards the stele is through the cytoplasm of connected cortical cells (the symplast) and not through the apoplast.

The ionic content within the symplast is regulated by the selective properties of the uptake mechanism and the selective accumulation of ions across the tonoplast into the vacuole. Whereas high levels of potassium are a feature of the symplast, sodium is usually preferentially accumulated in the cortical vacuoles and is thus partially removed from the symplast. Some ions within the symplast are utilized in a metabolic capacity, while the majority are moved radially across the root and into the stele. This radial ion movement is believed to be diffusional, the driving force being the negative gradient of concentration across the root, with cytoplasmic streaming aiding the intracellular transport (Tyree, 1970).

The ions which are transported radially across the root cross the endodermis through the symplast and are ultimately released within the stele for long-distance transport through the xylem. The mechanism of ion release within the stele has been the subject of some controversy, some investigators advocating a passive, thermodynamically downhill process, while others envisage an active ion efflux across the plasma membranes of the stelar parenchyma. Stelar cells are capable of accumulating ions when supplied through the symplast but are incapable of accumulation when isolated from the cortical tissue (Baker, 1973), and there are data which show that both anions and cations move into the xylem down a gradient of electrochemical potential (Dunlop and Bowling, 1971). Both of these observations are consistent with a passive leakage. However, the high accumulative capacity of xylem parenchyma and the inhibition of root pressure exudation by amino acid analogues, which do not inhibit ion uptake, are interpreted as support for the active efflux mechanism (Pitman, 1977).

In the more mature regions of the root there is a discrimination against calcium uptake which is correlated with the stage of development of the endodermis. In the mature regions a suberin layer is deposited over the whole endodermal cell wall so that the only accessible route into the endodermis for water and ions is through the plasmodesmata. This development of the endodermis coincides with the reduced ability to translocate calcium. Accessibility of the endodermal plasma membrane from the cortical free space thus appears to be a requisite for calcium transfer to the stele and it would appear that calcium cannot move readily in the symplast. The corollary of this is that potassium, which does not have its radial movement impaired by these developmental changes, is delivered to the endodermis through the cortical symplast (Clarkson and Robards, 1974).

An intact endodermis appears to be necessary for root pressure exudation, as neither isolated stele nor isolated cortex have been observed to exude (Baker, 1973). When the stele and cortex are separated the break occurs at the Casparian band, depriving both components of an apoplastic barrier. However, active movement of chloride across the symplast of isolated cortical sleeves has been observed, although potassium and sodium move passively across the cortex (Ginsburg and Ginsburg, 1970).

D. EFFECT OF TRANSPIRATION ON CATION UPTAKE

There is evidence that both the rate and the selectivity of cation absorption by intact plants may be influenced by transpiration. Although a large mass of often conflicting data exists regarding these effects of transpiration, in general, plants with a high salt status have a transpirationally-dependent component of cation uptake whereas plants with a low salt status do not (see Bowling, 1976). This influence of transpiration on cation uptake by intact plants has little or no effect on the accumulation of cations by the root, most of the cations taken up in response to increased transpiration being destined for the shoot (Bowling and Weatherley, 1965). Detailed discussions of this aspect of translocation can be found in the publications of Weatherley (1969) and Bowling (1976).

Evidence of an effect of transpiration on the selectivity of the plant to some cations has been reported for the uptake of potassium and sodium by barley plants (Pitman, 1965). Both the roots and the shoots showed a preference for potassium over sodium when the ions were presented together (Table IV). When the rate of transpiration was increased, however, the ratio of potassium to sodium was reduced in the shoot but not in the root. Thus, the increased water flux favoured sodium transport to

TABLE IV

Comparison of potassium:sodium in shoots of whole barley seedlings and exudate from detopped seedlings (after Pitman, 1965)

Concentration in solution (mM)		Ratio of K^+/Na^+			
K^+	Na^+	External solution	Roots	Shoots	Exudate
0·5	9·5	0·052	0·8	3·4	2·1
1·0	9·0	0·11	1·1	4·0	3·1
2·5	7·5	0·33	1·9	8·1	7·1
6·0	4·0	1·5	6·7	18	22
8·0	2·0	4·0	8·9	30	36

the shoot, possibly as a result of the change in water flux past the cation adsorption sites in the cell walls of the root cortex and xylem.

E. TRANSPORT IN THE XYLEM

The delivery of cations into the stele by radial transport across the root results in an accumulation within the xylem vessels. In a transpiring plant these ions are transported through the xylem in the transpiration stream in response to the hydrostatic tension developed in the xylem pathway. In the absence of this tension a positive pressure develops in the xylem, as a result of the continued accumulation of ions, causing an osmotic flow of water into the root. This positive pressure is manifested as an extrusion of fluid (guttation) in intact plants and as root pressure exudation in detopped root systems.

The composition and concentration of ions in xylem sap may be determined on samples extracted using a vacuum technique (Bollard, 1953) or by analysis of root pressure exudate. Although it has been argued that the sap obtained from exuding roots is not a representative sample of the transpiration stream, it does represent the composition of the xylem sap at zero transpiration. An analysis of root pressure exudate is presented in Table V where it can be seen that the major ions present in the external medium are concentrated in a selective manner in the xylem sap. This composition may be influenced by the balance of various cations and anions present in the external medium (Collins and Reilly, 1968). The concentration of a particular ion, such as potassium, in the xylem sap shows a consistent relationship to its concentration in the external medium indicating the operation of a complex interaction between the transport of water and ions across the root (Bowling and Weatherley, 1965; Baker and Weatherley, 1969).

Iron is taken up and transported more readily when supplied as a chelated complex, such as ferric ethylenediaminetetraacetate (FeEDTA) or as ferric diethylenetriaminepentaacetate (FeDTPA) (Wallace and North, 1953). There is some evidence that calcium may also be transported in a chelated form (Jacoby, 1966). Minor elements such as copper, manganese, and zinc are also present in the xylem sap as inorganic ions. These elements do not appear to form complexes in the xylem, copper, and zinc being less readily absorbed and transported when supplied in a chelated form (De Kock and Mitchell, 1957).

The transpiration stream supplies the leaves of the plant with ions selectively transported across the root and up the xylem. Some ions are selectively removed during this passage; sodium in particular may be accumulated by the root and stem tissues. In maize, for example, sodium

TABLE V

Concentrations of the major ions in the external solution (c_j^o) and root pressure exudate (c_j^i) from *Ricinus* (from Bowling *et al.*, 1966)

Ion	c_j^o (mM)	c_j^i (mM)
K^+	0·60	5·6
Na^+	0·33	0·74
Ca^{2+}	0·84	10·1
Mg^{2+}	0·45	3·9
NO_3^-	1·45	12·0
Cl^-	0·14	1·03
SO_4^{2-}	0·69	2·00
$H_2PO_4^-$	0·11	1·38
HPO_4^{2-}	0·13	0·56

levels in root pressure exudate are considerably lower when collected from longer lengths of root, whereas the sodium level of the root tissue increases with increasing length (Shone *et al.*, 1969). Electron probe microanalytical studies have indicated that sodium is accumulated in modified xylem parenchyma cells which may be responsible for the regulation of the ion content of the transpiration stream as it leaves the root (Yeo *et al.*, 1977). Sodium removal may also occur in stem tissues (Jacoby, 1965) contributing to the low sodium level in the leaves of many plant species. Lateral movement of some other ions out of the xylem also presumably takes place throughout the passage from the absorbing roots to the leaves, such ions often appearing in the phloem tissues (see next section). In trees considerable distances are involved and the xylem sap composition may be markedly altered before it reaches the leaves, in many cases transfer cells being involved in a selective extraction of ions (Gunning and Pate, 1969).

The walls of the xylem conduits, in common with other plant cell walls, contain numerous fixed negative charges which can retain cations. Divalent cations are preferentially bound and their movement through the xylem is thus retarded. The movement of calcium may be envisaged as a continuous displacement as in an ion-exchange column (Bell and Biddulph, 1963). When calcium is supplied as Ca EDTA, although the initial uptake is slower, the calcium absorbed in this form is more mobile as it is not adsorbed on the exchange sites in the xylem (Isermann, 1971).

As a result of the selective accumulation in the initial uptake followed by a further selective removal of ions in the xylem, the xylem sap, which ultimately reaches the mesophyll cells through the apoplast of the leaf, has a different composition to that experienced by the root cortical cells. The level of potassium may be considerably higher than that in the external medium, while sodium may be reduced to a very low level. This is reflected in the high ratio of potassium to sodium in the shoots of many plants.

On reaching the leaves the ions may once again undergo selective accumulation into the mesophyll and other leaf tissues by a process of absorption similar to that in other living plant tissues, except that the energy for uptake may be supplied by photosynthesis directly, as well as by respiration (Jeschke and Simonis, 1969; Nobel, 1969). As a result of this selective removal of ions the transpiration stream reaching the terminals of the leaf xylem will normally have a very different composition to that of the root pressure exudate. This difference is clearly demonstrated by some results presented by Clarkson (1974) (Table VI) where it can be seen that

TABLE VI

Changes in concentrations of ions during passage through the xylem of stems and leaves of marrow seedlings (from Clarkson, 1974)

Ion	Concentration (mM)			
	External solution	Xylem exudate	Petiole exudate	Guttation fluid
K^+	0·5	12·8	26·2	0·5
Na^+	0·2	0·3	0·6	0·7
Ca^{2+}	0·15	5·1	8·0	0·9
NO_3^-	1·0	7·1	12·9	0·4
$H_2PO_4^-$	0·01	0·2	0·5	0·2

the concentrations of calcium, potassium, and nitrate in the guttation fluid were substantially reduced. Sodium is of interest in that the concentration was higher in the guttation fluid than in the root pressure exudate, indicating that the leaf tissues of this plant discriminate against sodium absorption.

The continued delivery of ions via the transpiration stream ensures that the leaf is supplied with the mineral nutrients essential to its growth and development. However, the continued supply of ions and other solutes to a mature leaf can lead to excessively high levels within the leaf and may result in some toxicity. Some excess ions are accumulated into the vacuoles of the leaf mesophyll cells where they serve an osmotic function, but in many cases the surplus is exported from the leaf through the phloem pathway (see next section), such transfer of ions from xylem to phloem sometimes involving transfer cells.

Ions such as calcium, which are relatively immobile in the phloem, continue to accumulate within the leaves and often form precipitates. Crystals of calcium carbonate and calcium oxalate commonly occur in the vacuoles and cell walls of many species, while calcium phosphate is sometimes precipitated within mitochondria. These crystals increase in size and number as the leaf matures and are prominent in many deciduous plants just prior to leaf fall.

Leaching of solutes from leaves by rain or mist also contributes to the removal of excess levels of certain ions. In general cations are leached preferentially to anions and the leachate is slightly alkaline (Tukey, 1970). Leaching has its converse in the foliar feeding of certain nutrients which is widely practised in agriculture.

Under conditions of low transpiration, such as high humidity and darkness, the development of a positive root pressure in the xylem results in the extrusion of droplets of liquid from the leaf, the process of guttation. Often specialized pores, termed hydathodes, are the pathway for guttation. These pores are modified stomata, but do not show any opening or closing movements. Guttation may also occur through ordinary stomata. The guttation fluid, usually a dilute solution of inorganic ions and organic compounds, contributes to the overall loss of solutes from the leaf. The guttation water may evaporate, leaving an encrustation of solutes on the leaf, but any subsequent rainfall will leach these solutes from the system. The volume of fluid extruded by a guttating plant varies from a few microlitres on a grass blade overnight to the prodigious 200 ml recorded for a single leaf in one day of the Indian taro plant (*Colocasia antiquorum*).

Loss of solutes by leaching and guttation are somewhat fortuitous processes, depending upon rainfall and humidity, over which the plant can exercise no control. Certain halophytes have evolved mechanisms for the excretion of excess salts from specialized glands or for accumulation in salt hairs (see Stewart, this volume). However, for the majority of plants the removal of excess levels of certain ions takes place through the phloem, accompanying the export of the products of photosynthesis.

F. EXPORT OF CATIONS THROUGH THE PHLOEM

Analyses of phloem saps from a range of plants have demonstrated the presence of appreciable levels of certain ions (see Zeigler, 1975). The analysis of *Ricinus* phloem sap (Table VII) may be taken as representative in that among the cations there is a high level of potassium giving a high potassium:sodium ratio, and a low level of calcium giving a high magnesium:calcium ratio. These values indicate the relative mobility and/or ease of entry of these ions into the phloem pathway, suggesting the involvement of a regulatory membrane in the loading of these ions into the phloem. In many species the levels of ions and organic solutes present in the phloem sap increases markedly during leaf senescence, when they are exported from the leaf prior to abscission. In addition to the major inorganic ions presented in Table VII traces of molybdenum, copper, iron, manganese, and zinc have also been reported in phloem exudate (Tammes and van Die, 1966).

TABLE VII
The composition of phloem exudate obtained from *Ricinus* plants. Concentrations are expressed in mg ml^{-1} and also in meq or mM where relevant (from Hall and Baker, 1972)

	mg ml^{-1}	
Dry matter	100	
Sucrose	80–106	
Reducing sugars	Absent	
Protein	1·45–2·20	
Amino acids	5·2 (as glutamic acid)	35·2 mM
Keto acids	2·0–3·2 (as malic acid)	30–47 meq
Phosphate	0·35–0·55	7·4–11·4 meq
Sulphate	0·024–0·048	0·5–1·0 meq
Chloride	0·355–0·675	10–19 meq
Nitrate	Absent	
Bicarbonate	0·010	1·7 meq
Potassium	2·3–4·4	60–112 meq
Sodium	0·046–0·276	2–12 meq
Calcium	0·020–0·092	1·0–4·6 meq
Magnesium	0·109–0·122	9–10 meq
Ammonium	0·029	1·6 meq
ATP	0·24–0·36	0·40–0·60 mM
Auxin	10·5 × 10^{-6}	0·60 × 10^{-4} mM
Gibberellin	2·3 × 10^{-6}	0·67 × 10^{-5} mM
Cytokinin	10·8 × 10^{-6}	0·52 × 10^{-4} mM
Abscisic acid	105·7 × 10^{-6}	0·40 × 10^{-3} mM

A systematic study of the relative mobility of a large number of ions within the plant was conducted by Bukovac and Wittwer (1957) (Table VIII). A result of this mobility is that deficiencies of phloem-mobile elements occur in the older leaves first and those of non-mobile elements in the youngest leaves. Mobility is a relative term and among the so-called mobile elements some, such as potassium, are considerably more mobile than others.

The heavy metal micronutrients have an intermediate mobility, although this varies between species (see Rediske and Biddulph, 1953, and Brown *et al.*, 1965, for details of iron translocation; Uriu and Koch, 1964, for details of manganese and zinc translocation; Suida and Linck, 1963, and Wallihan and Heymann-Herschberg, 1956, for details of zinc translocation).

As calcium and boron are relatively immobile in the phloem, a continuous supply of these two elements through the xylem is critical for normal growth and development of the plant. Fruits and storage organs, primarily dependent on the phloem for their nutrient supply, are normally low in calcium and boron giving rise to deficiency disorders. Some evidence exists of limited calcium movement through the phloem (Ringoet

TABLE VIII
Mobility of ions in the phloem (after Bukovac and Wittwer, 1957)

Mobile	Partially mobile	Immobile
Rb	Zn	Ca
Na	Cu	Sr
K	Mn	Ba
P	Fe	
Cl	Mo	
S		

et al., 1968; Läuchli, 1968; Hall and Baker, 1972) but these findings do not alter the general conclusion that calcium mobility is very low. As pointed out by Clarkson (1974) calcium is not transported effectively in either the symplast or the phloem, even though its entry into cells forming part of the transport system is not prevented.

The high mobility of magnesium in the phloem is somewhat surprising for a divalent cation, and it is possible that it may be transported in a complexed form. Also the heavy metal micronutrients referred to above may require some complexing in order to ensure their mobility. Iron in particular is insoluble at the pH of phloem saps and must be transported in a non-ionic or chelated form.

III. Regulation of Ion Content in Whole Plants

The processes of ion transport and distribution within the plant result in certain regularities which can be related to various external and internal factors. A view shared by the majority of whole-plant physiologists is that the observed regularities reflect a control system operating at the whole organism level. Sadly this regulatory aspect of nutrient uptake has received only limited attention in the past (see Weatherley, 1969; Pitman and Cram, 1977). Undoubtedly the ion content of the leaves of the plant will vary as the degree of import and export of ions alters as the leaves age, the redistribution of these nutrients being an important process within the shoot. However, the supply of ions to the shoot would appear to be in balance with its overall requirement and is closely related to the growth of the plant (Pitman, 1975). The growth of the leaves needs to be limited by mineral nutrient availability when that availability is low, but the supply of mineral nutrients also needs to be reduced when availability is high, so that an excessive and potentially detrimental level of accumulation in the leaves may be prevented.

The observation of an integrated response in plant growth and long-distance nutrient transport may merely reflect an inbuilt mechanism which gives the appearance of a negative-feedback system, but is actually only a passive response. The evidence for integrated responses within the plant must be critically appraised in terms of this possible passive response before invoking the transmission of information from shoot to root.

The existence of such a negative feedback system has been considered; it could be involved in the regulation of the supply of mineral nutrients to the xylem, the loading of the phloem with assimilates and mineral nutrients, and the control of stomatal movement which will influence photosynthetic rates (Pitman and Cram, 1977). Such a regulation involves negative-feedback signals which may be hormonal, relating to specific internal parameters of the leaf cells such as turgor changes (see Cram, 1976; Raven, 1977).

The mechanisms of transmitting such a signal are relatively limited within the whole plant. There are two strongly favoured possibilities: either an hormonal signal or the supply of assimilates may provide, individually or together, a phloem-transmitted signal from shoot to root.

There is evidence that the endogenous plant growth hormones, auxin, gibberellins, cytokinins, and abscisic acid are present in the phloem sap (Table VII; Hall and Baker, 1972; Hoad, 1973; Allen and Baker, 1980) and may therefore be transmitted from shoot to root. Some effects have been reported whereby these hormones can alter the permeability of the roots to water and also induce changes in the ion uptake and transport to the xylem (see Pitman, 1975). However, these effects of hormones have been less specific than the integration of whole plant activity would seemingly require and their role in the regulation of ion content at this level of organization requires further, more detailed, elucidation.

The uptake and transport of ions by the root is an energy-dependent process which could conceivably be rate-limited by the supply of sugars and other assimilates to the root by the phloem. Indeed a reduction of sugar transport by stem cooling (Bowling, 1968) or by light and dark treatments (Hatrick and Bowling, 1973; Pitman and Cram, 1973) has been shown to reduce ion uptake. Studies on assimilate loading, which has been shown to be a co-transport of sugars with protons (see Baker, 1978), have implicated potassium as a possible regulator of proton extrusion and hence the availability of potassium would influence the supply of sugars to the root. Evidence in support of such a system is provided by the positive correlation between sucrose and potassium levels in the phloem sap of plants subjected to a range of potassium levels in the external medium (Fig. 1). However, the possible influence of endogenous phloem-transmitted hormones casts doubt as to whether such results were purely due to a reduced influx of energy in the root.

Pitman (1972) has demonstrated that the supply of sugar to the roots of plants growing at moderate to high growth rates does not appear to

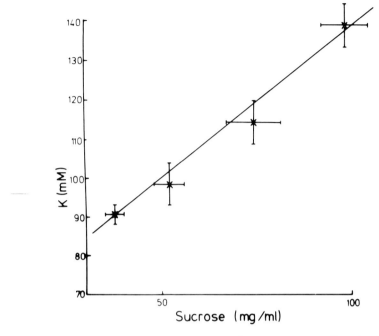

F<small>IG</small>. 1. The relationship between potassium and sucrose concentrations in phloem sap obtained from *Ricinus* plants subjected to a range of potassium concentrations in the external nutrient solution. Vertical and horizontal bars indicate ± SD.

regulate the rate of potassium uptake, this uptake being independent of the sugar level. Sugar is diverted into storage however, a fact which indicates that it is not limiting potassium transport. Such an effect may be more apparent at low growth rates and when potassium is poorly available, but to date the effect of sugars is even less specific than that of hormones and they are therefore weak candidates for the proposed signal mechanism.

This brief summary of the regulation of ion transport in whole plants is inevitably eclectic, and especially so because the researcher in this field is presented with such a variety of hypotheses; it should be the aim of future research to resolve these hypotheses into a coherent pattern. Although we have a mass, and in some areas perhaps an excess, of empirical data our knowledge of these important control processes is barely inchoate.

R<small>EFERENCES</small>

Allen, J. R. F. and Baker, D. A. (1980). *Planta* **148**, 69–74.
Asher, C. J. and Ozanne, P. G. (1967). *Soil Science* **103**, 155–161.

Baker, D. A. (1973). *Planta* **112**, 293–299.
Baker, D. A. (1978). *New Phytol.* **81**, 485–497.
Baker, D. A. and Weatherley, P. E. (1968). *J. exp. Bot.* **20**, 485–496.
Baker, D. A. and Hall, J. L. (1973). *New Phytol.* **72**, 1281–1291.
Bell, C. W. and Biddulph, D. (1963). *Plant Physiol.* **38**, 610–614.
Bollard, E. G. (1953). *New Zealand J. Sci. Tech.* **35A**, 15–18.
Bowling, D. J. F. (1968). *J. exp. Bot.* **19**, 381–388.
Bowling, D. J. F. (1976). "Uptake of Ions by Plants Roots", Chapman and Hall, London.
Bowling, D. J. F. and Weatherley, P. E. (1965). *J. exp. Bot.* **16**, 732–741.
Bowling, D. J. F., Macklon, A. E. S., and Spanswick, R. M. (1966). *J. exp. Bot.* **17**, 410–416.
Brown, A. L., Yamaguchi, S., and Leal-Diaz, J. (1965). *Plant Physiol.* **40**, 35–38.
Bukovac, M. J. and Wittwer, S. H. (1957). *Plant Physiol.* **32**, 428–435.
Clarkson, D. T. (1974). "Ion Transport and Cell Structure in Plants", McGraw Hill, London.
Clarkson, D. T. and Robards, A. W. (1974). *In* "Root Structure and Function", pp. 415–436, (J. Torrey and D. T. Clarkson, eds.), Academic Press, London and New York.
Collins, J. C. and Reilly, E. J. (1968). *Planta* **83**, 218–222.
Cram, W. J. (1976). *In* "Encyclopedia of Plant Physiology", New Series, Vol. 2, Transport in Plants, Part A, Cells, pp. 284–316, (U. Luttge and M. G. Pitman, eds.), Springer-Verlag, Berlin, Heidelberg, and New York.
DeKock, P. C. and Mitchell, R. L. (1957). *Soil Sci.* **84**, 55–62.
Dunlop, J. and Bowling, D. J. F. (1971). *J. exp. Bot.* **22**, 434–444.
Epstein, E. (1976), *In* "Encyclopedia of Plant Physiology", New Series, Vol. 2, Transport in Plants, Part B, Tissues and Organs, pp. 70–94. (U. Lüttge and M. G. Pitman, eds.), Springer-Verlag, Berlin, Heidelberg and New York.
Gerson, D. F. and Poole, R. J. (1971). *Plant Physiol.* **48**, 509–511.
Ginsburg, H. and Ginsburg, B. Z. (1974). *J. exp. Bot.* **25**, 667–669.
Gunning, B. E. S. and Pate, J. S. (1969). *Protoplasma* **68**, 107–133.
Hall, S. M. and Baker, D. A. (1972). *Planta* **106**, 131–140.
Hatrick, A. A. and Bowling, D. J. F. (1973). *J. exp. Bot.* **24**, 607–613.
Hewitt, E. J. and Smith, T. A. (1975). "Plant Mineral Nutrition", English Universities Press, London.
Hoad, G. V. (1973). *Planta* **113**, 367–372.
Isermann, K. (1971). *Z. Pflanzenr. Bodenkunde* **128**, 195–207.
Jacoby, B. (1965). *Physiol. Plant.* **18**, 730–739.
Jacoby, B. (1966). *Nature Lond.* **211**, 212.
Jeschke, W. D. and Simonis, W. (1969). *Planta* **88**, 157–171.
Laties, G. G. (1969), *Ann. Rev. Pl. Physiol.* **20**, 89–116.
Läuchli, A. (1968). *Vortage aus dem Gesamtgebiet der Botanik N.F.* **2**, 58–65.
Loneragan, J. F. and Snowball, K. (1969). *Aust. J. Agric. Res.* **20**, 465–467.
Nissen, P. (1974). *Ann. Rev. Pl. Physiol.* **25**, 53–79.
Nobel, P. S. (1969). *Plant and Cell Physiol.* **10**, 597–605.
Pitman, M. G. (1965). *Aust. J. biol. Sci.* **18**, 10–24.
Pitman, M. G. (1965). *Aust. J. Sci.* **18**, 987–998.
Pitman, M. G. (1972). *Aust. J. biol. Sci.* **25**, 905–919.
Pitman, M. G. (1975). *In* "Ion Transport in Plant Cells and Tissues", pp. 267–308, (D. A. Baker and J. L. Hall, eds.), Elsevier, New York.
Pitman, M. G. (1977). *Ann. Rev. Pl. Physiol.* **28**, 71–78.

Pitman, M. G. and Cram, W. J. (1973). *In*: "Ion Transport in Plants, (W. P. Anderson, ed.), Academic Press, London and New York.

Pitman, M. G. and Cram, W. J. (1977). *Symp. Soc. exp. Biol.* 391–424.

Raven, J. A. (1977). *Symp. Soc. exp. Biol.* **31**, 73–91.

Rediske, J. H. and Biddulph, O. (1953). *Plant Physiol.* **28**, 594–605.

Ringoet, A., Sauer, G., and Gielink, A. J. (1968). *Planta* **80**, 15–20.

Shone, M. G. T., Clarkson, D. T., and Sanderson, J. (1969). *Planta* **86**, 301–314.

Suida, T. W. and Linck, A. J. (1963). *Plant and Soil.* **19**, 249–254.

Sutcliffe, J. F. (1976). *In* "Encyclopedia of Plant Physiology", New Series, Vol. 2, Transport in Plants, Part B, Tissues and Organs, pp. 394–417, (U. Lüttge and M. G. Pitman, eds.), Springer-Verlag, Berlin, Heidelberg, and New York.

Tammes, P. M. L. and van Die, J. (1966). *Kon. Ned. Akad. Wetenschap. Proc. (C)* **69**, 656–659.

Thellier, M. (1973). *Ann. Bot.* **34**, 983–1009.

Tukey, H. B. (1970). *Ann. Rev. Pl. Physiol.* **21**, 305–324.

Tyree, M. T. (1970). *J. Theor. Biol.* **26**, 181–214.

Uriu, K. and Koch, E. C. (1964). *Proc. Am. Soc. Hort. Sci.* **84**, 25–31.

Van Iren, F. and Boers-van der Sluijs, P. (1980). *Planta* **148**, 130–137.

Wallace, A. and North, C. P. (1953). *California Agriculture* **7**, 10.

Wallihan, E. F. and Heymann-Herschberg, L. (1956). *Plant Physiol.* **31**, 294–299.

Weatherley, P. E. (1969). *In* "Ecological Aspects of the Mineral Nutrition of Plants", pp. 323–340, (I. H. Rorison, ed.), Blackwell, Oxford and Edinburgh.

Wyn Jones, R. G. (1975). *In* "Ion Transport in Plant Cells and Tissues", pp. 193–229, (D. A. Baker and J. L. Hall, eds.), Elsevier, New York.

Yeo, A. R., Kramer, D., Läuchli, A., and Gullasch, J. (1977). *J. exp. Bot.* **28**, 18–30.

Ziegler, H. (1975). *In* "Encyclopedia of Plant Physiology", New Series, Vol. 1, pp. 59–100, Phloem Transport, (M. H. Zimmerman and J. A. Milburn, eds.), Springer-Verlag, Berlin, Heidelberg, and New York.

CHAPTER 2

Mycorrhizal Fungi and Ion Uptake

P. B. TINKER AND A. GILDON

Rothamsted Experimental Station, Harpenden, Herts., England

I. Introduction

This chapter deals more with phosphorus than with metals, because this element occupies a very special position in relation to the mycorrhizal symbiosis, and by far the largest amount of published work relates to phosphorus. However, it is now known that trace metals are also involved. It is important to stress at the outset that mycorrhizal infection is not an abnormal or exotic occurrence. The overwhelmingly greater part of the world's vegetation is infected with mycorrhizal fungi, and the uninfected, solution-grown, excised root often used for laboratory experimentation is wholly unrepresentative of the real state of plants, even if it may be useful for investigating the basic processes of nutrient uptake.

II. Biology and Function of Mycorrhizas

There are two major classes of mycorrhizal fungi, the ectotrophic (ECM), and the vesicular-arbuscular (VAM). The ericaceous mycorrhizas are a

much smaller group, and the orchidaceous fungi have only small importance in practical terms despite the attention they received in early work.
Only the two major groups will be discussed here, general descriptions of
which can be found in Harley (1969), Marks and Koslowski (1973),
Sanders et al. (1975) and Mosse et al. (1981).

The ECM form a thick (40 μm) sheath of mycelium around roots, and
the mycelium also penetrates between the cortical cells, so that its total
mass may be comparable to that of the roots themselves. The external
mycelium extending into the soil has received relatively less attention, but
forms single hyphae, rhizomorphs, and hyphal strands on which fruiting
bodies may form. The form of this mycelium depends very much upon the
precise host–fungus combination. Ectomycorrhizas occur exclusively on
tree species, including all our major forest trees.

The vesicular-arbuscular mycorrhizas (VAM) are not obvious on inspec-
tion, consisting of internal and external mycelium, but no sheath. The
internal mycelium includes intracellular branched structures called arbuscules and intercellular vesicles and hyphae; the external mycelium carries
spores of differing size and may extend several centimetres into the
surrounding soil. They are found on most species except those with other
forms of mycorrhizas, though some species, mostly in the Chenopodiaceae
and Cruciferae, are not infected.

A number of the ECM fungi can be grown in pure culture and reproduce
sexually, so the taxonomy of many of them is fairly clear. The VAM fungi
are obligate symbionts and produce no known organs of sexual reproduc-
tion, so their taxonomy is poorly established and their genetics unknown.
This is probably the reason why work on ECM was much better developed
than that on VAM until relatively recently. However, the discovery of the
yield-enhancing effects of the latter, and their occurrence on many
important agricultural crops, have changed the emphasis greatly, and
VAM now probably receives more attention than ECM. The importance
of ECM for tree growth was established quite a long time ago, and the
artificial inoculation of forest nurseries with appropriate fungi had often
been shown to be highly effective (see Mikola, 1973). Yield increases
attributable to VAM infection were first reported in 1944 but significant
work did not start until the late 1950s (see Harley, 1969). Work on ECM
since the 1930s had implicated an increased nutrient supply in the
improvement of yields, but the equivalent deduction for VAM was only
drawn about 1960, when the strong interaction with phosphorus supply
became evident (see Mosse, 1973). It is now established that mycorrhizal
plants can grow satisfactorily on soils of low phosphorus status, on which
non-infected plants are severely deficient (see Tinker, 1980). There have
been continuing suggestions that nitrogen uptake is greater with ECM
plants, but this has never been very satisfactorily established, and no such

suggestions have been made for VAM. More positive evidence of enhanced uptake of N has been obtained with ericaceous mycorrhizas. There is now good evidence that VAM can supply Zn and Cu to the hosts and that ECM can do so for Zn (see below). It has been claimed that VAM can enhance uptake of K from soil (Powell, 1975) or from mica (Mojallalli and Weed, 1978) and that ECM can increase the uptake from K-containing minerals (Rosendahl, 1943). Edmonds *et al.* (1976) found that the large K uptake of beech mycorrhizas was not affected by the presence of Ca or Na. Above 20°C K was rapidly lost from the mycorrhizal tissues, but this could be prevented if glucose was supplied. However, none of this work has given convincing evidence of a *direct* increase in K uptake to the host via the fungal partner and, because improved growth due to an increased P supply could enhance the demand for all other nutrients, there is now doubt over this possibility. No work has been reported relating VAM or ECM to Mg or Ca effects on plant growth, but VAM infection increased the rate of uptake of ^{90}Sr by soyabeans (Jackson *et al.*, 1973) and Ca can be translocated by both types of mycorrhiza. The VAM can also aid in the uptake of S.

III. Phosphorus Uptake, Translocation, and Transfer Processes

The general principles of the supply of phosphorus by mycorrhizal fungi to the host are now clear, probably most so for VAM. It is assumed that similar principles apply to the other nutrients involved also. Early ideas concentrated on the properties and P content of the sheath of ECM, with the suggestion that this acted as a store of phosphorus for the plant. However, this suggestion is unlikely to be correct for VAM, where the total weight of fungus can probably not exceed 10% of the root weight (see Tinker, 1978). There is now ample proof that the external mycelium absorbs P from the soil outside the depletion zone which forms around any active root. The calculations of Sanders and Tinker (1971) indicated that a mycorrhizal root could absorb at a rate four times greater than the uninfected root and that this increased inflow could not have been transported to the root itself by diffusion, because of the low mobility of P in soil. Later direct tests showed uptake of P by hyphae several centimetres from the root (see Rhodes and Gerdemann, 1980), though Owusu-Bennoah and Wild (1979) could not detect any difference between the depletion zones beyond 2 mm away from roots or mycorrhizas. Measurements of the soil phosphorus L-value (size of the pool of soil phosphate which can be isotopically labelled) by several authors proved that no extra amount of unavailable soil P was "solubilized", or in any other way utilized by the fungus (see Tinker, 1978). Similarly clear evidence is not available for the ECM; their external mycelium can absorb P and transfer it, but in addition their proliferation in organic litter has led to repeated suggestions that they can hydrolyse organic phosphates in the soil (see Bowen, 1973).

The nutrition of mycorrhizal plants is thus very largely a question of the ability of the fungal symbiont to absorb, translocate, and transfer various elements, and these processes will be discussed in this order.

A. UPTAKE

Fungi absorb nutrients by processes believed to be closely similar to those of the higher plants. Measurements of uptake rates on VAM fungi are exceedingly difficult to make, and the only practical information is from Sanders and Tinker (1973), from which one may calculate a mean uptake rate for P of 2×10^{-15} mol. cm^{-1} hyphae s^{-1}, and from Cooper and Tinker (1978) who measured mean uptake rates per cm of hypha of the order of 10^{-17} mol. cm^{-1} s^{-1}. For hyphae of around 8μm diameter, the first value implies a flux of up to 10^{-12} mol. cm^{-2} s^{-1}, which is similar to P fluxes into roots.

Some interesting results have been produced by Cress et al. (1979), in which the classical Epstein-Hagen kinetic analysis was applied to the uptake of P by excised barley roots with and without VAM, using the usual equation for uptake rate $V = V_{max} C/(C + K_m)$, where C is solution concentration, and V_{max} and K_m are root parameters. Gray and Gerdemann (1969) earlier showed that mycorrhizal roots absorb P more rapidly than uninfected ones in solution culture, but Bowen et al. (1974) reported that the amounts of P taken up by excised roots or VA mycorrhizas from solution over short periods were not always very different, especially if they were taken from the same plants. All such studies of uptake in solution culture raise many questions, such as how much external mycelium remains on the root, how far internal fungal structures act as a strong local sink or source for P, and how far the stirring ensures that there are no gradients in the static layer of solution around each root surface. If we leave these uncertainties aside, the results of Cress et al. (1979) show that V_{max}, the maximum rate of uptake, was not changed by mycorrhizal infection. K_m was found to be two to three times smaller (about 1.6μM) in the mycorrhizal roots, implying a more efficient uptake mechanism in the fungus than in the root at low concentrations (but see page 27). However, all diffusion studies on P uptake by roots in P-deficient soils show that it is the diffusion impedance in the soil, rather than the physiology of the root, which is rate-limiting (Nye and Tinker 1977). So even if this difference in K_m is real, it is not likely to have a major effect on uptake rates in normal plants growing in soil. Cress *et al.* (1979) also tested the uptake of Zn and Mn in the same way, but were unable to detect any effect of infection on V_{max} or K_m for these elements.

The need to consider the whole soil-plant system before concluding that the presence of VAM is beneficial was stressed by work of Rhodes and Gerdemann (1978b, 1980). This confirmed uptake and translocation of S by VAM hyphae, but the authors concluded that this may be of no value to the host, because this element is so mobile in the soil that it can reach the root itself at an adequate rate.

Much work has been done with uptake of P, N, and K by the more easily handleable fungal sheath of ECM (see Harley, 1969, for a full discussion). The sheath can be dissected from the mycorrhiza and used in this work in a manner closely analogous to that of an excised root. Uptake has been shown to be metabolically controlled, temperature-sensitive, and oxygen-sensitive. Using excised tissues, it was shown that the sheath absorbed much more rapidly than uninfected roots on a dry weight basis, by a factor of up to 8. The uptake rates of the mycorrhizas from 32 mM phosphate solution was about 6×10^{-10} mol. g^{-1} fresh weight s^{-1} (Harley 1969, p. 140) which is a high but not exceptional value (Nye and Tinker, 1977, p. 118). From the rates of uptake by the mycorrhizal sheath, it was concluded that little P would reach the root proper by simple diffusion through the sheath when infected roots were absorbing from dilute solutions under normal growing conditions, and most P is indeed found in the sheath after short-term uptake experiments.

B. TRANSLOCATION

One of the most prominent properties of fungi is the ability to translocate materials. Around 1955 Melin and co-workers (see Harley, 1969, for references) showed that when isotopes of P, Ca, and N (as ammonium) were supplied to mycelium of mycorrhizal fungi they could then be detected several centimetres away. This work was probably not given sufficient attention then and was not expressed quantitatively, so it did not lead to immediate further developments.

Work on translocation in VAM has increased sharply over recent years. Rhodes and Gerdemann (1978a,b) showed that Ca, P, and S were translocated by hyphae, and Cooper and Tinker (1978) measured translocation rates of P, Zn and S. Sanders and Tinker (1973) calculated the maximum fluxes of P through the main VAM entry hyphae to the root to be a high value, of the order of 10^{-8} mol. cm^{-2} s^{-1}. Interest in translocation of P was greatly enhanced by the discovery of polyphosphate in the vacuoles of VAM hyphae and arbuscules (Cox *et al.*, 1975), which was largely concentrated in granules of less than 1μm in diameter within the small vacuoles. More recent work has substantiated this by measuring the P in the granules by electron beam microprobe (Cox *et al.*, 1979), and has shown that the measured fluxes of P in hyphae could be accounted for by protoplasmic streaming of the hyphal contents, carrying the vacuoles with polyphosphate, at reasonable rates of a few cm h^{-1}. Cooper and Tinker (1981) have shown that translocation of P in hyphae is stopped by the inhibitor of protoplasmic streaming, cytochalasin, and is also highly temperature-sensitive and dependent upon plant transpiration. A combined bulk flow and protoplasmic streaming mechanism, operating on P concentrations enhanced by the formation of polyphosphate granules, appears to offer a full explanation of the high rates of P translocation in VAM (Tinker, 1975).

Movement across the sheath of ECM is also a translocation process, in which a clear analogy with processes in VAM was established by the discovery of polyphosphate (Ling-Lee *et al.*, 1975). Harley and co-workers (Harley, 1969) have shown that there are large pools of phosphate in the sheath, with which recently absorbed P only slowly equilibrates, and part of this may probably now be identified with the polyphosphate.

C. TRANSFER

Very little is known about the processes of transfer of nutrients from the fungus to the host. In ECM it seems reasonable to assume that there is leakage of orthophosphate out of the fungal mycelium around the cortical cells of the root (the Hartig net), followed by normal uptake by the latter. For VAM, the possibilites are wider, because the arbuscules are constantly degenerating, with new ones being formed. It has therefore been suggested by several authors that the degeneration allowed arbuscular contents to be transferred directly into the host cell. Such a mechanism would imply that all the nutrient elements in the fungus could be transferred, with little specificity. However, Cox and Tinker (1976) calculated from the breakdown rate of the arbuscules that this could not supply the measured extra P inflow into the infected root, unless the P concentration in the arbuscules were increased to extremely high levels. The leakage of nutrient elements from the intact arbuscules and the internal mycelium, followed by uptake through the host plasmalemma, must therefore also be important.

In this case, we may regard the organisms as being arranged in series, with phosphate ions passing through the uptake mechanism of the fungus, and then later through the uptake mechanism of the host plant. The value of V_{max} for the host is that obtained by saturating its uptake mechanism, so it is not so surprising that Cress *et al.* (1979) found that V_{max} remains the same after infection. The formation of arbuscules, which are surrounded by host plasmalemma and cytoplasm, causes an increase in the plasmalemma area, but from the measurements of Cox and Tinker (1976) the difference is less than 20% in onion roots, so it can probably be neglected. It is in any case uncertain whether the plasmalemma area is rate-limiting in uptake.

The conclusion of Cress et al. that K_m was smaller in infected roots depended upon a very small number of measurements of unknown accuracy. Even if correct, it need not mean that this is the K_m value for the fungus. At low concentrations of the nutrient, it will not diffuse into the free space of the root rapidly enough to maintain uptake by the inner cortical cells, which may thus be almost ineffective (Nye and Tinker, 1977, p. 117). Any mechanism which injects extra phospate into the inner parts of the free space of the cortex, as we hypothesize that VA fungi do, will then increase uptake rate, and hence cause an apparent decrease in K_m. If substantiated, the conclusions of Cress et al. are thus entirely in accordance with expectation from current theory.

It is suspected that an alkaline phosphatase found in mycorrhizal roots is involved in the transfer step (Gianinazzi and Gianinazzi-Pearson, 1979). This enzyme occurs in the vacuoles of the fungal arbuscules and appears to reach maximum activity at the time when the growth response to mycorrhizal infection is starting to appear. However, it is not able to hydrolyse polyphosphates, and its function is presently obscure.

D. PHOSPHORUS NUTRITION OF HOST

The mechanisms described above indicate that an improvement in host plant nutrition, and possibly growth, should be caused by mycorrhizal infection whenever (1) the uptake rate of a nutrient by the simple root is restricted by transport processes in the soil (see Nye and Tinker, 1977) to below that required by the host to grow at the maximum rate allowed by the environment, and (2) the mycorrhizal fungus can absorb and translocate that nutrient. Yost and Fox (1979) have used field experiments with methylbromide-fumigated plots, to determine the soil-solution concentrations of P below which particular crops would respond to infection with VAM, but so far it is very difficult to predict how large such yield increases will be. Some reported growth responses have been very large, when the uninfected plants were very phosphorus-deficient. The responses to infection with VA fungi are often smaller than the maximum responses given by large dressings of soluble phosphates. The amount of soluble P fertilizer which gave the same growth response as VAM infection in a number of reported experiments was on average around 100 kg P ha^{-1}, but for a citrus crop it was as large as 500 kg ha^{-1} (Menge et al. 1978).

IV. UPTAKE OF TRACE METALS BY MYCORRHIZAL PLANTS

Much less is known about the effects of mycorrhiza formation on the uptake of the trace metals. There are certainly reasons why one could expect a mycorrhizal effect on trace metal uptake. It has been well established that metals such as Cu, Zn and Mn have very low mobility in soil, because of their strong adsorption by the soil colloids. The existence of well-defined depletion zones around roots has been proven for Zn (Wilkinson *et al.*, 1968). Mycorrhizal infection could thus possibly improve trace metal nutrition by a mechanism analogous to that for phosphorus, if the metals are absorbed and translocated in the mycelium. Since the metals are essential for fungal growth, they must be adsorbed by the mycelium. If as suggested, polyphosphate granules form the basis of the efficient P translocation mechanism, it is possible that trace metals could form complexes with the polyphosphate, and thus be translocated too. Effects of VAM inoculation on growth of Zn deficient crops have been demonstrated (Table I) for peaches (La Rue *et al.*, 1975; Gilmore, 1971), cotton

TABLE I

Effect of VAM inoculation on growth and on mineral content of leaves of 1-year old peach trees (after LaRue *et al.*, 1975)

Treatment	Height m	N %	P %	K %	Mn ppm	Zn ppm
Untreated	1·35	4·38	0·13	1·70	64	11
Fertilised (P + Zn)	1·93	3·42	0·13	1·75	29	13
Fertilised, Mycorrhizal	2.24	2.99	0.15	1.79	24	17
Mycorrhizal	2·42	2·99	0·16	1·80	25	19

(Wilhelm *et al.*, 1967) and apples (Benson and Covey, 1976). A very interesting situation can arise in fumigated citrus nurseries, in which both P and Zn are deficient, but adding more P suppresses re-infection by VAM, and this enhances the Zn deficiency. In an analogous way to P, high levels of Zn are toxic to the VAM fungi, although Gildon and Tinker (1981) found Zn- and Cd-tolerant fungal strains forming mycorrhizas on mine spoil.

One of the earliest detailed studies on trace metals was that of Bowen *et al.* (1974). They measured Zn uptake in short-term experiments on excised uninfected roots and mycorrhizas of *Pinus radiata* (ECM) and *Araucaria cunninghamii* (VAM). In all cases there was strong uptake at 2°C, assumed to be passive adsorption on cell walls, and further enhanced uptake at 20°C, the difference being regarded as the true uptake into the root cells. This enhanced uptake per unit surface area was 1·5 to 3 times greater for ectomycorrhizal than for uninfected roots. Uptake rates were much less for *Araucaria*, but mycorrhiza formation again increased the difference between uptake rates at 2°C and 20°C by about 2.5 times. These results certainly indicate a much increased uptake rate for mycorrhizae as compared to uninfected roots, but there must always be some doubt about exactly how one should interpret these short-term uptake experiments on excised roots in relation to the behaviour of intact plants. Cooper and Tinker (1978) showed that *Glomus mosseae* hyphae with a clover host absorbed and translocated Zn, though the rates of the processes were some two orders of magnitude less than those for P in the same conditions. There is thus no doubt about the ability of mycorrhizas to absorb Zn from soil.

There is now also clear evidence of the uptake of Cu by VAM. Timmer and Leyden (1980) showed that sour orange seedlings grew better and responded less to added copper when inoculated with the VAM fungus *Glomus fasciculatus*, but that the effect of inoculation on both growth and Cu concentration decreased as the P fertilizer level was increased, presumably because of the depressing effect of P on infection of the roots. Great care is needed in proving that VAM effects are mediated via elements other than P, for there is always the risk that the P status of the plant is altered, and that this may interact with the uptake of the element in question via effects on root or shoot growth. In our work (Gildon and Tinker, in preparation) using a factorially-designed experiment with Cu and P treatments, leeks were grown in a soil known to contain very little available Cu. The P treatment had no significant effect on growth or Cu concentration, but the Cu treatment gave a highly significant increase in growth and Cu concentration in the absence of VAM inoculation. When the leeks were infected the growth response to added Cu almost disappeared, and when no Cu was applied, VAM increased both growth and Cu concentration (Table II), whereas phosphorus concentrations were very little changed.

The interactions between Cu and P or Zn and P in plant nutrition are well known (e.g. Olsen, 1972), though the mechanism has always remained rather obscure. The effect is shown strikingly in work on corn by Safaya

TABLE II

Growth responses to copper (5 μg g^{-1} soil) of leeks (*Allium porrum*) with or without infection with *Glomus mosseae*

| Copper treatment | Cu_0 | | Cu_1 | | LSD |
Mycorrhiza	+	−	+	−	
Dry weight, g	1·33	0·79	1·31	1·41	0·16
Cu concentration, μg g^{-1}	5·1	3·6	5·7	5·5	1·5
P concentration %	0·13	0·12	0·13	0·13	0·03

(1976), when adding P reduced uptake rates per unit weight of root from 4·9 to 1·3 mg Zn g^{-1} fresh root day^{-1} and from 2·4 to 0·3 mg Cu g^{-1} fresh root day^{-1}. It seems likely that one of the mechanisms operating to produce this antagonism is that increased phosphorus tends to reduce VAM infection. Lambert et al. (1979) tested this idea with soyabean, and showed that adding P fertilizer significantly reduced both Zn and Cu concentrations in mycorrhizal plants, but had no effect when they were not infected. However, they did not apply Cu or Zn fertilizers, so the interaction between P and metals could not be tested properly in their experiments. A search of the literature indicated that this effect may be quite general.

Very recently (Lambert et al., 1980) there has been a suggestion that boron may be important in the rapid establishment of VAM infection in roots, and that a boron deficiency may induce trace metal deficiencies via this effect. There is no indication that boron is absorbed by the VAM fungus, and the effects were not very marked, so further evidence seems needed.

V. Conclusion

Studies on the trace metal nutrition of plants cannot ignore the mycorrhizal relationship, because it is quite certain that Cu and Zn supplies to plants can be markedly improved by mycorrhiza formation, and it may well be found that this applies to other trace elements also. Mycorrhizal effects may help to explain the very complex and confused state of the literature on trace metal uptake, especially where comparisons are made between solution or sand culture (where plants are likely to be non-mycorrhizal) and in soil cultures (where they probably are infected). Mycorrhizal infection will not always be relevant to the point at issue, but the possibility that it is cannot be neglected. Certainly, if there are interactions between metal uptake and phosphorus then mycorrhizal infection must always be considered.

References

Benson, N. R. and Covey, R. P. (1976). J. Amer. Soc. Hort. Sci. **11,** 252–253.
Bowen, G. D. (1973). In "Ectomycorrhizae", pp. 151–205, (G. C. Marks and T. T. Kozlowski, eds.), Academic Press, New York and London.
Bowen, G. D., Skinner, M. F., and Bevege, D. J. (1974). Soil Biol. Biochem. **6,** 141–144.
Cooper, K. M. and Tinker, P. B. (1978). New Phytol. **81,** 43–52.
Cooper, K. M. and Tinker, P. B. (1981). New Phytol. **88,** 327–339.
Cox, G. and Tinker, P. B. (1976). New Phytol. **77,** 371–378.

Cox, G. C., Moran, K. J., Sanders, F., Nockolds, C., and Tinker, P. B. (1979). *New Phytol.* **84**, 649–659.

Cox, G. C., Sanders, F. E., Tinker, P. B., and Wild, J. (1975). *In* "Endomycorrhizas", pp. 297–312, (F. E. Sanders, B. Mosse, and P. B. Tinker, eds.), Academic Press, London and New York.

Cress, W. A., Throneberry, G, O., and Lindsey, D. L. (1979). *Plant Physiol.* **64**, 484–487.

Edmonds, A. J., Wilson, J. M., and Harley, J. L. (1976). *New Phytol.* **76**, 307–316.

Gianinazzi, S. and Gianinazzi-Pearson, V. (1979). *New Phytol.* **82**, 127–132.

Gildon, A. and Tinker, P. B. (1981). *Trans. Brit. Mycol. Soc.* **77**, 648–649.

Gilmore, A. E. (1971). *J. Am. Soc. Hort. Sci.* **96**, 35–38.

Gray, L. E. and Gerdemann, J. W. (1969). *Plant Soil,* **30**, 415–422.

Harley, J. L. (1969). "The biology of mycorrhiza", 2nd Ed., Leonard Hill, London.

Jackson, N. E., Miller, R. H., and Franklin, R. E. (1973). *Soil Biol. Biochem.* **5**, 205–212.

Lambert, D. H., Baker, D. E., and Cole, H. (1979). *Soil Sci. Soc. Am. J.*, **43**, 976–980.

Lambert, D. H., Cole, H., and Baker, D. E. (1980). *Plant Soil* **57**, 431–438.

Ling-Lee, M., Chilvers, G. A., and Ashford, A. E. (1975). *New Phytol.* **75**, 551–554.

LaRue, J. H., McClellan, W. D., and Peacock, W. L. (1975) *California Agriculture* **29**, (5), 6–7.

Marks, G. C. and Kozlowski, T. T. (1973). "Ectomycorrhizae", Academic Press, London and New York.

Menge, J. A., Labanauskas, C. K., Johnson, E. L. U., and Platt, R. G. (1978). *Soil Sci. Soc. Amer. J.* **42**, 926–930.

Mikola, P. (1973). *In* "Ectomycorrhizae" pp. 383–412, (G. C. Marks and T. T. Kozlowski, eds.), Academic Press, New York and London.

Mojallali, H. and Weed, S. B. (1978). *Soil Sci. Soc. Am. J.* **42**, 367–372.

Mosse, B. (1973). *Ann. Rev. Phytopath.* **11**, 171–196.

Mosse, B., Stribley, D. P., and Le Tacon, F. (1981). "Advances in Microbial Ecology" (In press).

Nye, P. H. and Tinker, P. B. (1977). "Solute Movement in the Soil-root System", Blackwell, London.

Olsen, S. R. (1972). *In* "Micronutrients in Agriculture", pp. 243–264, (J. J. Mortvedt, P. M. Giordano, and W. L. Lindsay, eds.), Soil Sci. Soc. Amer., Madison.

Owusu-Bennoah, E. and Wild, A. (1979). *New Phytol.* **82**, 133–140.

Powell, C. Ll. (1975). *In* "Endomycorrhizas", pp. 461–468, (F. E. Sanders, B. Mosse, and P. B. Tinker, eds.), Academic Press, London and New York.

Rhodes, L. H. and Gerdemann, J. W. (1978a). *Soil Sci.* **126**, 125–126.

Rhodes, L. H. and Gerdemann, J. W. (1978b). *Soil Biol. Biochem.* **10**, 355–360.

Rhodes, L. H. and Gerdemann, J. W. (1980). *In* "Cellular Interactions in Symbiosis and Parasitisms", pp. 173–195, (C. B. Cook, P. W. Pappas and E. D. Rudolph, eds.), Ohio State Univ. Press., Columbus.

Rosendahl, R. O. (1943). *Soil Sci. Soc. Am. Proc.* **7**, 477–479.

Safaya, N. M. (1976). *Soil Sci. Soc. Am. J.* **40**, 719–722.

Sanders, F. E. and Tinker, P. B. (1971). *Nature (Lond.)* **233**, 278–279.

Sanders, F. E. and Tinker, P. B. (1973). *Pestic. Sci.* **4**, 385–395.

Sanders, F. E., Mosse, B., and Tinker, P. B. (1975). "Endomycorrhizas", Academic Press, London and New York.

Timmer, L. W. and Leyden, R. F. (1980). *New Phytol.* **85,** 15–23.

Tinker, P. B. (1975). *Symp. Soc. Exp. Biol.* **29,** 325–349.

Tinker, P. B. (1978). *Physiol. Veg.* **16,** 743–751.

Tinker, P. B. (1980). *In* "The Role of Phosphorus in Agriculture", pp. 617–653, (F. E. Kwasahneh, E. C. Sample, and E. J. Kamprath, eds.), Am. Soc. Agron., Madison.

Wilhelm, S., George, A., and Pendery, W. (1967). *Phytopath.* **57,** 103.

Wilkinson, H. F., Loneragan, J. F., and Quirk, J. P. (1968). *Soil Sci. Soc. Amer. Proc.* **32,** 831–833.

Yost, R. S. and Fox, R. L. (1979). *Agron. J.* **71,** 903–908.

CHAPTER 3

Adaptation to Salinity in Angiosperm Halophytes

G. R. STEWART* AND I. AHMAD†

*Department of Botany, Birkbeck College, University of London, England
†Department of Botany, University of Toronto, Canada

I. INTRODUCTION

But land that's salt
And bears a bitter name—a worthless land,
That neither bears nor hearkens to the plough.

Virgil's (The Georgics, Book II) comments are only too accurate when it comes to considering the response of most crop species to salinity. His dismissal of saline land as worthless overlooks the fact that in the natural vegetation of such soils we have, potentially at least, the possible key to the agricultural exploitation of saline soils. Increasing rates of salinization of agricultural land has prompted considerable interest in those characteristics of angiosperm halophytes which enable them to survive, grow, and reproduce in the presence of large amounts of sodium chloride. Knowledge of these characteristics is essential for any attempt to enhance the salt-tolerance of existing varieties of crop plants, particularly if the possibilities afforded by somatic cell hybridization and genetic engineering are to be ever fully exploited.

In this article we will consider the problems posed for plants growing in a saline environment and will go on to describe some of the characteristics which enable them to exploit such habitats. For the most part we will be concerned only with angiosperm halophytes; the sub-cellular structure and morphological organization of lower eukaryotes and prokaryotes makes comparisons of different salt tolerance strategies difficult and of limited value.

In considering salinity we will restrict our discussion to that associated with the prescence of large amounts of NaCl. The response to salinity of crop species (Maas and Hoffman, 1977) and the salt tolerance of other non-halophytes (Greenway and Munns, 1980) have been reviewed and will not be considered here.

II. Osmotic Stress and Adjustment

Saline soils can be defined quite simply as those having excessive concentrations of soluble salts (US Salinity Laboratory). In the coastal marshes of Britain and Europe salt concentrations (NaCl) around 500 mM are commonly encountered and for short periods of the year salinities in excess of 1 M may exist (Jeffries, Davy, and Rudmik, 1979). Although the ion species of particular saline soils can exert specific effects on plant growth these appear only rarely to be the most important cause of growth inhibition (Bernstein, 1975; Maas and Hoffman, 1977). It is the total ion concentration, rather than specific ions, which appears to influence plant growth in such soils.

The water potentials of saline soils fall in the range −0·2 to −5·0 M Pascals (Pa) while that of sea water is around −2·4 M Pa. In order that a net flow of water into the plant can occur, tissue water potentials must be lower than those of the external soil solution. The water relations of the plant can be represented by the familiar equation:

$$\psi = \psi_p + \psi_s + \psi_m \tag{1}$$

where,

ψ is the water potential of the cell, tissue organ, or plant,

ψ_p is the pressure (turgor) or hydrostatic potential of the cell,

ψ_m is the matrix potential, or potential of structurally-bound water,

ψ_s is the solute or osmotic potential of the cell contents.

A plant growing in saline soils is likely therefore to have tissue water potentials in the range −0·2 to −5·0 M Pa. It is worth noting at this stage that leaf water potentials lower than about −1·2 to −1·6 M Pa are found in severely wilted leaves of crop plants (see e.g. Hanson, Nelsen, and Everson, 1977), while well-watered, non-stressed leaf tissues generally have water potentials of −0·2 M Pa or greater. In the field, leaf water potentials of unwilted halophytes in the range −2·0–3·0 M Pa have been recorded (see e.g. Jeffries et al., 1979a). The capacity to effect osmotic adjustment in soils containing large amounts of NaCl is then an essential characteristic of the physiology of angiosperm halophytes.

Using the terminology developed in studies of osmoregulation in animals, Wyn Jones *et al.* (1977) have suggested there are two strategies of osmotic adjustment exhibited by angiosperm halophytes, that of osmoconformers and that of osmoregulators. Species in which a more or less constant gradient of osmotic potential is maintained between the plant and external solution are osmoconformers, while species in which a more or less constant plant osmotic potential is maintained over a wide range of external salinity are osmoregulators. The results presented in Fig. 1 show

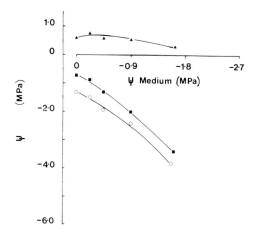

FIG. 1. Leaf water potential and component potentials of *Cochleria officinalis* grown at different salinities. Measurements of leaf ψ were made with a dew point hygrometer (Wescor Ltd, HR-33(T)) using a C-52 sample chamber. Leaf pieces were cut so as to fit the largest sample holder of the C-52. Expressed cell sap was obtained using a hydraulic press. A small volume of sap was transfered to a filter paper disc and ψ_s determined with the dew point hygrometer. ψ_p was determined from the difference between ψ_s and ψ. Details of plant growth were as described by Ahmad *et al.* (1981b). Leaf ψ = ■; ψ_s = ○; ψ_p = ▲.

that leaf osmotic potential of *Cochleria officinalis* decreases more or less linearly with an increase in ψ of the growth medium, suggesting this species could be regarded as an osmoconformer.

The response of *Atriplex littoralis* (Fig. 2) is somewhat different as more or less constant leaf ψ_s is maintained up to an external ψ of -0.9 M Pa, suggesting that this species is an osmoregulator. The ability of this species and possibly also *Avicennia nitida* to osmoregulate is limited; at external ψ less than -0.9 M Pa they behave more like osmoconformers. The behaviour of *Limonium vulgare* (Fig. 3) and *Plantago maritima* appears intermediate between that of osmoconformers and osmoregulators.

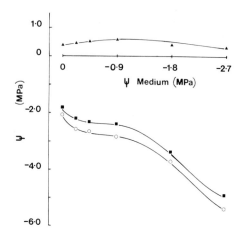

FIG. 2. Leaf water potential and component potentials of *Atriplex littoralis* grown at different salinities. Details as for Fig. 1. Leaf ψ = ■; ψ_s = ○; ψ_p = ▲.

Although the classification of halophytes into two distinct groups on the basis of their osmotic behaviour may be an oversimplification it is striking that the leaf ψ under non-saline conditions is in some species around $-2\cdot0$ M Pa, suggesting that they are already "adjusted" to low external ψ and that such species may most closely approach the status of an "obligate" halophyte.

It is evident from equation (1) that a change in leaf water potential can occur in a variety of ways; however, in most studies of osmotic adjustment in angiosperm halophytes emphasis has been placed on changes in ψ_s being

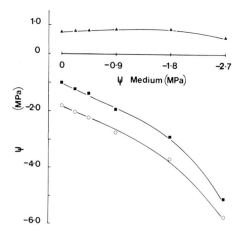

FIG. 3. Leaf water potential and component potentials of *Limonium vulgare* grown at different salinities. Details as for Fig. 1. Leaf ψ = ■; ψ_s = ○; ψ_p = ▲.

the component potential effecting changes in leaf ψ. The results in Figs 1–3 show that changes in leaf ψ are accompanied by very similar changes in ψ_s. This is readily seen with leaf tissue of *Limonium vulgare* (Fig. 3) where leaf ψ and ψ_s decrease in parallel over a change in external ψ from near zero to −2·7 M Pa. The other major component potential, ψ_p, remains more or less constant up to −1·8 M Pa. The behaviour of these halophytes contrasts with that of water-stressed plants where decreases in ψ_p are often more pronouned than decreases in ψ_s and are responsible for a major component of the decrease in leaf ψ under conditions of drought (see e.g. Gardner and Ehlig, 1965).

A change in solute or osmotic potential can occur in several ways: by an accumulation of solutes taken up from the external solution; by the synthesis of solutes; or by a reduction in cell water content, that is by cell dehydration. Only the first two processes can be regarded as being mechanisms of osmotic adjustment or compensation. Many studies of the response of plants to salinity (particularly those with glycophytes) have failed to distinguish adequately between physiologically active changes in solute content and those occurring passively as the result of cell dehydration (see Janes, 1966). It is evident from the results in Figs 1–3 that ψ_p often decreases in high salinities. This is usually associated with a reduction in water content and growth rate.

The process of osmotic adjustment in angiosperm halophytes is generally thought to occur primarily by the uptake and accumulation of NaCl (see Flowers *et al.*, 1977). The results in Table I confirm that in leaf tissue sodium and chloride ions make a major contribution to the decrease in leaf ψ_s which occurs under saline conditions. In species such as *Atriplex littoralis*, *Plantago maritima* and *Suaeda maritima* over 70% of ψ_s can be accounted for by $Na^+ + Cl^-$. In other species, such as *Cochleria*, $Na^+ + Cl^-$ accounts for only 46–50% of ψ_s. These results and those

TABLE I
The contribution of NaCl to leaf osmotic potentials

Species	Leaf ψ_s MPa	ψ $Na^+ + Cl^-$ MPa	% of leaf ψ_s
Atriplex littoralis	−2·82	−2·0	70
Avicennia nitida	−3·29	−1·88	57
Cochleria dancia	−2·28	−1·05	46
Cochleria officinalis	−2·42	−1·24	50
Limonium vulgare	−2·70	−1·56	57
Plantago maritima	−2·04	−1·38	68
Suaeda maritima	−2·63	−1·93	73

Plants were grown on 200 mM NaCl ($\psi \equiv 0·9$ MPa). See Fig. 1 for other details.
ψ $Na^+ + Cl^-$ is the calculated contribution of Na^+ and Cl^- to the measured leaf ψ_s.

obtained in many other studies (see e.g. Wallace and Kleinkop, 1974; Storey, Ahmad, and Wyn Jones, 1977; Jeffries, Rudmik, and Dillon, 1979) indicate that sodium and chloride ions are major sources for osmotic adjustment in angiosperm halophytes. There are, however, some species in which NaCl does not appear to be a major osmotic solute. For example, the salt marsh ecotype of *Agrostis stolonifera* accumulates considerably smaller concentrations of NaCl but larger amounts of organic solutes than the less tolerant spray-zone and inland ecotypes (Ahmad, Wainwright, and Stewart, 1981).

III. METABOLISM AT LOW WATER POTENTIALS

It has been suggested above that the capacity to adjust osmotically by the accumulation of NaCl is a critical attribute of angiosperm halophytes and it follows that the metabolic process of these plants must be able to proceed at internal water potentials in the range $-2\cdot0$ to $-4\cdot0$ M Pa. Furthermore it would seem from analysis of the solutes contributing to ψ_s that this implies an ability to maintain metabolic activity in the presence of large amounts of NaCl. However, studies carried out in the early 1970s demonstrated that enzymes such as malate dehydrogenase, glucose-6-phosphate dehydrogenase, isocitrate dehydrogenase, and RUBP-carboxylase from salt-tolerant plants such as *Atriplex spongiosa* (Greenway and Osmond, 1972; Osmond and Greenway, 1972) and *Suaeda maritima* (Flowers, 1972), were markedly inhibited by concentrations of NaCl which were within the range of concentrations found in leaf tissues of halophytes. Subsequent studies have for the most part confirmed the salt-sensitivity of enzymes from angiosperm halophytes (see Flowers *et al.*, 1977; Greenway and Munns, 1980). Generally the salt inhibition of enzymes is fairly non-specific, a wide variety of neutral salts will inhibit their order of effectiveness following the Hofmeister (lyotropic) series. Kinetic analysis of salt inhibition suggests that NaCl behaves as a mixed type inhibitor, reducing maximum reaction rate (V_{max}) and increasing K_m, thus decreasing the apparent affinity (Greenway and Sims, 1974; Ahmad *et al.*, 1979a). The mechanistic basis for salt inhibition is uncertain, although salt-induced changes in the orientation of key residues at the active site or domain is a popular explanation (see e.g. Von Hippel *et al.*, 1973). Somero, Neubauer, and Low (1977) have argued that such an explanation cannot account for the influence of salts on both V and the catalytic activation volume ($\Delta V^{\mp}$). They argue that salts affect the ability of an enzyme molecule as a whole, to take part in the conformational changes which are essential for catalysis. Such an explanation is based on the energy and volume changes resulting from transfers of protein groups between the hydrophobic interior and the protein-solvent interface which occur during catalytic conformational changes. Thus salt is suggested to influence the transfer free energies of these processes of hydration and dehydration.

Large salt concentrations can also affect protein stability as well as catalytic activity. Chloroplastic glutamine synthetase partially purified from the halophyte *Triglochin maritima* is unstable in the presence of NaCl and other neutral salts (Fig. 4).

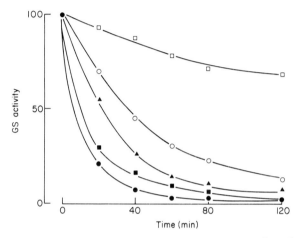

FIG. 4. Influence of NaCl on the stability of *Triglochin maritima* glutamine synthetase. Tris-acetate pH 8·4 alone = □; + 200 mM NaCl = ○; + 600 mM NaCl = ●; + 200 mM KCl = ▲; + 200 mM KNO₃ = ■.

Multi-enzyme processes of halophytes also exhibit salt sensitivity. Oxygen uptake and phosphorylation by mitochondria isolated from the halophyte *Suaeda maritima* were found to be inhibited by NaCl (Flowers, 1974). Moreover, Hall and Flowers (1973) have shown protein synthesis in *Suaeda* to be salt sensitive.

The general view which emerges from such studies is that the metabolic processes of angiosperm halophytes are in no way especially adapted to function in the presence of large amounts of NaCl, indeed they show little or no difference with respect to salt sensitivity from those of glycophytes.

The only groups of organisms which have been shown to exhibit salt-adaptation at the macro-molecular level are the extreme halophilic bacteria of the Archaebacteria. These bacteria, which grow in environments with salt concentrations in excess of 2 M NaCl, accumulate internally large amounts of K^+ (3–4 M) and have macromolecules whose structure and function is obligately dependent on the presence of at least 1 M salt (Larsen, 1967; Lanyi, 1974). The catalytic and structural proteins are characterized by a large excess of acidic amino acid residues and it has been suggested (Baxter, 1959; Soo-Hoo and Brown, 1967; Kushner *et al.*, 1964) that the salt requirement arises from the necessity to shield these negative charges in the protein. In the absence of salt it is assumed that charge repulsion would cause unfolding of the polypeptide chain with subsequent destabilization and loss of function. However, this charge-screening hypothesis has many weaknesses, in particular screening should be maximal at molarities at least an order of magnitude below those required for the stabilization of proteins from those extremely halophilic bacteria (Lanyi and Stevenson, 1970). The requirement of unusually large concentrations of NaCl or KCl for maintaining the optimal conformation of halophilic protein may reside more in new hydrophobic interactions which occur in the presence of salt rather than simply through the electrostatic shielding of charged groups (see Lanyi, 1974). It is conceivable however that charge screening may be a phenomena worth considering in relation to salt inhibition of angiosperm enzymes.

IV. Sub-cellular Ion Localization

The salt sensitivity of enzymes and organelles from angiosperm halophytes suggests that growth and development would be severely reduced and disorganized if ion concentrations at the sites of metabolic activity resembled those measured in whole leaf analyses. The resolution of this paradox lies in the structure of the angiosperm cell with its large central vacuole comprising as much as 95% of cell volume, and it is in this vacuole that the bulk of the accumulated ions are thought to be located (Flowers *et al.*, 1977). Direct evidence for the sub-cellular localization of Na^+ and Cl^- has however been difficult to obtain. A variety of techniques have been employed but until recently few have yielded unequivocal results. The isolation of subcellular components by non-aqueous cell fractionation has yielded some data on chloroplast ion concentrations (Larkum and Hill, 1970; Harvey and Flowers, 1978). Relatively large amounts of chloride ($\simeq$ 400 mM) have been measured in chloroplasts of *Limonium* (Larkum and Hill, 1970) and *Suaeda* (Harvey and Flowers, 1978) and similar concentrations of sodium were found in *Suaeda* chloroplasts. This relatively large accumulation of salt in chloroplasts poses some problems with respect to the functioning of chloroplastic enzymes such as RUBP-carboxylase (Osmond and Greenway, 1972), glutamine synthetase and acetolactate synthetase (Stewart and Lee, 1974) and malic enzyme (Greenway and Sims, 1974) all of which exhibit a salt-sensitivity similar to that of other non-chloroplast enzymes.

Another widely used approach has been that of ion precipitation; but in general the results obtained using this approach are difficult to interpret since large losses of ions are encountered during sample preparation (Harvey, Flowers, and Hall, 1976; Harvey, 1978).

Recent work in which leaf tissue has been prepared by freeze-substitution and examined using transmission analytical electron microscopy has yielded very interesting results concerning the ion concentrations of cytoplasm, chloroplasts, vacuoles, and cell walls (Harvey et al., 1981). The results in Table II are taken from this study and show the very marked

TABLE II
Na⁺ and Cl⁻ concentrations in Mesophyll Cells of *Suaeda maritima*

Cell type Frequency of occurrence	Compartment	mol m^{-3}	
		Na$^+$	Cl$^-$
60% of cells	Vacuole	565	388
	Chloroplasts	93	85
	Cytoplasm/Cell Wall	109	21
	Intercellular space/Cell Wall	132	36
30% of cells	Vacuole	422	301
	Chloroplasts	75	86
	Cytoplasm/Cell Wall	148	112
	Cell Wall	381	368
	Intercellular space	22	38
10% of cells	Vacuole	286	284
	Chloroplasts	257	212
	Cytoplasm/Cell Wall	71	134
	Intercellular Space/Cell Wall	0	58

Ion concentration determined by transmission analytical electron microscopy, data from Harvey et al. (1981).

differences in Na⁺ and Cl⁻ concentrations in different compartments of the cell. The vacuolar concentrations of Na⁺ and Cl⁻ in the bulk of the mesophyll cells of *Suaeda* are at least five times greater than those of the cytoplasm or chloroplast. It is particularly striking that chloroplast chloride concentrations determined in this way vary considerably less than those determined on chloroplasts prepared by non-aqueous extraction (cf. Harvey and Flowers, 1978). Another interesting feature of these results are the differences in ion distribution within different cells of the mesophyll; in 10% of the mesophyll cells little differential ion compartmentation is evident. A similar heterogeneity in chloride distribution has been reported in leaf cells of the mangrove *Aegiceras corniculatum* (Von Steveninck et al., 1976) and those of *Salicornia pacifica* (Hess, Hansen, and Weber, 1975).

These data provide the most direct evidence available at present for the spatial separation of potentially disruptive salt and the salt sensitive metabolic sites of halophytes. Thus, while these halophytes accumulate large amounts of salt in order to provide a source of solute for osmotic adjustment to the low external osmotic potentials of saline environments, they protect their salt-sensitive metabolic activities by sequestering the bulk of the accumulated salt in the cell vacuole.

V. Intracellular Osmotic Adjustment

The existence of cytoplasmic ion concentrations which are considerably lower than those of the vacuole necessitate a lowering of cytoplasmic water potential in order to maintain the water potential equilibrium between cytoplasm and vacuole. The use of inorganic ions other than Na^+ and Cl^- would seem to be precluded in view of the rather non-specific nature of the salt inhibition exhibited by the enzymes of angiosperm halophytes. One way in which water potentials of cytoplasm and vacuole may be kept in balance is by the accumulation in the cytoplasm of "compatible" organic solutes.

A remarkable feature of many angiosperm halophytes is that they accumulate very large amounts of certain metabolites including the imino acid, proline (Goas, 1965) and the methylated quaternary ammonium compound, glycine betaine (Storey and Wyn Jones, 1975). Both of these compounds have been suggested to function as compatible cytoplasmic solutes (Stewart and Lee, 1974; Storey and Wyn Jones, 1975). Wyn Jones and his colleagues have carried out extensive studies of the ionic and solute relations of many species and from such studies a close correlation has emerged between glycine betaine concentration and sap osmotic potential (Wyn Jones et al., 1977; Storey and Wyn Jones, 1979), and also between the capacity to accumulate glycine betaine and the salt tolerance of different species (Storey and Wyn Jones, 1977). A general model of cytoplasmic osmoregulation has been proposed in which glycine betaine synthesis is used to generate low cytoplasmic osmotic potentials (Wyn Jones et al., 1977).

Similar studies with proline-accumulating species have shown a similar correlation between tissue salt concentrations and proline content (Treichel, 1975) and between the salt tolerance of ecotypes and their capacity to accumulate proline (Stewart and Lee, 1974; Ahmad et al., 1981a).

Large amounts of other methylated onium compounds besides glycine betaine have been found in some angiosperm halophytes; β-alanine betaine has been shown in the tissues of salt tolerant species of the

Plumbaginaceae (Larher and Hamelin, 1975; Larher, 1976) and related sulphonium compounds are present in *Spartina anglica* (Larher, Hamelin, and Stewart, 1977) and *Diplotaxis tenuifolia tennis* (Larher and Hamelin, 1979). β-alanine betaine is present in large amounts in plants of *Limonium vulgare* grown under non-saline conditions and increases some two to three fold at 600 mM NaCl (Stewart *et al.*, 1979). This pattern of response is similar to that found for glycine betaine in *Spartina anglica* and *Suaeda maritima* (Stewart *et al.*, 1979). The concentration of dimethyl propiothetin (dimethyl sulphonio propionate) in *Spartina anglica* was found to be unaffected by low water potentials (Stewart *et al.*, 1979).

The occurrence and distribution of proline and methylated onium compounds in temperate halophytes is now well documented (Storey *et al.*, 1977; Stewart *et al.*, 1979; Gorham, Hughes, and Wyn Jones, 1980). One group of species is recognizable in which proline appears to be the major organic solute accumulated under saline conditions, a second group comprises species in which the major organic solute is glycine betaine, and in a third group both methylated onium compounds and proline are accumulated. A final group consists of species which, although halophytic, appear to accumulate neither proline nor methylated onium compounds. One species in this groups, *Plantago maritima*, accumulates the polyol sorbitol (Ahmad, Larher, and Stewart, 1979; Jeffries *et al.*, 1979b). The concentration of sorbitol in the tissues of *P. maritima* follows very closely that of NaCl (Ahmad *et al.*, 1979b) and in many respects there are close similarities between the behaviour of sorbitol in *P. maritima* and that of proline in proline-accumulators such as *Puccinellia maritima* (Ahmad, Larher, and Stewart, 1981b) and *Agrostis stolinifera* (Ahmad·*et al.*, 1981a).

Several other species of *Plantago* accumulate sorbitol when subject to low water potential whether these arise from the presence of large amounts of NaCl or a non-penetrating osmoticant such as polyethylene glycol (Table III). Sorbitol accumulation, like that of proline and glycine betaine seems then to occur in response to drought as well as salinity (see Stewart and Larher, 1980).

Although these various groups of halophytes can be recognized with respect to their capacity to accumulate different organic compounds, it should be emphasized that, in some, many metabolites alter in response to increased salinity. In leaf tissue of *Puccinellia maritima* the amides asparagine and glutamine, soluble carbohydrates and the non-protein amino acid Δ'-acetyl ornithine increase along with proline in response to increased salinity (Ahmad *et al.*, 1981b). The salt marsh ecotype of *Agrostis stolinifera* shows accumulations of proline, asparagine, glutamine, serine, and glycine under saline conditions (Fig. 5; see also Ahmad *et al.*, 1981a) and the mangrove *Avicennia nitida* accumulates betaine, and the ureides, allantoin, and allantoic acid (Fig. 6). Many proline accumulating

TABLE III
Influence of salinity and osmotic stress on sorbitol accumulation in *Plantago*

Species		ψ leaf MPa	Sorbitol μmol gH_2O^{-1}
Plantago albicans	Control	0·85	5·3
	200 mM NaCl	4·62	59·3
	15% PEG	1·45	41·3
Plantago alpina	Control	0·91	6·0
	200 mM NaCl	1·82	38·9
	15% PEG	1·23	37·8
Plantago cynops	Control	0·84	6·1
	200 mM NaCl	4·73	62·7
	15% PEG	1·12	27·5
Plantago maritima	Control	0·77	5·1
	200 mM NaCl	1·19	26·9
	15% PEG	0·88	20·1

Details for plant growth and analysis were as described by Ahmad *et al.* (1981b).

species accumulate other amino acids such as pipecolic acid and 5′-hydroxy pipecolic acid (Goas, Larher, and Goas, 1970). There is little doubt that yet other organic solutes are involved in intracellular osmotic adjustment since, in a number of species, no major source of organic solute has as yet been identified (see Stewart *et al.*, 1979).

Studies of the effect of proline (Stewart and Lee, 1974), glycine betaine (Flowers, Hall, and Ward, 1978; Pollard and Wyn Jones, 1979) and sorbitol (Ahmad *et al.*, 1979b) on the activity of various enzymes indicate

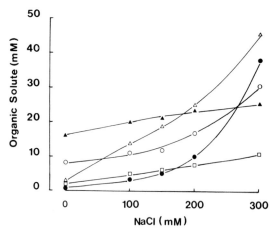

FIG. 5. Organic Solute Accumulation in *Agrostis stolonifera* (salt marsh ecotype) (See Ahmed *et al.*, 1981a for details.) Asparagine = △; Proline = ●; Serine = ○; Glycine betaine = ▲; Glutamine = □.

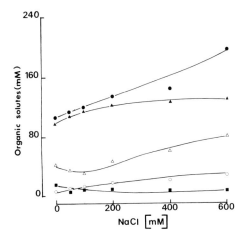

FIG. 6. Organic Solute Accumulation in *Avicennia nitida*. Glycine betaine = ●; amino compounds = △; ureides = ○; organic acids = ■; sugars = ▲. (Analyses were performed as described by Ahmad *et al.*, 1979a).

that large concentrations (up to 1 M) have little or no inhibitory effect. Indeed in one instance, glycine betaine was found to decrease the inhibition of malic enzyme by NaCl (Pollard and Wyn Jones, 1979). The basis for this apparently protective effect is uncertain although Greenway and Munns (1980) have analysed this data and suggest that glycine betaine increases the affinity of malic enzyme for its substrate malate, thereby reducing the inhibition by NaCl. Glycine betaine, proline, and sorbitol have been found to stabilize glutamine synthetase from *Triglochin* (Table IV) and sorbitol in particular is very effective in stabilizing the enzyme in

TABLE IV

The influence of osmotic solutes on the stability of glutamine synthetase from *Triglochin maritima*

Treatment	Decay constant kd (h^{-1})
Tris-acetate buffer, pH 8·4	0·16
+ 300 mM Glycine betaine	0·03
+ 300 mM Proline	0·05
+ 300 mM Sorbitol	0
+ 300 mM NaCl	1·77
+ 300 mM Glycine betaine + 300 mM NaCl	1·07
+ 300 mM Proline + 300 mM NaCl	1·08
+ 300 mM Sorbitol + 300 mM NaCl	0·35

Kd, the first order decay constant for glutamine synthetase activity was determined from plots of 1n residual activity against time at 30° for partially purified enzyme treated as above.

the presence of NaCl. The physiological significance of these protective effects of organic solutes on enzyme stability and functioning are uncertain, although Schobert (1977) has argued that compounds such as proline and glycine betaine function as modifiers of water structure which enhance protein hydration at low water potentials and that this rather than osmotic adjustment is their primary function (Schobert and Tzchesche, 1978). Any compound which accumulates in the cytoplasm to levels where it comprises over 60% of the total solute potential must, of course, be active in the osmotic relations of a cell. Irrespective of a somewhat speculative function as water structure regulators the evidence discussed above is consistent with the view that these organic solutes are "compatible" with the metabolic processes of halophytes.

Evidence regarding the sub-cellular localization of these solutes is sparse. Wyn Jones et al. (1977) have found evidence for at least a partially cytoplasmic localization of glycine betaine in beetroot storage cells. Recently, however, histochemical staining applied to freeze-substituted material of Suaeda maritima has provided the best evidence for a cytoplasmic localization for glycine betaine (Hall, Harvey, and Flowers, 1978). This staining procedure is based on the formation of a complex between methylated quarternary ammonium compounds and platinum halogenates, and electron dense deposits containing platinum and iodine (determined by X-ray microprobe analysis) were found in the cytoplasm but not the vacuole of saltgrown plants of Suaeda maritima. It is surprising, however, that little staining was observed with plants grown in the absence of NaCl since their glycine betaine content was 25–30% of that in saltgrown plants (Hall et al., 1978).

Thus compounds such as glycine betaine, proline, and sorbitol increase in response to low external water potentials and the intracellular accumulation of NaCl, they do not inhibit the metabolic processes of halophytes and, in the case of glycine betaine at least, accumulate in the cytoplasm. This evidence is for the most part consistent with them playing a role as compatible cytoplasmic solutes in maintaining the water potential equilibrium between cytoplasm and vacuole.

An exclusively cytoplasmic localization for all of the organic solutes which increase in response to salinity seems, however, unlikely. Plants of, e.g. Puccinellia maritima grown on 600 mM NaCl exhibited an increase of approximately 240 mM organic solutes while the Na^+ and Cl^- concentration was 800 mM, suggesting that if the cytoplasm represented 5–10% of the total cell volume the osmotic potential of the cytoplasm would be considerably lower than that of the vacuole (Ahmad et al., 1981b). Variation in ion content of individual cells (Hess et al., 1975; Van Steveninck et al., 1976; Harvey et al., 1981) implies that a similar variation must exist in the organic solute content of different cells. Harvey et al.

(1981) point out "as cells differ significantly in the concentrations of ions they accumulate in the various sub-cellular compartments, then they may also vary in the extent to which they synthesize compatible solutes such as glycine betaine". Non-vacuolated meristematic cells with little potential for the spatial separation of ions from sensitive metabolic sites must osmotically adjust by a synthesis of compatible solutes. In the apices and expanding leaves of water stressed wheat, very large amounts of proline (amounting to the equivalent of -0.6 M Pa) are accumulated (Munns, Brady, and Barlow, 1979). Similarly Gorham et al. (1980) have shown the florets of *Aster tripolium* to have a high ratio of organic solutes to inorganic salts compared to leaf tissue.

VI. Concluding Remarks

Lawton, Todd, and Naidoo (1981) have recently suggested that studies of halophytes "have failed to correlate any one aspect of the physiology of halophytes in general with a reasonable hypothesis of salt tolerance". However, as is clear from the discussion above, the work carried out in the past ten years, principally by Flowers and his colleagues at the University of Sussex and by Wyn Jones and his colleagues at the University College of North Wales, Bangor, have provided the experimental data for a generalized model of salt tolerance in angiosperm halophytes. The key features of this model are firstly, that osmotic adjustment to the low water potentials of saline soils occurs primarily by an accumulation of sodium and chloride ions. Secondly, the potential disruption of salt-sensitive metabolic activities is circumvented by a sequestering of Na^+ and Cl^- in the cell vacuole. Thirdly, intracellular osmotic adjustment is effected through the synthesis and accumulation of compatible solutes in the cytoplasm. It may be argued that as yet we do not understand the molecular mechanisms which underlie these three key characteristics, but certainly the phenomenological basis of salt tolerance can, in general terms, be stated.

Flowers et al. (1977) have discussed the characteristics of salt uptake in halophytes and it is pertinent to point out that while these plants do accumulate massive amounts of NaCl, the ability to regulate salt accumulation is as important in them as it is in non-halophytes which appear to tolerate low salinities by excluding NaCl. In halophytes growing at the upper limit of their salt tolerance an apparently uncontrolled uptake of salt results in massive increases in salt content of leaf tissue (Fig. 7). This accumulation of salt in species such as *Atriplex patula* and *Hibiscus tiliaceus* occurs almost as if some barrier to salt entry had been removed. The large concentrations of salt in such plants cannot be solely accounted for by a decrease in tissue water content although this does occur. Such plants lose

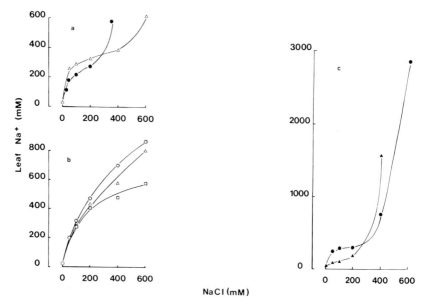

Fig. 7. Pattern of sodium accumulation in leaf tissue of different halophytes. (Details for growth and analyses are given in Ahmad *et al.*, 1979). *Atriplex patula* = ● (c); *Avicennia nitida* = □ (b); *Cochleria officinalis* = ● (a); *Hibiscus tiliaceus* = ▲ (c); *Limonium vulgare* = △ (a); *Plantago maritima* = △ (b); *Suaeda maritima* = ○ (b).

water even though there appears to be more then sufficient salt to effect osmotic adjustment. Similar behaviour has been observed in non-tolerant ecotypes of *Agrostis stolonifera* (Ahmad *et al.*, 1981a) and a possible explanation for this behaviour is that the plants are no longer able to maintain intracellular salt accumulation. Consequently, there is a build-up of salt in the cell walls of leaves,, and the accumulation of large amounts of ions in the cell walls results in osmotic imbalance and loss of turgor (see Bingham, Fenn, and Oertli, 1968; Oertli, 1966; 1968).

Given that the plant can regulate the uptake and accumulation of salt, some means of maintaining a differential localization of the accumulated salt must operate. One possibility is the continuous operation of ion pumps on the tonoplast which would maintain low cytoplasmic concentrations of Na^+ and Cl^-. Such a system would seem to require the expenditure of considerable energy and a more feasible means of maintaining the differential ion concentrations across the tonoplast would be that the tonoplast has a low permeability to NaCl. Once pumped into the vacuole there would be little leakage back into the cytoplasm. Consistent with this view are observations that Na^+ and Cl^- are not readily re-translocated in halophytes (Greenway, Gunn, and Thomas, 1966).

At present there is little evidence concerning mechanisms through which the synthesis of organic solutes are coupled to salt accumulation and compartmentation. The metabolism of proline and methylated onium compounds appears essentially similar in both glycophytes and halophytes (see Stewart and Larher, 1980) and there is little evidence of "salt stimulated" reactions. Gutknecht and Bisson (1977) have discussed the relationship between ion transport and osmotic regulation of giant algal cells and suggest that constant turgor in such cells is effected through a turgor-sensitive ion-carrier system and that changes in turgor pressure are the primary signal for turgor regulation.

Thus, while evidence available at present is consistent with the Flowers–Wyn Jones model of salt tolerance, further work is necessary to elucidate the molecular mechanisms underlying the three key components of the model, namely the regulation of salt accumulation, the compartmentation of salt, and intracellular osmotic adjustment.

References

Ahmad, I. and Wainwright, S. J. (1977). *New Phytol.* **76**, 361.

Ahmad, I., Larher, F., and Stewart, G. R. (1979a). *New Phytol.* **82**, 671.

Ahmad, I., Larher, F., Rhodes, D., and Stewart, G. R. (1979b). *In* "Nitrogen Assimilation in Plants", pp. 653–657, (E. J. Hewitt and C. V. Cutting, eds.), Academic Press, New York and London.

Ahmad, I., Wainwright, S. J., and Stewart, G. R. (1981a). New Phytol. **87**, 615.

Ahmad, I., Larher, F., and Stewart, G. R. (1981b). *Phytochem.* **20**, 1507.

Baxter, R. M. (1959), *Can. J. Microbiol.* **5**, 47.

Bernstein, L. (1975). *Ann. Rev. Phytopath.* **13**, 295.

Bingham, F. T., Fenn, L. B., and Oertli, J. (1968), *Proc. Soil. Sci. Am.* **32**, 249.

Flowers, T. J. (1972). *Phytochem.* **11**, 1881.

Flowers, T. J. (1973). *J. Expt. Bot.* **25**, 101.

Flowers, T. J., Troke, P. F., and Yeo, A. R. (1977). *Ann. Rev. Plant Physiol.* **28**, 89.

Flowers, T. J., Hall, J. L., and Ward, M. E. (1978). *Ann. Bot.* **42**, 1065.

Gardner, W. R., and Ehlig, C. F. (1965). *Plant Physiol.* **40**, 705.

Goas, M. (1965). *Bull. Soc. Fr. Physiol. veg.* **11**, 309.

Goas, M., Larher, F., and Goas, G. (1970). *C.R. Acad. Sci. Paris* **271**, 1368.

Gorham, J., Hughes, L., and Wyn Jones, R. G. (1980). *Plant Cell Environ.* **3**, 309.

Greenway, H. and Osmond, C. B. (1972). *Plant Physiol.* **49**, 256.

Greenway, H. and Sims, A. P. (1974). *Aust. J. Plant. Physiol.* **1**, 15.

Greenway, H., and Munns (1980). *Ann. Rev. Plant Physiol.* **31**, 149.

Greenway, H., Gunn, A., and Thomas, D., (1966). *Aust. J. Biol. Sci.* **19**, 741.

Gutknecht, J. and Bisson, M. A. (1977). *J. Membr. Biol.* **37**, 85.

Hall, J. L. and Flowers, T. J. (1973). *Z. Planzenphysiol.* **71**, 200.

Hall, J. L., Harvey, D. M. R., and Flowers, T. J. (1978). *Planta* **140**, 59.

Hanson, A. D., Nelsen, C. E., and Everson, E. H. (1977), *Crop Sci.* **17**, 720.

Harvey, D. M. R. (1978). Ph.D. thesis, University of Sussex, England.

Harvey, D. M. R. and Flowers, T. J. (1978). *Protoplasma* **97**, 337.

Harvey, D. M. R., Flowers, T. J., and Hall, J. L. (1974). *New Phytol.* **77**, 319.
Harvey, D. M. R., Hall, J. L., Flowers, T. J., and Kent, B. (1981). *Planta* **151**, 555.
Hess, W. M., Hansen, D. J., and Weber, D. J. (1975). *Can. J. Bot.* **53**, 1176.
Janes, B. E. (1966). *Soil. Sci.* **101**, 180–188.
Jeffries, R. L., Davy, A. J., and Rudmik, T. (1979a). *In* "Ecological processes in Coastal Environments", pp. 243–268, (R. J. Jeffries and A. J. Davy, eds.), Blackwell Scientific Publications, Oxford.
Jeffries, R. L., Rudmik, T., and Dillon, E. M. (1979b). *Plant Physiol.* **64**, 989.
Kushner, D. J., Bayley, S. T., Boring, J., Kates, M., and Gibbons, N. E. (1964). *Can. J. Microbiol.* **10**, 483.
Lanyi, J. K. (1974). *Bact. Rev.* **38**, 272.
Lanyi, J. K. and Stevenson, J. (1970). *J. Biol. Chem.* **245**, 4070.
Larher, F. (1976). Thèse Doct. Sci. Université de Rennes.
Larher, F. and Hamelin, J., (1975). *Phytochem.* **14**, 205.
Larher, F. and Hamelin, J. (1979). *Phytochem.* **18**, 1396.
Larher, F., Hamelin, J., and Stewart, G. R. (1977). *Phytochem.* **16**, 2019.
Larkum, A. W. D. and Hill, A. E. (1970). *Biochim. biophys. acta.* **203**, 133.
Larsen, H., (1967). *Adv. Microbiol. Physiol.* **1**, 97.
Lawton, J. R., Todd, A., and Naidoo, D. K. (1981). *New Phytol.* **88**, 713.
Maas, E. V. and Hoffman, G. J. (1977). *A.S.C.E. J. Irrig. Drain. Div.* **103**, 115.
Munns, R., Brady, C. J., and Barlow, E. W. R. (1979). *Aust. J. Plant Physiol.* **6**, 379.
Oertli, J. J. (1966). *Soil. Sci.* **102**, 258.
Oertli, J. J. (1968). *Agrochimica* **12**, 461.
Osmond, C. B. and Greenway, H. (1972). *Plant Physiol.* **49**, 260.
Pollard, A. and Wyn Jones, R. G. (1979). *Planta* **144**, 291.
Schobert, B. (1977). *J. Theor. Biol.* **68**, 17.
Schobert, B. and Tzchesche, H., (1978). *Biochem. biophy. acta* **541**, 270.
Somero, G. N., Neubauer, M., and Low, P. S. (1977). *Arch. Biochem. Biophys.* **181**, 438.
Soo-Hoo, T. S. and Brown, A. D. (1967). *Biochim. biophys. acta* **135**, 164–166.
Stewart, G. R. and Lee, J. A. (1974), *Planta* **120**, 279.
Stewart, G. R. and Larher, F. (1980). *In* "Biochemistry of Plants" Vol. 5, pp. 609–635, (B. J. Miflin, ed.), Academic Press, New York and London.
Stewart, G. R., Larher, F., Ahmad, I., and Lee, J. A. (1979). *In* "Ecological Processes in Coastal Environments", pp. 211–227, (R. L. Jeffries and A. J. Davy, eds.), Blackwell Scientific Publications, Oxford.
Storey, R. and Wyn Jones, R. G. (1975). *Plant Sci. Lett.* **4**, 161.
Storey, R. and Wyn Jones, R. G. (1977). *Phytochem.* **16**, 447.
Storey, R. and Wyn Jones, R. G. (1979). *Plant Physiol.* **63**, 156.
Storey, R., Ahmad, N., and Wyn Jones, R. G. (1977). *Oecolgia* **27**, 319.
Treichel, S. (1975). *Z. Planzen physiol.* **76**, 56.
Wallace, A. and Kleinkopf, G. E. (1974). *Plant Sci. Lett.* **3**, 251.
Van Steveninck, R. F. M., Armstrong, W. D., Peters, P. D., and Hall, T. A. (1976). *Aust. J. Plant. Physiol.* **3**, 367.
Von Hippel, P. H., Peticolas, V., Schack, L., and Karlson, L., (1973). *Biochem.* **12**, 1256.
Wyn Jones, R. G., Storey, R., Leigh, R. A., Ahmad, N., and Pollard, A. (1977). *In* "Regulation of Cell Membrane Activities in Plants", pp. 121–136, (E. Marré and O. Ciferri, eds), North Holland, Amsterdam.

CHAPTER 4

Adaptation to Toxic Metals

P. J. PETERSON

*Department of Botany and Biochemistry, Westfield College, University of London, England**

I. Introduction

The evolutionary adaptations of plants to metalliferous soils has long been recognized and the resulting concept of geobotanical indicators has proved to be a useful approach in mineral exploration in certain regions. More recent ecological studies of these sites have revealed the presence of specific communities or floras which have developed on soils containing elevated levels of, for example, selenium, zinc, copper, nickel, and chromium. Some species are restricted to particular metalliferous soils and are thus endemics, whilst others growing on these soils may occur in other phytogeographically distinct areas uncontaminated by metals. Specific floras have also developed over soils deficient in one or more elements, such as the Chimanimani Mountain Floras of Zimbabwe, Shale Barrens etc., but this review will be restricted to plant responses to metalliferous soils only.

Toxic metal sites, at least those in the tropics, have probably been

* Present address: Department of Biological Sciences, Chelsea College, University of London, London SW 10.

producing specialized soil conditions since the Precambrian and long before the evolutionary development of angiosperms in the Cretaceous (Wild and Bradshaw, 1977). In these soils, therefore, the origin of at least some of the endemics must have closely followed angiosperm development. It is surprising to find that although the floras are impoverished, relatively few of the species are endemic. This must indicate that many of the plants growing on such soils today are relatively recent in origin, and that many of the endemics presumably have not survived past climatic changes.

In temperate regions however, the pleistocene ice age will have eliminated endemic floras and present-day plants will be of recent origin. Nevertheless, isolation processes and population selection have produced significant physiological and some morphological changes in plants on derelict mine and smelter waste in a very small number of generations. The reason why many of the grasses have adapted to such conditions, and most legumes have not, has not been adequately explained.

There is much evidence in the literature which suggests that plants growing on metalliferous soils contain the element, or elements, and hence it can be said that they cannot prevent metal uptake, they can only restrict it. Plants which contain elevated levels of an element have been termed "accumulators", or in certain cases "hyperaccumulators" if they exceed $1000 \mu g$ metal g^{-1} dry weight of plant (Brooks, Lee, Reeves and Jaffré, 1977); this latter term will not however be used in this review for the concept divides accumulators on an arbitrary concentration basis.

The presence of plants on metalliferous soils necessarily implies that they are tolerant of the toxic metal, especially if they also accumulate the metal to high cellular concentrations. Before going on to discuss possible mechanisms of adaptation to toxic metals, three basic concepts should be mentioned. Firstly, metal accumulation and hence tolerance is metal specific. This seems to be very well established although there is an early report that plants tolerant of copper were also tolerant of nickel, nickel was apparently absent from the soil (Gregory and Bradshaw, 1965). Recent reports confirm that a number of plants may indeed have multiple-metal tolerance and some a co-tolerance (Hutchinson and Kuja, 1979; Cox and Hutchinson, 1979). Secondly, with all endemic species, tolerance is an absolute phenomenon, whereas with metal tolerant ecotypes, races, or physiotypes, a continuous gradation between weak and strong degrees of tolerance can be measured. Thirdly, metal tolerance is an inherited characteristic, usually dominant, although a large number of genes may be involved.

For convenience, tolerant plants can be divided into three possible basic types: excluders, or plants with a restricted transport; index plants which reflect soil concentrations; and accumulator species.

II. Metal Exclusion and Restrictions in Transport

The concept of metal exclusion from plants is not well established although there are several examples where metals are restricted to varying degrees in the roots. Lead is normally considered to be accumulated in roots with little transported to the leaves (Peterson, 1978). Indeed, the data of Nicolls, Proven, Cole, and Tooms (1965) for the grass *Triodia pungens* reveals that the concentration of lead in the leaves was constant despite a range of soil lead concentrations of several orders of magnitude. Similarly, Barry, and Clark (1978) reported that there was little increase in foliar lead in *Agrostis tenuis*, *Festuca ovina*, and *Minuartia verna* over the range 500–20 000 μg lead g^{-1} in the soil around abandoned lead-mines.

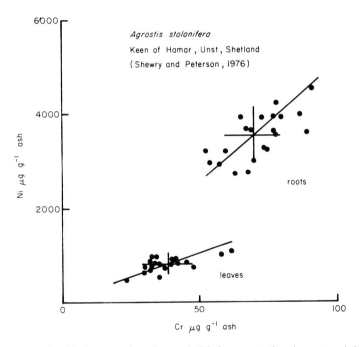

FIG. 1. The relationship between chromium and nickel concentrations in roots and shoots of plants of *Agrostis stolonifera* collected from a serpentine site. Regression lines are drawn and the arithmetic means are marked with crosses, the limits of which show the standard deviations.

Further information on species which respond with a restriction in transport can be gained from an examination of the data of Reilly and Reilly (1973) for copper in some Zambian plants (Table I). The data shows that the two grasses *Trachypogon spicatus* and *Stereochlaena cameronii* contain most of the copper in their roots, whereas the other two species *Becium homblei* and *Combretum psidioides* contain much of the copper in their leaves. Similarly, the roots of *Agrostis stolonifera* contained much of the nickel and chromium in the plants collected from a serpentine site (Shewry and Peterson, 1976) (Fig. 1) although other species accumulate these elements in their leaves (Peterson, 1979). Despite these and other observations, the physiological basis for differential transport has not been closely examined.

TABLE I
Copper leaf/root ratios for Zambian plants
(Reilly and Reilly, 1973)

Species	Ratio
Trachypogon spicatus (Gramineae)	0·003
Stereochlaena cameronii (Gramineae)	0·05
Becium homblei (Labiatae)	1·9
Combretum psidioides (Combretaceae)	2·4

III. Index Plants

Over areas of similar geochemical environments, differences in metal accumulation are apparent for many plant species and for most, if not all, elements, but the physiological differences which help explain these observations are largely lacking. Figure 2 shows species differences in their mean nickel and chromium concentrations when collected from the same soil. Similarly, plants collected from a range of similar geochemical environments also indicate that metal concentration is determined both by the plant species and the amount of metal in the soil. This concept is illustrated in Fig. 3 with a range of pasture species growing on soils derived from the Black Shales. Such plants are often called "indicator plants", but it seems more appropriate to refer to them as "index plants" for they provide an estimation of the soil metal concentration. The term "indicator plant" seems best used in the historical geobotanical context as indicating the presence of ores rather than providing a quantitative assessment of soil metal levels.

Irrespective of the nomenclature of this group of plants the foliar metal content of many species reflects soil concentration and provides the basis for biogeochemical exploration. Presumably in these cases the plants do not control their metal uptake and transport processes, they merely reflect soil concentrations. Yet species differences are considerable and differences in the behaviour of essential and non-essential elements within

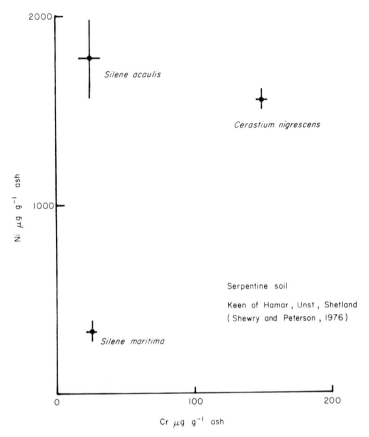

FIG. 2. The relationships between concentrations of chromium and nickel in the shoots of three species of plants collected from a serpentine site. The arithmetic means are marked with crosses, the limits of which show the standard deviations.

species have also been observed (Timperly *et al.*, 1970). Various plants can apparently accumulate metals to concentrations which would normally be considered toxic to other plants e.g. accumulation of nickel and chromium (Lyon *et al.*, 1968). Presumably mechanisms of adaptation must exist for such plants but have been largely neglected in studies on the development of tolerance.

IV. ACCUMULATOR SPECIES

Many lists have been compiled of the elemental concentrations in a wide variety of plant species collected from a range of geochemical and polluted

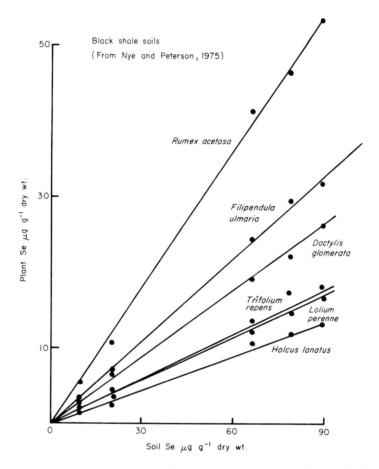

FIG. 3. The relationships between total soil selenium concentration and plant selenium for samples collected from the one geochemical environment.

landscapes and the striking observation is one of a wide diversity between species. Some species can be defined as accumulators yet others growing nearby lack this facility. An accumulator can be defined as "a species whose metal concentration exceeds that present in the soil or, alternatively, as "a species whose metal concentration exceeds the normal values for metal concentrations in plants on a particular soil". Undoubtedly, "index plants" growing on soils containing high concentrations of an element may not be described as "accumulators" under the first definition but some species could be if the second definition were used.

Irrespective of how accumulator species may be defined, their high metal concentration must place the plant under physiological, biochemical, and presumably genetical stress. Plants must, therefore, have evolved a tolerance of these conditions to allow them to survive and compete under what can only be described as adverse growth conditions. Unusual element accumulation as a taxonomic character has been discussed elsewhere (Peterson, 1982) and will not be considered further. This review is concerned with the wide variety of adaptive mechanisms which presumably enable the plant to survive its own accumulated element concentration or organo-metal concentrations should the element be metabolized by the plant. Possible mechanisms can be grouped into three major sections which are considered below.

A. METABOLIC ADAPTATIONS

Phytochemical, biochemical, and analytical studies have revealed the presence in accumulator plants of high concentrations of some inorganic ions, low molecular weight organo-metallic compounds, as well as high molecular weight metal-containing biopolymers, although the significance of these observations in terms of physiological and biochemical tolerance mechanisms has scarcely been examined.

1. Low Molecular Weight Compounds

(a) Seleno-amino acids. Selenium-rich soils with their characteristic floras have been recognized for many years from parts of the USA as being particularly toxic to livestock (Rosenfeld and Beath, 1964). Similar disorders in grazing animals have been reported from the Republic of Ireland, Australia, S. Africa, and elsewhere, while in S. America seeds of the accumulator *Lecythis ollaria* are toxic to man (Kerdel-Vegas *et al.*, 1965). Seleniferous floras in the Great Plains area of the USA contain many selenium accumulators including several dozen species of the legume genus *Astragalus* e.g. *A. pectinatus*, *A. bisulcatus* as well as members of the Cruciferae (*Stanleya pinnata*), Compositae (*Oonopsis condensata*) and many others where the selenium concentration may, especially in mature seeds, exceed 1%. In all cases seleno-amino acids comprise most of the selenium present in the plant, although volatile dimethylselenides have also been detected. Because the selenium-accumulators belong to a variety of unrelated genera it seems likely that selenium-accumulation arose through convergent evolution rather than evolving from a single selenium-accumulating ancestor.

Selenocystathionine and Se-methylselenocysteine are the two principal amino acids and their occurrences are outlined in Table II. These amino acids may also occur in seeds as the γ-glutamyl derivative with both isomeric γ-glutamyl derivatives of selenocystathionine being isolated from *A. pectinatus* (Nigam and McConnell, 1976). Non-accumulator and selenium-sensitive plants on the other hand, synthesize predominantly selenomethionine and to a lesser extent selenocysteine, both amino acids being incorporated into proteins via the usual aminoacyl-tRNA synthetase enzymes.

The nature of the biochemical differences responsible for tolerance towards selenium are twofold. Firstly, accumulators synthesize large amounts of the non-protein seleno-amino acids, whereas these are scarcely

TABLE II
Seleno-amino acids

Compound	Species	Reference	
Selenocystathionine	Neptunia amplexicaulis	Peterson and Robinson,	1972
	Morinda reticulata	Peterson and Butler,	1971
	Lecythis ollaria	Kerdel-Vegas *et al.*	1965
	Astragalus species	Shrift and Virupaksha,	1965
	Stanleya pinnata	Shrift and Virupaksha,	1965
Se-methylselenocysteine	Astragalus species	Shrift and Virupaksha,	1965
Selenomethionine	Many species	Peterson and Butler,	1962
Selenocysteine	Vigna radiata	Brown and Shrift,	1980

detectable in non-accumulators; and secondly, the proteins of accumulators contain very little selenium and less than selenium-sensitive plants (Peterson and Butler, 1967). Confirmation of this exclusion mechanism comes from more recent work on selenium analysis of proteins of three *Astragalus* accumulators and three non-accumulators (Brown and Shrift, 1981; Anderson, this vol. Thus in *N. amplexicaulis* and in some other *Astragalus* species selenium tolerance is associated with exclusion of selenium from proteins and, of course, enzymes. A more detailed investigation of the mechanism of protein synthesis has revealed that the cysteinyl-tRNA synthetase from the accumulator *A. bisulcatus* did not link selenocysteine to tRNA, but that the equivalent preparation from a selenium-sensitive *Astragalus* was able to form the amino acyl-tRNA (Burnell and Shrift, 1979). However, other exclusion mechanisms also exist, for the cysteinyl-tRNA synthetase from *N. amplexicaulis* was able to use selenocysteine as a substrate yet the amino acid was not found in protein (Burnell, 1981).

Discrimination during protein synthesis, therefore, appears to be likely for the exclusion of selenocysteine from proteins, but selenomethionine is presumably not synthesized in accumulators for its presence in them has not been reported.

(b) Nickel-organic acids. The nickel concentration in plants usually varies around a mean of $0.05\,\mu g$ nickel g^{-1} dry weight tissue but for plants growing on ultrabasic soils, the nickel concentration may exceed $1000\,\mu g$ g^{-1} (Table III). Many nickel accumulators have now been described from countries including Italy e.g. *Alyssum bertolonii* (Minguzzi and Vergnano, 1948), New Caledonia e.g. *Hybanthus austro-caledonicus* (Brooks, Lee

TABLE III
Nickel in Plants (per cent ash weight)

Species	Plant	Soil	Ratio	Reference	
Hybanthus floribundus (Violaceae)	13	0·08	162·5	Severne and Brooks,	1972
H. austro-caledonicus (Violaceae)	27	0·50	54·0	Brooks *et al.*,	1974
Pearsonia metallifera (Papilionaceae)	15·3	0·55	27·8	Wild,	1974
Alyssum serpyllifolium (Cruciferae)	10·3	0·40	25·8	Menezes de Sequeira,	1968
Dicoma niccolifera (Asteraceae)	2·8	0·70	4·0	Wild,	1970
Pimelia suteri (Thymelaeaceae)	0·59	0·33	1·8	Lyon *et al.*,	1971
Silene acaulis (Caryophyllaceae)	0·18	0·33	0·5	Shrewry and Peterson,	1976

and Jaffré, 1974), Australia e.g. *H. floribundus* (Severne and Brooks, 1972), Zimbabwe e.g. *Pearsonia metallifera* (Wild 1974), and more recently from Canada e.g. *Arenaria humifusa* (Roberts, 1980) and USA e.g. *Arenaria rubella* (Samiullah, Kruckeberg, and Peterson, unpublished data).

An assessment of nickel accumulation in relation to the evolutionary status based on the Advancement Index of Sporne (1969) indicates that nickel accumulation occurs especially in "primitive" families. Aluminium accumulation is also correlated with "primitive" characters (Chenery and Sporne, 1976) whereas phylogenetic considerations suggest that manganese accumulation is a later evolutionary development (Jaffré, 1979). These relationships are discussed in greater detail in Peterson (1982).

Pelosi *et al.* (1976) have indicated that nickel is complexed with malic and malonic acids in the Italian nickel accumulator *Alyssum bertolonii* and later more detailed studies by Lee *et al.* (1978) have shown that these organic acids are also important complexing agents in the Portuguese *A. serpyllifolium ssp. lusitanicum* and in the Zimbabwean *Pearsonia metallifera*.

In many other nickel accumulators such as *Hybanthus austro-caledonicus*
H. caledonicus, *Sebertia acuminata*, *Psychotria douarrei*, *Geissois pruni-*
nosa, *Homalium francii*, and other *Homalium* spp., the citrato-nickel (II)
complex (see Hewitt, this vol.), predominates (Lee *et al.*, 1977,78).
Indeed a strong correlation was reported between the concentration of
nickel and citric acid in the leaves of seventeen New Caledonian accumulators. High concentrations of the hydrated cation $Ni(H_2O)_6^{2+}$ were also
recorded in some species e.g. *S. acuminata*.

As an anionic complex of nickel with similar electrophoretic behaviour
to nickel-citrate has also been reported to occur in various non-
accumulator plants (Tiffin, 1971), perhaps this element is transported
largely as the citrate complex. Nevertheless, no clear evidence has been
presented to establish whether this complex is formed metabolically or
non-metabolically. In any event the occurrence of iron as the citrate
complex in xylem sap of various plants (Tiffin, 1971) provides further
indication of the important role of organic acids in the chelation of metals
within plants.

(c) Chromium-organic acids. Few chromium accumulators have been
discovered. In all cases the plant/soil ratio was less than one (Table IV)
which contrasts strongly with the nickel-accumulators where values greater
than one are relatively commonplace.

TABLE IV
Chromium in plants (μg g^{-1} ash weight)

Species	Plant	Soil	Ratio	Reference	
Sutera fodina (Scrophulariaceae)	48 000	125 000	0·38	Wild,	1974
Leptospermum scoparium (Rubiaceae)	2 470	8 950	0·28	Lyon *et al.*,	1971
Dicoma niccolifera (Asteraceae)	30 000	115 000	0·24	Wild,	1974
Cerastium nigrescens (Caryophyllaceae)	147	1 800	0·08	Shewry and Peterson,	1976
Silene maritima (Caryophyllaceae)	22	1 800	0·01	Shewry and Peterson,	1976

Three Zimbabwean species, *Sutera fodina*, *Pearsonia metallifera*, and
Dicoma niccolifera have been reported to contain high concentrations of
chromium in their ash — up to 5% (Wild, 1974). while *Leptospermum*
scoparium from a disused chromite mine in New Zealand contained up to
one per cent (Lyon *et al.*, 1971). The biochemical form of much of the
soluble chromium in leaf tissue of *L. scoparium* has been characterized as

the trioxalatochromate (III) ion, $[Cr(C_2O_4)_3]^{3-}$, (Lyon *et al.*, 1969a,b). Chromium was, however, transported in the xylem sap as chromate, indicating that metabolism to the complex took place within the leaf tissues. The form of chromium in the other accumulators has not been described. Presumably the function of the chromium-organic acid complex would be to reduce the cytoplasmic toxicity associated with Cr^{3+} and chromate ions.

(d) Gold cyanide. An examination of the gold concentrations in a range of plants has revealed that *Phacelia sericea* has the ability to accumulate considerably more gold than many other species (Girling, Peterson, and Warren, 1979). Chemical tests have shown that this species is a cyanogenic plant which can liberate free cyanide into the soil solution with the resultant dissolution of gold (Girling and Peterson, 1978). The gold-cyanide complex is readily absorbed by the plant and translocated to its leaves where gold concentrations can increase to values higher than are found when gold is supplied as an inorganic salt. In this latter case, gold accumulates in the root and shoot cell wall with little soluble gold remaining in solution. Gold in the form of the cyanide complex binds only slightly to soluble proteins and nucleic acids but gold supplied as the chloride rapidly binds to these cell constituents. Presumably gold-cyanide formation is a mechanism which enables the plant to avoid the more toxic cation and permits significant gold accumulation to take place.

(e) Copper-proline. Reilly (1972) has described the occurrence of copper-amino acid complexes in both the copper accumulating *Becium homblei* and in a non-tolerant ecotype, although he did not consider that the complexes were of importance in the development of copper tolerance. A specific copper-proline complex has been isolated from the roots of *Armeria maritima* (Farago and Mullen, 1979) but again the relevance of the complex to tolerance is uncertain, for high concentrations of copper were also stored in the cell walls associated with the carbohydrate fraction (Farago *et al.*, 1980). It is of interest that Farago and Mullen (1979) detected high concentrations of proline in the tissues of their tolerant plants. Since high concentrations of proline are produced by plants under arid conditions (Hegarty and Peterson, 1973) and plants on tailings undoubtedly grow under physiological water-stress, then the high proline values may not be a direct response to metal toxicity after all.

(f) Zinc malate. Zinc-tolerant *Agrostis tenuis* has been shown to accumulate higher levels of malate than non-tolerant clones, which has led to the view that zinc-malate may be involved in zinc transfer and perhaps tolerance (Ernst, Mathys, and Jamiesch, 1975; Mathys, 1977). An examination of organic acid concentrations by Brookes *et al.* (1981) working with zinc tolerant *Deschampsia caespitosa*, revealed that the malate levels were considerably elevated in the tolerant clone compared with the non-tolerant individuals. Whether or not this stimulation in malate production is directly related to zinc accumulation or tolerance is not known.

(g) Fluoroacetate. Fluorine-accumulator plants reported from various geographical regions of the world belong to diverse families (Peters, 1960; Oelrichs and McEwan, 1961). Plants like *Gastrolobium grandiflorum* convert the absorbed inorganic fluoride to fluoroacetate which can reach 120 μg fluorine g^{-1} dry weight (McEwan, 1964). Species which do not metabolize fluoride are apparently unable to accumulate it to high levels. The common tea plant *Camellia sinensis*, although containing elevated levels of fluorine compared with many agronomic plants, is not an effective accumulator and contains mainly the fluoride ion. It is of interest that *C. sinensis* has also been reported to be an aluminium accumulator (Sivasubramanium and Talibudeen, 1971). The significance of the production of fluoroacetate by accumulator plants remains obscure; this compound is toxic to animals because it is converted to fluorocitrate, a potent inhibitor of the tricarboxylic acid cycle.

2. High Molecular Weight Compounds

(a) Metallothionein. Biochemical studies some 25 years ago led to the isolation of cadmium and zinc containing mammalian metallo-proteins rich in cysteine residues which were subsequently termed metallothioneins, i.e. a metallo-derivative of the sulphur-rich protein thionein. Their isolation and possible biological functions have been closely examined throughout the animal kingdom but only within the past few years have studies on plants been successfully undertaken.

Premakumar *et al.* (1975) isolated from yeast and mung bean a copper-binding protein low in sulphur, "a copper chelatin", that was produced in response to Cu^{2+} exposure but Bremner and Young (1976) suggested that it was an artifact of isolation. Casterline and Barnett (1977) first reported the presence of a cadmium-thionein in tissue homogenates of soyabeans exposed to cadmium, while Bartolf, Brennan, and Price (1980) have isolated a similar metallothionein from the roots of cadmium-treated tomato plants. The only reports of metallo-proteins in accumulator plants are by Curvetto and Rauser (1979) and Rauser and Curvetto (1980), who isolated a copper-thionein from the roots of copper-tolerant *Agrostis gigantea*. Although evidence is accumulating that metallothionein-like proteins exist in plants, conclusive evidence that they are indeed metallothioneins has yet to be presented. Despite the occurrence of these proteins in plants, their importance as a tolerance mechanism or cytoplasmic metal scavenger remains to be ascertained.

B. CELL WALL AND MEMBRANE ADAPTATIONS

The reduced transport of copper from roots to shoots of some grasses has already been mentioned (Table I) but this restriction is not confined to just this element. A number of studies have shown that zinc-tolerant as well as other copper-tolerant grasses and *Silene maritima* also restrict the transport of these elements to leafy tissues (Antonovics *et al.*, 1971; Baker, 1974, 1978). Studies with *S. maritima* demonstrated that the edaphic adaptation was close, for the root/shoot concentration ratio was correlated with the total soil zinc (Fig. 4). Immobilization of zinc and copper in the roots poses the question of the cellular location of these metals.

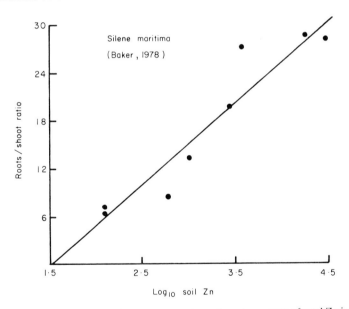

FIG. 4. Linear regression of root/shoot concentration ratio onto amount of total Zn in the soil, expressed logarithmically, for population samples of *Silene maritima* grown in solution cultures.

1. Root Cell Wall Binding

Differential solvent extraction of root tissue homogenates has revealed the washed cell wall fraction as the major site of zinc or copper localization in tolerant ecotypes of *A. tenuis* and *A. stolonifera* (Peterson, 1969; Turner, 1970; Turner and Marshall, 1971,72). Furthermore, the extracted residues from tolerant plant roots contained a higher percentage of zinc than those of the non-tolerant plants (Peterson, 1969). Indeed the greater the tolerance to zinc, the greater the amount of zinc in the cell wall (Fig. 5).

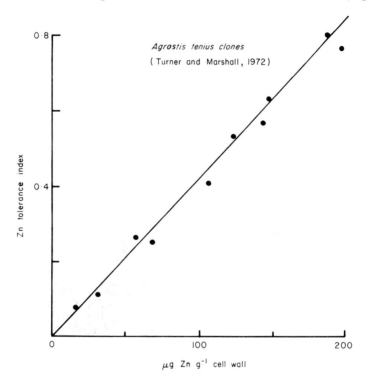

FIG. 5. The relationship between the zinc-tolerance index measured by rooting studies and the concentration of zinc in the cell wall fraction from *Agrostis tenuis* clones collected from a range of sites exhibiting different zinc concentrations in the soil.

The zinc distribution in the copper-tolerant plants was similar to that in the non-tolerant plants, indicating that the tolerance mechanism for the two elements are different. Chemical fractionation revealed that much of the zinc (greater than 50%) was released into solution when the wall-pectates were solubilized, but less than 20% was solubilized following digestion with the proteolytic enzyme pronase. Wyn Jones *et al.* (1971), although finding that trypsin released significant quantities of zinc from cell walls, reported that 60% of the bound zinc was released by cellulase digestion and that this fraction chromatographed on Sephadex G25 predominantly as a single band.

Comparable solvent extraction experiments by Farago *et al.* (1980) have also shown that much of the copper in the tolerant *Armeria maritima* is associated with the pectates and other carbohydrates in the roots. Their results are in agreement with the earlier conclusion that tolerance to copper is linked somehow with metal storage in the cell walls.

These observations lead to the hypothesis that in some copper- and zinc-tolerant plants, metal-binding in root cell walls is an important process in the development of a tolerance mechanism, but the exact nature of the mechanism has not been elucidated. Further studies are required on the molecular structures of the cell walls which appear to confer the observed specificity for metal binding. Moreover, accumulation by root cell walls is not a universal feature of metal tolerance and other mechanisms also operate in plants.

2. Cell Membrane Permeability

In view of the number of membrane-bound enzymes and the importance of membrane integrity for cellular processes, it is surprising that so little work has been directed towards the study of the relationships between accumulation and membrane protection from high metal concentrations. The most relevant study is a series of experiments performed by Wainwright and Woolhouse (1975) designed to measure the leakage of K^+ ions from roots of copper and zinc-tolerant *A. tenuis* plants. Increasing the concentration of copper caused increasing leakage of K^+ from non-tolerant and zinc-tolerant plants but much less leakage from the copper-tolerant root tips. These results were interpreted as meaning that zinc ions did not damage the plasmalemma whereas copper ions did, suggesting that plasmalemma modifications are involved in copper-tolerance but not in zinc-tolerance. A similar study by Benson and Peterson (unpublished) on copper and arsenic tolerant races of *A. tenuis* revealed that increasing arsenic concentrations did not affect the leakage of K^+ ions but increasing copper had the same effect as that described earlier by Wainwright and Woolhouse (1975). These results again indicate that plants have responded differentially towards copper, zinc, and other elements during the evolutionary development of their accumulation and tolerance mechanisms.

C. ENZYME ADAPTATIONS

Enzyme adaptations conferring resistance to an element might be expected to provide an excellent biochemical basis for the evolution of accumulator plants, or at least metal-resistant ecotypes. However, in *in vitro* experiments with zinc-, copper- and non-tolerant populations of *Silene cucu-*

balus, no resistant nitrate reductase, malate dehydrogenase, isocitrate dehydrogenase or glucose-6-phosphate dehydrogenase enzymes could be detected (Mathys, 1975). Likewise, there was no difference in sensitivity to zinc of malate and glutamate dehyrogenases extracted from the roots of zinc-tolerant clones of *Anthoxanthum odoratum*. Perhaps this is due to the fact that tolerant plants accumulate zinc in their vacuoles or form zinc-complexes as already mentioned above.

The studies of Wainwright and Woolhouse (1975) upon copper tolerant *Agrostis tenuis* revealed that grasses growing on contaminated soils possess cell wall acid phosphatases which form less stable metal-complexes than plants from uncontaminated sites. Cox and Thurman (1978) working with zinc-tolerant and non-tolerant clones of *Anthoxanthum odoratum* concluded that the tolerant clones possessed soluble and cell wall acid phosphatases which had significantly higher inhibition constants than those of the non-tolerant clones. The overall correlation of metal sensitivity of both types of acid phosphatases with the index of tolerance is suggested, but there was no correlation between the sensitivity of certain isozymes and the index of tolerance. This leads to the view that enzyme adaptations conferring tolerance on plants may not be a widespread phenomenon and perhaps will be found to apply only to root cell wall enzymes which come into contact with high metal concentrations at the soil/root interface.

V. DISCUSSION

It is quite clear from the foregoing sections that the mechanisms involved in metal tolerance are distinctive adaptations at cellular and sub-cellular levels although much detailed biochemical research is required before their precise natures are established. In addition to the formation of a range of organo-metal complexes already outlined, a number of other complexes have been suggested but sound experimental data is lacking. Calcium vanadate has been described from *Astragalus spp.* (Cannon, 1952) and aluminium succinate from *Orites excelsa* (Hutchinson, 1945) while the predominant form of uranium in *Coprosma australis* collected from mineralized soil was in the form of an RNA complex (Whitehead *et al.*, 1971).

Few studies have described the intra-cellular distributions of accumulated metals other than those typified by a differential centrifugation approach (Peterson, 1969, Turner, 1970; Turner and Marshall 1971, 1972). Brookes *et al.* (1981), however, carried out compartmental flux analysis using ^{65}Zn and found that only the zinc-tolerant clone of *Deschampsia caespitosa* was capable of actively pumping zinc from the cytoplasm of root cells across the tonoplast into the vacuoles when exposed to concentrations of up to 1 mN. This process was inhibited above 0·1 mM zinc in the non-tolerant clone. These results suggest that zinc-tolerance may be an active process in this plant rather than the more passive process of organo-metal complex formation and cell-wall binding reported for some species.

Studies of cytoplasmic sensitivity to metals which measure metal effects on specific enzymes have yielded few promising results despite reports of the apparent cytoplasmic resistance towards nickel in the epidermal cells of nickel-tolerant *Indigofera setiflora* (Ernst, 1972). Cytoplasmic resistance was also noted towards copper and zinc in copper and zinc-tolerant *I. setiflora*, but there was no tolerance towards nickel in these plants. Populations from normal soil had no cytoplasmic resistance towards these metals.

More recent studies on cytoplasmic resistant epidermal cells have confirmed that the nickel accumulator *Alyssum bertolonii* has greater resistance than in the nickel-sensitive *A. argenteum* (Vergnano-Gambi, pers. comm.) The implications of these results to enzymatic sensitivity remain to be ascertained.

Much of this review has been concerned with the concept of the evolutionary development of a specific physiological or biochemical mechanism which would enable the plant to survive the toxicity associated with accumulated metals. But there is some evidence for the existence of "botanical" mechanisms. For example, frequent bush fires may remove the foliage and hence copper in plants like *Becium homblei*. Plants such as zinc-tolerant *Thlaspi cepeaefolium* have been reported to concentrate zinc in their deciduous aerial parts (Rascio, 1977). Leaf fall in other deciduous plants would also reduce metal concentrations to a lower level.

In conclusion, it can be stated that the physiological, biochemical, and "botanical" strategies adopted by plants to avoid, restrict or alleviate potential metal toxicity are many and varied; and, moreover, much additional research will be required to explain present observations and recognize any unifying adaptive principles.

REFERENCES

Antonovics, J., Bradshaw, A. D., and Turner, R. G. (1971). *In*, "Advances in Ecological Research", pp. 1–85, (J. B. Cragg, ed), Academic Press, New York and London.
Baker, A. J. M. (1974). Heavy metal tolerance and population differentiation in *Silene maritima* With. PhD thesis, University of London.
Baker, A. J. M. (1978). *New Phytol.* **80**, 635–642.
Barry, S. A. S. and Clark, S. C. (1978). *New Phytol.* **81**, 773–783.
Bartolf, M., Brennan, E., and Price, C. A. (1980). *Plant Physiol.* **66**, 438–441.

Bremner, I. and Young, B. S. (1976). *Biochem. J.* **157**, 517–520.
Brookes, A., Collins, J. C., and Thurman, D. A. (1981). *J. Plant Nutr.* **3**, 695–705.
Brooks, R. R., Lee, J., and Jaffré, T. (1974). *J. Ecol.* **62**, 493–499.
Brooks, R. R., Lee, J., Reeves, R. D., and Jaffré, T. (1977). *J. Geochem. Explor.* **7**, 49–57.
Brown, T. A. and Shrift, A. (1980). *Plant Physiol.* **66**, 758–761.
Brown, T. A. and Shrift, A. (1981). *Plant Physiol.* **67**, 1051–1053.
Burnell, J. N. (1981). *Plant Physiol.* **67**, 316–324.
Burnell, J. N. and Shrift, A. (1979). *Plant Physiol.* **63**, 1095–1097.
Casterline, J. L. and Barnett, N. M. (1977). *Plant Physiol.* **59**, S–124.
Chenery, E. M. and Sporne, K. R. (1976). *New Phytol.* **76**, 551–554.
Cox, R. M. and Hutchinson, T. C. (1979). *Nature (Lond.)* **279**, 231–233.
Cox, R. M. and Thurman, D. A. (1978). *New Phytol.* **80**, 17–22.
Curvetto, N. R. and Rauser, W. E. (1979). *Plant Physiol.* **63**, S–59.
Cannon, H. L. (1952). *Amer. J. Sci.* **250**, 735–737.
Ernst, W. (1972). *Kirkia.* **8**, 125–145.
Ernst, W., Mathys, W., and Janiesch, P. (1975). *J. Plant Nutr.* **3**, 695–705.
Farago, M. E. and Mullen, W. A. (1979). *Inorg. Chim. Acta Letters.* **32**, L93–94.
Farago, M. E., Mullen, W. A., Cole, M. M., and Smith, R. F. (1980). *Environ. Pollut.* **A21**, 225–244.
Girling, C. A. and Peterson, P. J. (1978). *Trace Subst. Environ. Health*, **12**, 105–118.
Girling, C. A., Peterson, P. J., and Warren, H. V. (1979). *Econ. Geol.* **74**, 902–907.
Gregory, R. P. G. and Bradshaw, A. D. (1965). *New Phytol.* **64**, 131–143.
Hegarty, M. P. and Peterson, P. J. (1973). *In* "Chemistry and Biochemisty of Herbage", pp. 1–62, (G. W. Butler and R. W. Bailey, eds.), Academic Press, New York and London.
Hutchinson, G. E. (1945). *Soil Sci.* **60**, 29–39.
Hutchinson, T. C. and Kuja, A. (1979). *In* "Heavy Metals in the Environment", pp. 191–197, (R. Perry, ed.), C.E.P. Consultants, Edinburgh.
Jaffré, T. (1979). *C.R. Acad. Sci. (Paris)*, **289D**, 425–428.
Kerdel-Vegas, F., Wagner F., Russell, P. B., Grant, N. H., Alburn, H. E., Clark, D. E., and Miller, J. A. (1965). *Nature (Lond.)* **205**, 1186–1187.
Lee, J., Reeves, R. D., Brooks, R. R., and Jaffré, T. (1977). *Phytochem.* **16**, 1503–1505.
Lee, J., Reeves, R. D., Brooks, R. R., and Jaffré, T. (1978). *Phytochem.* **17**, 1033–1035.
Lyon, G. L., Brooks, R. R., Peterson, P. J., and Butler, G. W. (1968). *Plant Soil.* **291**, 225–240.
Lyon, G. L., Peterson, P. J., and Brooks R. R. (1969a). *Planta (Berl.)* **88**, 282–287.
Lyon, G. L., Peterson, P. J., and Brooks, R. R. (1969b). *N.Z. J.Sci.* **12**, 541–545.
Lyon, G. L., Peterson, P. J., Brooks, R. R., and Butler, G. W. (1971). *J. Ecol.* **59**, 421–429.
Mathys, W. (1975). *Physiol. Plant.* **33**, 161–165.
Mathys, W. (1977). *Physiol. Plant.* **40**, 130–140.
McEwan, T. (1964). *Nature (Lond.)* **201**, 827–828.
Menezes de Sequeira, E. (1968). *Agronomia Lusit.* **30**, 115–154.
Minguzzi, C. and Vergnano, D. (1948). *Mem. Soc. Tosc. Sci. Nat.* **55**, 49–74.
Nicholls, O. W., Provan, D. M. J., Cole, M. M., and Tooms, J. S. (1965). *Trans. Inst. Min. Metall.* **74**, 695–799.

Nigam, S. N. and McConnell, W. B. (1976). *Biochim. Biophys. Acta.* **437**, 116–121.

Nye, S. M. and Peterson, P. J. (1975). *Trace Subst. Environ. Health.* **9**, 113–121.

Oelrichs, P. B. and McEwan, T. (1961). *Nature (Lond.)* **190**, 808–809.

Pelosi, P., Fiorentini, R., and Galoppini, C. (1976). *Agric. Biol. Chem.* **40**, 1641–1649.

Peters, R. (1960) *Biochem. J.* **76**, 32P.

Peterson, P. J. (1969). *J. Exptl. Bot.* **20**, 863–875.

Peterson, P. J. (1978). *In* "The Biogeochemistry of Lead in the Environment", pp. 355–384, (J. O. Nriagu, ed.), Elsevier, Amsterdam.

Peterson, P. J. (1979). *Phil. Trans. R. Soc. Lond.* **B228**, 167–177.

Peterson, P. J. (1982). *In* "Anatomy of the Dicotyledon: Leaves, Stems and Wood in Relation to Taxonomy, (C. R. Metcalfe and L. Chalk, eds.), 2nd ed., pp. 167–173, Clarendon Press, Oxford.

Peterson, P. J. and Butler, G. W. (1962). *Austr. J. Biol. Sci.* **15**, 126–146.

Peterson, P. J. and Butler, G. W. (1967). *Nature (Lond.)* **213**, 599–600.

Peterson, P. J. and Butler, G. W. (1971). *Austr. J. Biol. Sci.* **24**, 175–177.

Peterson, P. J. and Robinson, P. J. (1972). *Phytochem.* **11**, 1837–1839.

Premakumar, R., Winge, D. R., Wiley, R. D., and Rajagopalan, K. V. (1975). *Arch. Biochem. Biophys.* **170**, 278–288.

Rascio, N. (1977). *Oikos.* **29**, 250–253.

Rauser, W. E. and Curvetto, N. R. (1980). *Nature (Lond.)* **287**, 563–564.

Reilly, C. (1972). *Z. Pflanzenphysiol.* **66**, 294–296.

Reilly, A. and Reilly, C. (1973). *New Phytol.* **72**, 1041–1046.

Rosenfeld, I. and Beath, O. A. (1964). "Selenium". Academic Press, New York and London.

Roberts, B. A. (1980). *Bot. Soc. Amer. Abstr., Publ.* No. **158**, 21.

Severne, B. C. and Brooks, R. R. (1972). *Planta. (Berl.)* **103**, 91–94.

Shewry, P. R. and Peterson, P. J. (1976). *J. Ecol.* **64**, 195–212.

Shrift, A. and Virupaksha, T. K. (1965). *Biochim. Biophys. Acta.* **100**, 65–75.

Sivasubramaniam, S. and Talibudeen, O. (1971). *J. Sci. Food Agric.* **22**, 325–329.

Sporne, K. R. (1969). *New Phytol.* **68**, 555–566.

Tiffin, L. O. (1971). *Plant Physiol.* **48**, 273–277.

Timperley, M. H., Brooks, R. R., and Peterson, P. J. (1970). *J. Appl. Ecol.* **7**, 429–439.

Turner, R. G. (1970). *New Phytol.* **69**, 725–731.

Turner, R. G. and Marshall, C. (1971). *New Phytol.* **70**, 539–545.

Turner, R. G. and Marshall, C. (1972). *New Phytol.* **71**, 671–676.

Wainwright, S. J. and Woolhouse, H. W. (1975). *In* "The Ecology of Resource Degredation and Renewal", pp. 231–257, (M. J. Chadwick and G. T. Goodman, eds.), Blackwell, London.

Whitehead, N. E., Brooks, R. R., and Peterson, P. J. (1971). *Austr. J. Biol. Sci.* **24**, 67–73.

Wild, H. (1970). *Kirkia* **7**, Suppl. 1–62.

Wild, H. (1974). *Kirkia* **9**, 209–232.

Wild, H. and Bradshaw, A. D. (1977). *Evolution* **31**, 282–293.

Wyn Jones, R. G., Sutcliffe, M., and Marshall, C. (1971). *In* "Recent Advances in Plant Nutrition", (R. M. Samish, ed.), pp. 575–585, Gordon and Breach, New York.

CHAPTER 5

Potassium Transport and the Plasma Membrane-ATPase in Plants

R. T. LEONARD

*Department of Botany and Plant Sciences,
University of California, USA*

I. Introduction

Potassium is essential for the normal growth and development of virtually all living organisms. In higher plants, potassium is the most abundant cation and the fifth most abundant element behind carbon, hydrogen, oxygen, and nitrogen. However, unlike these elements, potassium does not occur as part of any stable organic compounds, but rather it is present as the free monovalent cation. K^+ is also the most mobile nutrient element and is found in phloem sap at high concentrations. This is why the necrosis in the leaf margin that is characteristic of potassium deficiency often develops first in older leaves and later in the youngest leaves of the growing shoot.

   K^+ has three major types of nutritional roles in plants (Clarkson and Hanson, 1980). Firstly, K^+ is an activator of many enzymes and hence is biochemically important in various metabolic processes including photosynthesis, respiration, and protein synthesis. Secondly, K^+ is the most important inorganic atom involved in the control of osmotic potential. So it is essential for turgor-driven cell growth and for other turgor-generated plant responses including stomatal and leaf movements. For further information on both of these aspects the reader is referred to an earlier article in this series (Wyn Jones *et al.*, 1979). Thirdly, K^+ is the most common cation involved in neutralization of charge. This has particular significance for the synthesis and translocation of organic anions in plants and for membrane transport processes such as hydrogen ion extrusion involved in the regulation of cytoplasmic pH. Under conditions of potassium deficiency plants respond by accumulating basic amines which probably substitute for K^+ in neutralizing charge (Murty *et al.*, 1971).

It is clear that the level of K^+ nutrition in the plant has far-ranging effects on cell and tissue physiology. The process of K^+ transport through cell membranes is a fundamental aspect of cellular activity in all plant tissues.

The exact mechanism of K^+ transport through cell membranes of higher plants is not known (Poole, 1978, for references). The evidence available suggests that K^+ transport through the plasma membrane is mediated by a transport protein which functions to conduct the hydrophilic K^+ ion through the hydrophobic lipid bilayer. Such transport proteins distinguish K^+ from chemically similar ions and utilize metabolic energy to accumulate K^+ in cells at concentrations which are much greater than those in the external medium (Epstein, 1973; Hodges, 1973). The energy required for the selective accumulation of K^+ is supplied by respiration in the form of ATP (e.g. Petraglia and Poole, 1980). ATP is utilized at the plasma membrane by an ATP phosphohydrolase (ATPase). There is debate over whether or not the plasma membrane-ATPase directly or indirectly energizes K^+ transport (Poole, 1978; Leonard, 1982). It is also not known if there is a similar ATPase on the tonoplast membrane of the vacuole which could be involved in K^+ transport into that cellular compartment (e.g. Briskin and Leonard, 1980; Walker and Leigh, 1981). In this paper, the role of ATPase in the coupling of energy in ATP to K^+ transport through the plasma membrane will be discussed. The question of the mechanism and energetics of K^+ transport through the tonoplast membrane will not be considered. Before discussing the mechanism of K^+ transport through the plasma membrane of higher plant cells, it may be helpful to review what is known about K^+ transport in some other eukaryotic cells.

II. K^+ Transport in Mammalian and Fungal Cells

The plasma membrane of most mammalian cells contains a pump which functions to replace intracellular sodium ions with extracellular potassium ions (see Post, 1979; Robinson and Flashner, 1979 for references). This pump produces a large concentration gradient of each ion across the plasma membrane and these gradients are utilized for a variety of cell functions including electrical signaling and transport of amino acids or sugars. The pump transports 3 Na^+ ions outward and 2 K^+ ions inward across the plasma membrane per molecule of ATP hydrolyzed. The extra positive charge transported outward is not completely compensated for by other ion movements and hence the pump is electrogenic. As such, it generates an outward electric current which contributes to the electrical potential difference, cytoplasm negative, across the plasma membrane. Cardioactive glycosides such as ouabain are specific inhibitors of this pump. The plasma membrane can be isolated from broken cells and shown to contain an ATPase activity which is synergistically stimulated by the combination of Na^+ and K^+, and the stimulation by these ions is specifically inhibited by ouabain. This ATPase has been the subject of intensive research during the past twenty years. The enzyme has been partially solubilized, purified to about 95% homogeneity, and reconstituted into phospholipid vesicles to restore ATP-dependent, coupled Na^+ and K^+ transport (Dixon and Hokin, 1980). This is viewed as unequivocal proof that the $(Na^+ + K^+)$-ATPase activity represents the action of the Na^+ and K^+ pump.

The $(Na^+ + K^+)$-ATPase is a large protein complex which spans the membrane and is intimately associated with membrane lipids. The functional complex has a molecular weight of about 280 000 daltons and consists of two identical catalytic subunits of about 100 000 daltons each and at least one (probably two) glycopeptides of about 40 000 daltons (Peterson and Hokin, 1981). The large subunits may form an aqueous channel through the lipo-protein complex from one side of the membrane to the other. The large subunit is phosphorylated by ATP in the presence of Mg^{2+} and Na^+, and dephosphorylated by K^+. It appears that the phosphorylation/dephosphorylation cycle is associated with conformational changes which lead to the transport event involving a site or "gate" in the aqueous channel between the two large polypeptides. That is, Na^+ can approach a site from the inside of the cell and phosphorylation of the protein leads to a change in orientation of the site such that Na^+ ion can leave the site towards the outside. K^+ on the outside can now approach a site and dephosphorylation would return the site to an alternative conformation allowing K^+ access to the inside. The reaction mechanism for the enzyme in relation to the transport event is the subject of ongoing research and it will be some time before the details are fully known.

In the fungus, *Neurospora*, ATP also provides the direct source of energy for K^+ transport but, in contrast to the situation in mammalian cells, the plasma membrane ATPase may not directly pump K^+ (Bowman et al., 1980; Scarborough, 1980). It appears that the ATPase in Neurospora is an electrogenic proton pump (Scarborough, 1980) which functions to establish and maintain both an electrical potential difference and a pH gradient across the plasma membrane. The energy conserved in this proton-motive force is thought to be utilized to drive a variety of secondary transport systems, including one for K^+. Hence, K^+ transport is energized by ATP, but may occur at a transport site which is separated from the ATPase. It should be stressed, however, that while there is compelling evidence that the Neurospora ATPase is an electrogenic proton pump, it is still possible that future research will show that the ATPase also pumps K^+ ion. There are indications that the plasma membrane ATPase in yeast tightly controls fluxes of both H^+ and K^+ (Dufour et al., 1980; Malpartida and Serrano, 1981).

The plasma membrane ATPases of Neurospora and yeast have been partially solubilized and substantially purified (Bowman et al., 1980; Amory et al., 1980; Malpartida and Serrano, 1981). The enzyme is composed of a major polypeptide of at least 100 000 daltons which appears to be phosphorylated by $Mg \cdot ATP$ (Dame and Scarborough, 1980; Amory et al., 1980). This is in marked contrast to primary proton-pumping ATPases of mitochondria, chloroplasts, and certain bacteria which have not been shown to form a covalent, phosphoryl-enzyme intermediate. The fungal ATPase is not inhibited by ouabain, but like all other ATPases which form a phosphorylated intermediate, it is inhibited by orthovanadate. As yet, it has not been possible to reconstitute a purified preparation of the Neurospora ATPase into tight lipid vesicles so that the ATP driven electrogenic H^+ pumping observed in plasma membrane vesicles (Scarborough, 1980) can be verified. A reconstituted preparation of the yeast plasma membrane ATPase appears to function as an electrogenic H^+, K^+-pump with variable stoichiometry (Malpartida and Serrano, 1981).

To summarize, K^+ transport through the plasma membrane of mammalian cells is directly mediated by the $(Na^+ + K^+)$-ATPase, a transmembrane lipo-protein complex which couples metabolic energy in ATP to energetically uphill ion movements by conformational changes associated with a phosphorylation-dephosphorylation cycle. In fungal cells, K^+ transport may be energized by an ATPase-generated, proton motive force, and therefore K^+ may not be directly pumped by the plasma membrane ATPase, although the possibility that the proton pumping ATPase is also a K^+ pump cannot be completely ruled out.

III. K$^+$ Transport in Higher Plant Cells

A. GENERAL PROPERTIES

The plasma membrane of all plant cells functions to establish and maintain a large gradient in K$^+$ concentration between the cytoplasm and the external medium (Leonard and Hodges, 1980). The accumulation of K$^+$ in cells is a selective, carrier-mediated process (Epstein, 1973) which, at least under certain conditions, can be shown to occur against the electrochemical potential gradient for K$^+$ (Poole, 1978; Mercier and Poole, 1980). The energy required for the selective accumulation of K$^+$ is provided by respiration or by photosynthesis in green tissues and the direct source of energy is ATP (Petraglia and Poole, 1980). The way in which the energy in ATP is utilized to drive K$^+$ transport has been and continues to be the subject of much research. The results available are consistent with the view that K$^+$ influx is closely associated with H$^+$ efflux and that K$^+$ uptake may be energetically coupled to an electrogenic H$^+$ efflux pump on the plasma membrane.

The extrusion of H$^+$ appears to be a fundamental characteristic of all plant cells. The synthesis of organic acids is the major source of the extruded H$^+$, and it appears that H$^+$ extrusion is an essential component of the mechanism for maintaining a stable cytoplasmic pH (Smith and Raven, 1979). The efflux of H$^+$ into the extracellular solution (at pH 9 or below) is against the electrochemical potential gradient for H$^+$ and therefore is an active, energy requiring process (Smith and Raven, 1979).

Net H$^+$ efflux is usually associated with the accumulation of K$^+$ or other cations (Chastain *et al.*, 1981; Rasi-Caldogno *et al.*, 1980; Cheeseman *et al.*, 1980, for references). This is necessary to maintain charge balance, otherwise H$^+$ efflux would be limited by a build-up of a membrane potential difference, cytoplasm negative, across the plasma membrane. If cation influx lags slightly behind H$^+$ efflux, then the H$^+$ efflux will carry a net positive charge out of the cell and contribute to the resting electrical potential difference across the plasma membrane. There is a widespread belief that the energy-dependent H$^+$ efflux essential to regulation of cytoplasmic pH is mediated by an electrogenic pump which carries a net positive charge out of the cell and thereby contributes to the membrane potential difference (Poole, 1978).

The foundation for the concept of an electrogenic pump in plant cells comes from the pioneering work of Noe Higinbotham and colleagues (Higinbotham, 1973), who demonstrated that the electrical potential difference measured from vacuole to outside of the cell is to a large extent dependent on energy from respiration. Higinbotham postulated that there must be an electrogenic pump which creates the energy-dependent portion

of the membrane potential difference. In recent years it has been possible
to show that most, if not all, of the electrogenic membrane potential and
hence the electrogenic pump is associated with the plasma membrane and
not with the tonoplast membrane of the vacuole (Fischer *et al.*, 1976;
Goldsmith and Cleland, 1978; Rona *et al.*, 1980). If, as discussed below,
the plasma membrane ATPase is an electrogenic H^+, K^+ exchange pump,
then the absence of an electrogenic potential difference across the tonoplast membrane implies that the vacuole may have a different type of
ATPase-dependent transport system.

ATP appears to be the direct source of energy for both H^+ efflux and the
electrogenic component of the membrane potential (e.g. Mercier and
Poole, 1980), although this point may not be fully resolved (Löppert,
1981). Because of this, and the work described above for *Neurospora*, it
has been suggested that the plasma membrane contains an ATPase which
functions as an electrogenic proton-extrusion pump (see Poole, 1978, for
references). The energy conserved in the proton-motive force created by
the pump can be used to drive a variety of secondary transport systems.
Furthermore, because of the close relationship between K^+ influx and
electrogenic H^+ efflux, K^+ may be transported in response to the
ATPase-generated proton-motive force but not by the ATPase itself.
However, some recent reports (Mercier and Poole, 1980; Löppert, 1981)
suggest that there is not a simple relationship between K^+ influx and
electrogenic H^+ efflux. These researchers have presented data which show
an increase in the ATP-dependent K^+ influx rate under conditions where
there is little effect of ATP on the membrane potential difference. These
results suggest that ATP-dependent K^+ influx can occur by a mechanism
which is not dependent on the ATP-generated, proton-motive force.
Alternatively, the ATPase which pumps H^+ also directly pumps K^+, but
the stoichiometry of the H^+/K^+ exchange can be varied depending on
cellular requirements (Mercier and Poole, 1980).

To summarize this section, it appears that K^+ transport, through the
plasma membrane of higher plant cells is an energy-dependent, carrier-
mediated process. The direct source of energy is ATP, but there is debate
over whether or not the plasma membrane contains a primary K^+ pump
which directly utilizes ATP or a secondary K^+ transport system which
utilizes energy conserved in an ATP-generated proton motive force.

B. THE PLASMA MEMBRANE-ATPASE

The first convincing evidence for the existence of a membrane-associated
ATPase activity in homogenates of plant cells which might function in K^+
transport was provided by Hodges and colleagues in 1970 (see Hodges,

1976, for review). They measured the total membrane associated, K^+ (or Rb^+)-stimulated ATPase activity at pH 7·2 in roots of corn, wheat, oats, and barley. These activities were compared to K^+ (or Rb^+) influx rates into excised roots of these same species. The two parameters showed a strong correlation suggesting that K^+-stimulated ATPase activity may be representative of the action of a K^+ influx pump. This result also implied that ATP was the direct source of energy for K^+ transport in root tissues.

The next important step was to show that the K^+-stimulated ATPase activity was associated with the major permeability barrier of the cell, the plasma membrane. Hodges attempted to purify the ATPase and determine the identity of the membrane system with which it was associated. This task was confounded by two major difficulties. Firstly, plant cells contain a variety of enzymes which can hydrolyze ATP, and therefore can be mistaken for a "transport ATPase". Some of these activities, such as those associated with mitochondria and chloroplasts, can be readily identified. But others, such as those associated with endomembranes, nuclei, and vacuoles, are more difficult to distinguish from a putative plasma membrane, transport ATPase. Secondly, there is a paucity of marker enzymes which can be used to distinguish plasma membrane vesicles from other smooth vesicles such as those generated from tonoplast, endoplasmic reticulum, Golgi apparatus, and broken organelles in homogenates of plant tissues. Hence, when a cell fraction, rich in ATPase, is obtained, it is difficult to identify unequivocally the major membranes in the fraction.

In 1972, Hodges and co-workers (Hodges et al., 1972) isolated a membrane fraction from oat roots which was rich in K^+-stimulated ATPase activity, contained a high quantity of smooth vesicles, and was relatively free from recognizable organelles, such as mitochondria. A high proportion (at least 75%) of the smooth vesicles gave a positive reaction with an electron microscope staining procedure which is relatively specific for the plasma membrane (see Leonard and Hodges, 1980, for discussion). It was concluded that K^+-stimulated ATPase activity was associated with the plasma membrane. This basic observation has since been confirmed by several workers (see Pierce and Hendrix, 1979; Leonard and Hodges, 1980; Travis and Berkowitz, 1980, for references).

There are two other lines of investigation that support the view that an ATPase activity is associated with the plasma membrane. One, cytochemical localization of ATPase activity with lead nitrate, although controversial (see Hall et al., 1980, for discussion), shows activity in several regions of the cell, with intense activity at the plasma membrane (e.g. Gilder and Cronshaw, 1974; Bentwood and Cronshaw, 1978). Two, when isolated plant protoplasts are surface-labeled with a radioactive molecule, ATPase activity co-purifies with the radioactively labeled plasma membrane (Galbraith and Northcote, 1977; Perlin and Spanswick, 1980).

At present, the assumption is made that there is only one major transport ATPase associated with the plasma membrane. There are indications that a calmodulin-stimulated ATPase activity representing the action of a Ca^{2+} efflux pump may also be associated with the plasma membrane of plant cells (Dieter and Marme, 1981). However, in this article the K^+-stimulated ATPase activity will be treated as representing one pumping activity, but the reader should be aware that the ATPase activities measured in membrane fractions may represent contributions from more than one pump.

The ATPase activity measured in fractions rich in plasma membrane shows a clear preference for ATP as the substrate, but there is often nucleoside diphosphatase activity present depending on the plant species used for the source of plasma membranes. For example, the oat plasma membrane fraction is low in IDPase activity (Leonard et al., 1973) while barley is extremely high (Nagahashi et al., 1978), presumably because of greater contamination with Golgi membrane vesicles in the fraction from barley. It should also be noted that ATP is a good substrate for phosphatase (e.g. Mettler and Leonard, 1979), and sometimes even a better substrate than p-nitrophenol phosphate which is commonly used for assay of acid phosphatase activity. Hence, a preference for ATP as the substrate does not necessarily distinguish an energy transducing ATPase from a phosphatase. However, lack of substrate specificity for ATP is a clear indication of non-specific phosphatase and questionable relevance to "transport ATPase". This point has been overlooked far too often in studies on transport ATPase in plants (e.g. Benson and Tipton, 1978; Cross et al., 1978; Tognoli et al., 1979).

The plasma membrane-associated ATPase requires the presence of a divalent cation in concentration equal to ATP, and Mg^{2+} is always among those most preferred (Leonard and Hodges, 1973; Leonard and Hotchkiss, 1976). The reason for the Mg^{2+} requirement is not known for sure, but it is likely that $Mg^{2+} \cdot ATP$ is the true substrate for the enzyme (Balke and Hodges, 1975). Ca^{2+} (0·1 mM) is an effective inhibitor of ATPase activity, presumably because it interferes with the Mg·ATP complex. Transport ATPases are characteristically inhibited by mM concentrations of Ca^{2+}. In contrast, non-specific phosphatase activity is often enhanced in the presence of Ca^{2+} (e.g. Nagahashi et al., 1978; Mettler and Leonard, 1979).

The activity of the ATPase in the presence of Mg·ATP is further increased by addition of KCl (Table I). This increment of activity is referred to as K^+-stimulated ATPase activity and it may represent activity induced by actual transport of K^+ through the ATPase complex. However, it should be noted that K^+-stimulated ATPase activity is often a relatively small proportion of the total ATPase activity in the plasma membrane fraction (e.g. compare data for oat roots and pea epicotyl in Table I). It is

TABLE I

Effect of monovalent cations on plasma membrane ATPase activity of various plants

Addition	ATPase Activity[a] (μmol Pi mg^{-1} protein h^{-1})			
		Corn		
	Oat Root[b]	Root[c]	Leaf[d]	Pea Epicotyl[e]
Mg^{2+} alone	30·1	13·4	9·2	58·1
Mg^{2+} plus				
KCl	60·0 (100)[f]	19·7 (100)	21·5 (100)	69·7 (100)
RbCl	56·1 (87)	19·0 (89)	19·4 (83)	68·4 (89)
NaCl	54·9 (83)	18·6 (83)	18·0 (72)	67·7 (83)
CsCl	53·4 (78)	18·3 (78)	15·0 (47)	64·8 (58)
LiCl	42·6 (42)	16·4 (47)	13·7 (37)	58·7 (5)

[a] The assay conditions varied somewhat, but generally contained 3 mM ATP, 30–35 mM Tris-MES buffer (pH 6·0–6·5), and, when added, 3 mM divalent cation and 50 mM monovalent salt. See references listed below for exact assay conditions.
[b] From Leonard and Hodges, 1973
[c] From Leonard and Hotchkiss, 1976
[d] From Perlin and Spanswick, 1981
[e] From Hendrix and Pierce, 1980
[f] Numbers in parentheses represent the increase in ATPase activity produced by addition of monovalent salt expressed relative to that produced by KCl.

possible that the enhancement of ATPase activity produced by addition of KCl is of a general nature and not related to K$^+$ transport. The facts that K$^+$ salts give greater stimulation of ATPase than salts of the other alkali cations (Table I), and that the magnitude of K$^+$ stimulation is independent of the nature of the anion (e.g. Leonard and Hodges, 1973; Perlin and Spanswick, 1981) support the view that the enzyme is specifically affected by cations. This would not be expected if the enzyme was simply responding to a change in ionic strength produced by addition of KCl. However, it is still quite possible that K$^+$ stimulation of ATPase activity is a specific enzyme activation but, not directly related to K$^+$ transport.

There are two observations which provide evidence in support of the view that K$^+$-stimulated ATPase activity is, in fact, related to K$^+$ transport. One, there is a correlation between the ability of the alkali cations to stimulate ATPase and the specificty of alkali cation transport into root cells (Sze and Hodges, 1977). Two, the effect of K$^+$ concentration on the kinetics of the K$^+$-stimulated ATPase activity is complex, but fundamentally similar to the kinetic data for K$^+$ transport (Leonard and Hodges, 1973; Leonard and Hotchkiss, 1976). However, neither correlation is a perfect one (see Sze and Hodges, 1977; DuPont et al., 1981), and it may turn out that the correlations observed are not related to a direct role of the ATPase in K$^+$ transport.

ATPase activity of the plasma membrane fraction shows simple Michaelis-Menten enzyme kinetics with respect to Mg·ATP concentration (Balke and Hodges, 1975; Leonard and Hotchkiss, 1976). The apparent K_m for Mg·ATP is about 1 mM. The major effect of K^+ in stimulating ATPase activity is to increase V_{max}, but the apparent K_m for Mg·ATP is lowered in the presence of K^+.

The pH optimum for the plasma membrane ATPase is about 6·5, and it is clearly distinct from the alkaline pH optimum characteristic of the mitochondrial F_1-type ATPase. The mitochondrial ATPase inhibitor, N, N'-dicyclohexylcarbodiimide, does inhibit plasma membrane-ATPase activity, but other mitochondrial ATPase inhibitors such as oligomycin and azide do not (Leonard and Hodges, 1973; Leonard and Hotchkiss, 1976; Gallagher and Leonard, unpublished results). Plasma membrane-ATPase activity is sensitive to diethylstilbestrol, octylguanidine, orthovanadate, and various sulfhydryl inhibitors (Balke and Hodges, 1979a; DuPont et al., 1981, for references). Ouabain, a specific inhibitor of the Na^+, K^+-pump in animal cells, is not an inhibitor of the plant ATPase. This is not surprising since there is no evidence for a ouabain-sensitive, coupled transport of Na^+ and K^+ in higher plants and no indication of a synergistic stimulation of the plant ATPase by Na^+ and K^+. Progress in studying the plant ATPase has been limited by the failure to discover an inhibitor, analogous to ouabain, which is specific enough to be useful for correlating reduction in K^+ transport with effects on K^+-stimulated ATPase activity. Of the inhibitors listed above, diethylstilbestrol has proved to be helpful (Balke and Hodges, 1979b), and orthovanadate appears to be specific enough to be useful for research on ATPase function.

It appears that the majority of studies on the plasma membrane ATPase from plants have been conducted using membrane vesicles which are not sealed. That is, ions and other substances including ATP, can freely penetrate into the interior of the vesicle during the ATPase assay. If one assumes that (1) the catalytic site for ATP hydrolysis is exclusively on the cytoplasmic surface of the plasma membrane, and (2) that enzyme activity requires ion movement through the ATPase complex, then the following conclusions can be drawn. One, plasma membrane vesicles as isolated by most reserachers to date are probably a mixture of right- and wrong-side-out orientations, but this will have little effect on the ATPase assay since ATP can freely penetrate the vesicle to reach the catalytic site. Two, charge or concentration gradients normally generated by ion pumping do not develop across the leaky vesicle membranes, and hence do not restrict ATPase activity. Recently, Sze (1980) reported that ATPase activity associated with sealed vesicles (possibly of plasma membrane origin) from tobacco callus microsomes is stimulated by the H^+, K^+-ionophore, nigericin, or by the K^+-ionophore, valinomycin in combination with a protonophore. This is consistent with the idea that ATPase activity of sealed vesicles is limited by ion gradients presumably generated by ATPase action. Addition of an ionophore collapses the gradient and allows the enzyme to turn over at a higher rate. The fact that the stimulation of ATPase activity required an ionophore that conducts both K^+ and H^+ suggests that the plant ATPase is a K^+, H^+-pump. Furthermore, Sze and Churchill (1981) have demonstrated that the sealed vesicle fraction exhibits ATP-dependent thiocyanate accumulation. Thiocyanate is a permanent anion which will passively distribute across the vesicle membrane in response to an electrical potential difference. This observation together with her earlier results (Sze, 1980) implies that the plant ATPase may function as an electrogenic H^+, K^+-pump.

It is important to solubilize and purify the ATPase so that its enzymatic structure can be determined. Furthermore, it should be possible to incorporate the purified enzyme into tight lipid vesicle so that its function can be studied. Relatively few attempts have been made to solubilize and purify the plant ATPase, and all have met with limited success (see DuPont and Leonard, 1980, for references and discussion). There are no reports in the literature on reconstitution of the enzyme into lipid vesicles. The K^+-stimulated ATPase of corn root plasma membranes has been solubilized with the detergent, octyl-β-D-glucopyranoside (octylglucoside), but the solubilized enzyme loses activity rapidly (DuPont and Leonard, 1980). The loss of activity is presumably related to extensive removal of lipids which destabilizes the enzyme complex. Enzyme activity can be maintained if the solubilized ATPase complex is induced to form lipid vesicles by treatment with ammonium sulfate. About a four-fold purification is also achieved by this procedure and the ATPase retains most of the enzymatic characteristics observed for the original plasma membrane preparation (DuPont et al., 1981). The only significant difference between the plasma membrane-associated enzyme and the partially purified preparation is that the effect of K^+ concentration on the kinetics of K^+-stimulated ATPase activity changes from complex to near Michaelis-Menten (DuPont et al., 1981). This result does not necessarily diminish the significance of the correlation, discussed above, between the kinetics of K^+-ATPase and K^+ transport. The lipid vesicle associated with the partially purified ATPase is likely to be quite different from the plasma membrane vesicle and this may result in altered kinetic properties of the enzyme.

While the attempt to purify the ATPase has not, as yet, been completely successful, there is an important conclusion that can be drawn from the work conducted so far. The K^+-stimulated ATPase of corn roots is a large, intrinsic membrane protein or protein complex which is intimately associated with membrane lipids and retains activity only when associated with lipids. This type of structure has been observed for all the cation-transporting ATPase studied in animal cells, and is fundamentally different from the large, extrinsic F_1-ATPases of mitochondria and chloroplasts.

Another fundamental property of ATPases which function directly in ion transport is that they contain at least one catalytic subunit of about 100 000 dalton in size which is phosphorylated during ATPase action (Table II). The phosphorylation induces a conformational change which is associated with the ion pumping action of the enzyme. Research just completed (Briskin and Leonard, 1982) shows that the plasma membrane-ATPase of corn roots is specifically phosphorylated by ATP, and the polypeptide which is phosphorylated has a molecular weight of about 100 000 daltons. Interestingly, K^+ and other monovalent cations reduce the level of the phosphorylated intermediate, and the sequence observed among the monovalent cations ($K^+ > Rb^+ > Na^+ > Li^+$) is identical to that observed for stimulation of ATPase activity.

In summary, the plasma membrane of plant cells contains a Mg^{2+} requiring, K^+-stimulated ATPase activity. There is debate over the transport significance of the K^+-stimulated component of enzyme activity. It is possible that the ATPase functions exclusively as an electrogenic proton pump, but it is equally likely that the same ATPase also directly pumps K^+. Progress in determining the function of the ATPase has been slowed by the lack of a specific inhibitor. Ongoing research utilizing "sealed" vesicles offers much promise for resolving many questions on the function of the ATPase, but the unequivocal elucidation of the role of the ATPase in energy coupling to K^+ transport may depend on successful reconstitution of the purified pump.

C. A PROPOSAL FOR THE STRUCTURE AND FUNCTION OF THE PLASMA MEMBRANE-ATPase

I would now like to speculate on the structure and function of the plasma membrane-ATPase and its role in K^+ transport in plants. I view the ATPase as a large, multi-subunit enzyme complex which is an intrinsic membrane protein (Fig. 1). The complex is likely to be composed of at least two large polypeptides with a molecular weight of about 100 000 daltons each and several smaller polypeptides. The large subunits form one or more aqueous channels through the ATPase complex from one surface to the other surface of the plasma membrane. The other subunits in the complex are probably involved in the regulation of pump activity. The complex is situated in the membrane with the catalytic site for ATP hydrolysis exclusively exposed to the cytoplasmic surface, and with ion binding sites ("gates") in the aqueous channels. Ions (H^+, K^+, other

TABLE II
Some characteristics of ATPases and transport proteins

Enzyme	Function	Est. MW of Complex	MW of Large Subunit	Phosphorylated?	References
$(Na^+ + K^+)$-ATPase	$Na^+ + K^+$-Pump Mammalian cells	280 000	100 000	Yes	Peterson and Hokin, 1981; Post, 1979; Robinson and Flashner, 1979
Ca^{2+}-ATPase	Ca^{2+}-Pump Sarcoplasmic Reticulum	200 000	100 000	Yes	Klip and Maclennan, 1978; Inesi et al., 1978
$(Ca^{2+} + Mg^{2+})$-ATPase	Ca^{++} Transport Red blood cells	—	150 000	Yes	Sarkadi, 1980
H^+, K^+-ATPase	H^+-Transport Gastric mucosa	—	100 000	Yes	Sachs et al., 1979; Wallmark et al., 1980
H^+-ATPase	H^+-Pump Yeast, fungi	—	100 000	Yes	Bowman et al., 1980; Dame and Scarborough, 1980; Malpartida and Serrano, 1981
K^+-ATPase	K^+/H^+ Pump Plant cells	—	100 000	Yes	Briskin and Leonard, 1982
Anion Transporter	Anion-transport Mammalian cells	>200 000	100 000	N/A	Wilson, 1978
Mitochondrial-F_1-ATPase Chloroplast-CF_1-ATPase Bacterial-F_1-ATPase	ATP Synthesis and H^+-Pump	350 000	60 000	No	McCarty, 1978; Foster and Fillingame, 1979

N/A = not applicable

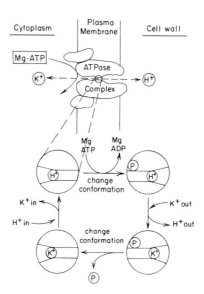

Fig. 1. A proposal for the structure and function of the plasma membrane-ATPase of plant cells.

monovalent cations) may approach the gates from either side of the membrane.

The transport of K+ may occur as follows (Fig. 1): H+ binding to sites on the large catalytic subunits induces Mg·ATP-dependent phosphorylation of the subunits. Phosphorylation causes a conformational change which allows H+, originally in the cytoplasm, access to the external solution. K+ binding to a site (possibly a similar site for H+ binding, but not necessarily the same site) causes dephosphorylation, and a return to the original conformation. Hence, K+ can leave the aqueous channel towards the cytoplasmic surface of the complex.

As described here, the ATPase is a pump which utilizes ATP to exchange H+ on the inside of the cell with K+ on the outside. The stoichiometry of the H+–K+ exchange may be electroneutral or electrogenic, depending on cellular requirements.

In short, the plasma membrane ATPase of plant cells is likely to be an electrogenic H+, K+-exchange pump analogous in structures and function to the (Na+–K+)–ATPase of mammalian cells, except that in plants, H+ rather than Na+ is, in most cases, exchanged for K+.

REFERENCES

Amory, A., Foury, F., and Goffeau, A. (1980). *J. Biol. Chem.* **255**, 9353–9357.
Balke, N. E. and Hodges, T. K. (1975). *Plant Physiol.* **55**, 83–86.

Balke, N. E. and Hodges, T. K. (1979a). *Plant Physiol.* **63**, 48–52.
Balke, N. E. and Hodges, T. K. (1979b). *Plant Physiol.* **63**, 53–56.
Benson, M. J. and Tipton, C. L. (1978). *Plant Physiol.* **62**, 165–172.
Bentwood, B. J. and Cronshaw, J. (1978). *Planta* **140**, 111–120.
Bowman, B. J., Blasco, F., Allen, K. E., and Slayman, C. W. (1980). *In* "Plant Membrane Transport: Current Conceptual Issues", pp. 195–206, (R. M. Spanswick, W. J. Lucas and J. Dainty, eds), Elsevier/North-Holland Biomedical Press, New York.
Briskin, D. P. and Leonard, R. T. (1980). *Plant Physiol.* **66**, 684–687.
Briskin, D. P. and Leonard, R. T. (1982). *Proc. Natl. Acad. Sci. USA* **79** (In Press).
Chastain, C. J., LaFayette, P. R. and Hanson, J. B. (1981). *Plant Physiol.* **67**, 832–835.
Cheeseman, J. M., LaFayette, P. R., Gronewald, J. W., and Hanson, J. B. (1980). *Plant Physiol.* **65**, 1139–1145.
Clarkson, D. T. and Hanson, J. B. (1980). *Annu. Rev. Plant Physiol.* **31**, 239–298.
Cross, J. W., Briggs, W. R., Dohrmann, U. C., and Ray, P. M. (1978). *Plant Physiol.* **61**, 581–584.
Dame, J. B. and Scarborough, G. A. (1980). *Biochemistry* **19**, 2931–2927.
Dieter, P. and Marmé, D. (1981). *FEBS Letters* **125**, 245–248.
Dixon, J. F. and Hokin, L. E. (1980). *J. Biol. Chem.* **225**, 10681–10686.
Dufour, J-P., Boutry, M., and Goffeau, A. (1980). *J. Biol. Chem.* **255**, 5735–5741.
DuPont, F. M. and Leonard, R. T. (1980). *Plant Physiol.* **65**, 931–938.
DuPont, F. M., Burke, L. L., and Spanswick, R. M. (1981). *Plant Physiol.* **67**, 59–63.
Epstein, E. (1973). *Int. Rev. Cytol.* **34**, 123–168.
Fischer, E., Lüttge, U., and Higinbotham, N. (1976). *Plant Physiol.* **58**, 240–241.
Foster, D. L. and Fillingame, R. H. (1979). *J. Biol. Chem.* **254**, 8230–8236.
Galbraith, D. W. and Northcote, D. H. (1977). *J. Cell Sci.* **24**, 295–310.
Gilder, J. and Cronshaw, J. (1974). *J. Cell Biol.* **60**, 221–235.
Goldsmith, M-H. M. and Cleland, R. E. (1978). *Planta* **143**, 261–265.
Hall, J. L., Browning, A. J., and Harvey, D. M. R. (1980). *Protoplasma* **104**, 193–200.
Hendrix, D. L. and Pierce, W. S. (1980). *Plant Sci. Letters* **18**, 365–373.
Higinbotham, N. (1973). *Annu. Rev. Plant Physiol.* **24**, 25–46.
Hodges, T. K. (1973). *Adv. Agron.* **251**, 163–207.
Hodges, T. K. (1976). *In* "Encyclopedia of Plant Physiology", New Series, Vol 2, Transport in Plants, Part A, Cells, pp. 260–283, (U. Lüttge and M. G. Pitman, eds.), Springer-Verlag, New York.
Hodges, T. K., Leonard, R. T., Bracker, C. E., and Keenan, T. W. (1972). *Proc. Nat. Acad. Sci. USA* **69**, 3307–3311.
Inesi, G., Coan, C., Verjovsic-Almeido, S., Kurznack, M. and Lewis, D. E. (1978). *In* "Frontiers of Biological Energetics: Electrons to Tissues", Vol. II, pp. 1129–1136, (P. L. Dutton, J. S. Leigh and A. Scarpa, eds.), Academic Press, New York and London.
Klip, A. and Maclennan, D. H. (1978). *In* "Frontiers of Biological Energetics: Electrons to Tissues", Vol. II, pp. 1137–1147, (P. L. Dutton, J. S. Leigh and A. Scarpa, eds.), Academic Press, New York and London.
Leonard, R. T. (1982). *In* "Membranes and Transport", Vol 2, pp. 633–637, (A. N. Martonosi, ed.), Plenum Press, New York.

Leonard, R. T. and Hotchkiss, C. W. (1976). *Plant Physiol.* **58**, 331–335.
Leonard, R. T. and Hodges, T. K. (1973). *Plant Physiol.* **52**, 6–12.
Leonard, R. T. and Hodges, T. K. (1980). *In* "Biochemistry of Plants, A Comprehensive Treatise", Vol. 1, pp. 163–182, (P. K. Stumpf and E. E. Conn, eds.), Academic Press, New York and London.
Leonard, R. T., Hansen, D., and Hodges, T. K. (1973). *Plant Physiol.* **51**, 749–754.
Löppert, H. (1981). *Planta* **151**, 293–297.
MacRobbie, E. A. C. (1977). *In* "International Review of Plant Biochemistry II", Vol. 13, pp. 211–247, (D. H. Northcote, ed.), University Park Press, Baltimore.
Malpartida, F. and Serrano, R. (1981). *J. Biol. Chem.* **256**, 4175–4177.
McCarty, R. E. (1978). *Current Topics in Bioenergetics* **7**, 245–278.
Mercier, A. J. and Poole, R. J. (1980). *J. Membrane Biol.* **55**, 165–174.
Mettler, I. J. and Leonard, R. T. (1979). *Plant Physiol.* **64**, 1114–1120.
Murty, K. S., Smith, T. A., and Bould, C. (1971). *Ann. Bot.* **36**, 687–695.
Nagahashi, G., Leonard, R. T., and Thomson, W. W. (1978). *Plant Physiol.* **61**, 993–999.
Perlin, D. S. and Spanswick, R. M. (1980). *Plant Physiol.* **65**, 1053–1057.
Perlin, D. S. and Spanswick, R. M. (1981). *Plant Physiol.* **68**, 521–526.
Peterson, G. L. and Hokin, L. E. (1981). *J. Biol. Chem.* **256**, 3751–3761.
Petraglia, T. and Poole, R. J. (1980). *Plant Physiol.* **65**, 969–972.
Pierce, W. S. and Hendrix, D. L. (1979). *Planta* **146**, 161–169.
Poole, R. J. (1978). *Annu. Rev. Plant Physiol.* **29**, 437–460.
Post, R. L. (1979). *In* "Cation Flux Across Biomembranes", pp. 3–19, (Y. Mukohata and L. Packer, eds.), Academic Press, New York and London.
Rasi-Caldogno, F., Cerana, R., and Pugliarello, M. C. (1980). *Plant Physiol.* **66**, 1095–1098.
Robinson, J. D. and Flashner, M. S. (1979). *Biochim. Biophys. Acta* **549**, 146–176.
Rona, J-P., Pitman, M. G., Lüttge, U., and Ball, E. (1980). *J. Membrane Biol.* **57**, 25–35.
Sachs, G., Rabon, E., and Saccomanni, G. (1979). *In* "Cation Flux Across Biomembranes", pp. 53–66, (Y. Mukohata and L. Packer, eds.), Academic Press, New York and London.
Sarkadi, B. (1980). *Biochim, Biophys, Acta* **604**, 159–190.
Scarborough, G. A. (1980). *Biochemistry* **19**, 2925–2931.
Smith, E. A. and Raven, J. A. (1979). *Annu. Rev. Plant Physiol.* **30**, 289–311.
Sze, H. (1980). *Proc. Natl. Acad. Sci.* **77**, 5904–5908.
Sze, H. and Churchill, K. A. (1981). *Proc. Natl. Acad. Sci.* **78**, 5578–5582.
Sze, H. and Hodges, T. K. (1977). *Plant Physiol.* **59**, 641–646.
Tognoli, L., Biffagna, N., Pesci, P., and Marre, E. (1979). *Plant Sci. Letters* **16**, 1–14.
Travis, R. L. and Berkowitz, R. L. (1980). *Plant Physiol.* **65**, 871–879.
Walker, R. R. and Leigh, R. A. (1981). *Planta* **153**, 140–149.
Wallmark, B., Steward, H. B., Rabon, E., Saccomani, G., and Sachs, G. (1980). *J. Biol. Chem.* **255**, 5313–5319.
Wilson, D. B. (1978). *Annu. Rev. Biochem.* **47**, 933–965.
Wyn Jones, R. G., Brady, C. J., and Speirs, J. (1979). *In* "Recent Advances in the Biochemistry of Cereals", pp. 63–103, (D. L. Laidman and R. G. Wyn Jones, eds.), Academic Press, London and New York.

CHAPTER 6

Boron and Membrane Function in Plants

A. J. PARR[*] AND B. C. LOUGHMAN

Department of Agricultural Science, University of Oxford, England

I. INTRODUCTION

The special interest for biochemists in the role of boron is that it is the only element known to be required for higher plants for which there is no role in animals. Attention has been centred on those aspects of plant function that have no corresponding role in animals and those systems thought to be under the control of specific plant hormones. The fact that fungi do not require boron perhaps draws attention away from the cell wall, and the absence of requirement in most algae suggests that there is no specific role in photosynthesis. Some lower plants require it and work by Lewin (1966) has also shown this to be true for some diatoms. Monocots, dicots, and gymnosperms have all been shown to require boron but within a particular group wide differences occur. This requirement is normally shown by the demonstration of specific symptoms a few days after withdrawal of boron

* Present address. Botany School, Downing Street, Cambridge, England.

from the growth medium, such as inhibition of root growth, breakdown of root and shoot apical meristems and malfunction of reproductive systems. The longer term effects are characterized by breakdown of cell walls and the presence of brown or black slime, as in the deficiency disease of swedes known as "brown-heart". Most diseases of this type on low boron soils can be prevented or alleviated by the application of boric acid at the rate of a few kilos per hectare. A recent article (Lewis 1980) has reviewed the theories concerning the role of boron and this paper will concentrate on more recent experimental evidence.

At normal cell pH the weak acid H_3BO_3 and the monovalent borate ion $B(OH)_4^-$ are the major constituent forms (Mengel and Kirkby, 1978) and presumably are responsible for any interaction with physiological and biochemical systems. The well-known ability of boric acid/borate mixtures to form complexes (I–III) with sugars or other compounds with *cis* hydroxyls such as diphenols has focused attention on a specific role for boron in carbohydrate metabolism but little definite information has emerged. Neales (1960) has pointed out that a supply of boron is required throughout the growth and development of the plant and B may therefore have a structural role in which binding with diols is involved. Phenylboronic acid (IV) also complexes with diols and can substitute for boric acid in the diatom *Cylindrotheca fusformis*, whereas derivatives such as tetraphenyl borate (V) do not substitute (Neales, 1967).

The structure of the only biologically synthesised metabolite known to contain boron, the antibiotic boromycin produced by *Streptomyces antibioticus*, shows the boron is bound to four oxy groups as in III (Dunitz *et al.*, 1971).

II. Possible Roles of Boron

The problem of pinpointing the primary role of boron in plants is particularly difficult for a number of reasons not the least of which is the minute amount involved (1–$100 \, \mu g \, g^{-1}$ dry wt.). The wide range of biochemical symptoms associated with deficiency arise over extended periods and represents secondary effects rather than the primary action of the element. Comparison of normal and deficient tissues is difficult because of differences in physiological age. An experimental approach designed to assess the primary role of boron during the onset of deficiency appears to us to be most useful in order to eradicate to a major extent these secondary effects. The major experimental approaches are summarized in Table I.

The wide range of types of measurement indicated in Table I confirms that, although the requirement for boron is established, the mechanisms of its action are far from agreed. The one aspect that could underlie all the types of responses shown is that of impaired membrane function—thus affecting transport of all types of metabolites required for normal growth and development. It is worth mentioning at this point that impaired transport capacity can be demonstrated within a few hours of withholding boron from young maize seedlings at a time when it is impossible to distinguish any diminution of growth or change in fine structure with the electron microscope. Consequently, it has been our aim to work with tissues of this kind in an attempt to isolate the primary role of boron in plants.

Perhaps the most widely held view on the role of boron is concerned in some way with the build up of phenolic compounds and the way in which this theory has developed from studies of changes in enzyme levels is worthy of attention. Dugger and Humphries (1960) measured the levels of some glycolytic enzymes in a number of species and showed that relatively high concentrations of borate are inhibitory. Phosphoglucomutase is the most sensitive of the glycolytic enzymes in germinating peas and a role for boron as a possible *in vivo* regulator of this enzyme was proposed by Loughman (1961) because 50% inhibition of the enzyme could be achieved *in vitro* by $0\cdot2$ mM boric acid. If the enzyme activity is regulated by the endogenous level of boron, then under conditions of boron deficiency diversion of glucose from sucrose formation to polysaccharide synthesis could occur. The specific nature of the boron effect is shown in Fig. 1 where the effect of boron on phosphohexoisomerase, the next enzyme in the glycolytic chain, is seen to be negligible when compared with the effect on phosphoglucomutase in a partially purified enzyme preparation from pea seeds containing both enzymes but no other glycolytic enzymes.

The possible role of boron as a regulator of carbohydrate metabolism

TABLE I

Experimental Systems Used in Studies on the Role of Boron

Biochemical System	Authors	Approach and Species
1. Sugar Transport	Gauch and Dugger (1953)	^{14}C sucrose movement in intact plants; tomato, snap bean.
	Turnowska-Starck (1960)	^{14}C sucrose movement in excised leaves; French bean.
2. Cell Wall Synthesis	Wilson (1961)	Analysis of wall material; tobacco.
	Slack and Whittington (1964)	^{14}C glucose incorporation into wall material; field bean.
	Dugger and Palmer (1980)	Glucan formation from UDPG; cotton (ovule).
3. Lignification	McIlrath and Skok (1964)	Lignin determination; sunflower, tobacco.
	Parish (1969)	Peroxidase levels and distribution; field bean, wheat.
4. Cell wall structure	Kouchi and Kumazawa (1976)	Electron microscopy; tomato.
5. Carbohydrate metabolism	Dugger and Humphreys (1960)	Effect of borate on sucrose synthesis *in vitro*.
	Loughman (1961)	Effect of borate on phosphoglucomutase.
	Lee and Aronoff (1967)	Effect of borate on 6 phospho-gluconate dehydrogenase.
6. RNA metabolism	Shkol'nik and Kositsyn (1962)	Incorporation of labelled precursors.
	Chapman and Jackson (1974)	Incorporation of labelled precursors; mung bean.

	Birnbaum, Dugger, and Beasley (1977)	Production of B deficiency-like symptoms by base analogues; cotton.
7. Respiration	Maevskaya and Alekseeva (1966)	Measurement of ATP levels; sunflower.
	Robertson and Loughman (1974a)	^{32}P incorporation into organic phosphates; field bean.
8. IAA metabolism	Coke and Whittington (1968)	Effects on IAA degradation, bioassay of IAA levels; field bean.
	Robertson and Loughman (1974b)	Anatomical and biochemical comparison of effects of IAA and B deficiency; field bean.
	Rajaratnam and Lowry (1974)	Measurement of IAA levels; oil palm.
	Shkol'nik, Krupnikova, and Dmitrieva (1964)	Measurement of IAA oxidase activity; sunflower.
	Bohnsack and Albert (1977)	Measurement of IAA oxidase activity; squash.
9. Phenol metabolism	Lee and Arnoff (1967)	Effect of borate on an enzyme of the pentose phosphate pathway.
	Shkol'nik (1974)	Determination of phenol levels, comparison of effects of phenols and B deficiency.
10. Membranes	Tanada (1974)	Effects on bioelectric fields; mung bean.
	Pollard, Parr, and Loughman (1977)	Ion transport and membrane enzyme studies; maize.
	Hirsch and Torrey (1980)	Ultrastructural studies; sunflower.

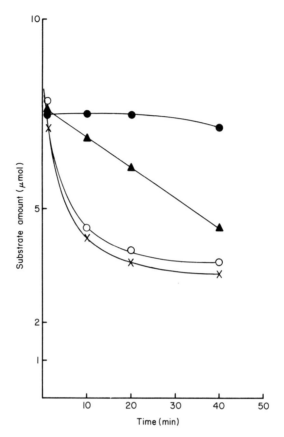

Fig. 1. The effect of sodium borate on phosphohexoisomerase and phosphoglucomutase activity of a purified enzyme preparation from germinating pea seads. 0·20 ml enzyme; 8 mM glucose-1-phosphate or fructose-6-phosphate; 5 mM-MgCl₂; 0·1 M glyoxaline-HCl buffer pH 7·2; additions to a final volume of 1·30 ml 25°C.

Phosphoglucomutase. ●——● 2 mM-Na₂B₄O₇ ▲——▲ No addition
Phosphophexoisomerase. ○——○ 2 mM-Na₂B₄O₇ ×——× No addition

leading to diversion of glucose *via* different pathways was explored by Lee and Aronoff in 1967. They showed that borate complexes with 6-phosphogluconate, inhibiting 6-phosphogluconate dehydrogenase and thereby decreasing the flow of glucose through the pentose phosphate pathway. It was proposed that deficiency of boron released the restraint on this enzyme with the result that increased activity of the pathway leads *via* shikimic acid to the production of phenolic compounds. The inhibition by 0·6 mM borate, however, amounted to about 30% with the enzyme *in vitro* and it is likely that the endogenous level of boron is significantly lower. Clearly, as other control points exist in the pathway from glucose to phenolic compounds it is unlikely that inhibitory effects of this magnitude would lead to a large change in the flux through the pathway under normal cellular conditions. While Shkol'nik (1974) has shown the build-up of a range of phenolic compounds under conditions of severe deficiency, the molecular basis of this accumulation remains uncertain.

III. Effects on Membrane Function

An argument used against the phenolic theory is contained in a paper by Shkol'nik *et al.* (1972) where in the early stages of deficiency in maize leaves there is little change or even a decrease in phenol content. It has been shown that at this stage of deficiency, the efficiency of ion transport is impaired and our attention has recently been focused on this aspect of the problem (Robertson and Loughman, 1974a, Pollard, Parr and Loughman, 1977).

A. ION TRANSPORT

Electrochemical studies also support the view that boron plays a role in membrane transport, e.g. the hyperpolarization of the plasmalemma in the presence of K^+ and the depolarization in the presence of Na^+ brought about by borate in *Nitella syncarpa* suggests that boron varies the selective permeability of Na^+/K^+ relative to Cl^- without altering the cation discrimination (Vorob'ev and Pelkhanov, 1973). Tanada (1974) showed that addition of $0.5\ \mu M\ H_3BO_3$ to boron deficient excised mung bean hypocotyls for 10 minutes altered the bio-electric field generated by red light. There is an obvious need for proper functioning of membranes in order to maintain optimal pH, ion content, osmolarity and a controlled intracellular environment. If, in fact, boron contributes to the integrity of membrane function then this could well be its primary role, from which symptoms of deficiency develop during B deprivation as a result of alteration in the normal flux of metabolites.

Experimental observations relating to the earlier section illustrate our approach. Coke and Whittington (1968), Timashov (1969), and our group in the early seventies all reported impaired membrane transport during prolonged B deficiency. The evidence presented here is the result of experiments designed to investigate membrane function during the early stages of deficiency before morphological changes can be seen and before the later changes in ultrastructure occur. In our view a reduced capacity for ion transport is the earliest symptom of deficiency demonstrable in maize, amounting to 25% within 15 h (Fig. 2). Smyth and Dugger (1979) have confirmed this in a diatom, reporting that Rb^+ absorption is inhibited by

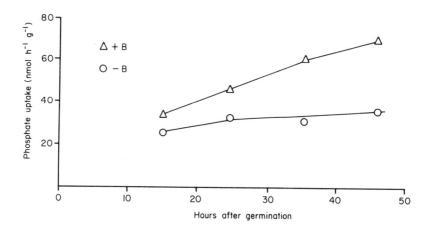

Fig. 2. The development of impaired phosphate transport during early B deficiency in maize. Seeds were germinated then transferred at time = 0 to B deficient or B sufficient growth solutions. Phosphate uptake capacity was then assayed after varying periods. Absorption from 10 μM KH$_2$PO$_4$ at pH 6·0 for 30 min.

20% after removal of boron from the culture solution for as little as 5 h although anion uptake remained unaffected. The work of Robertson and Loughman (1974a) showed that the reduced capacity for phosphate absorption observed in the early stages of B deficiency in field beans could be restored almost completely by supplying 10 μM H$_3$BO$_3$ for 1 h. This work was followed up by Pollard *et al.* (1977) and data for a range of ions in three species are shown in Table II. An intriguing aspect of these

TABLE II
The effect of boron deficiency on absorption of ions by roots of three species

Species	Ion	Absorption by B-deficient Root Tips (% of B sufficient rate)
Maize 4 days deficient	Cl$^-$ H$_2$PO$_4^-$	75 ± 3 51 ± 5
Sunflower 40 h deficient	Cl$^-$ H$_2$PO$_4^-$	47 69
Maize 4 days deficient	Rb$^+$	64
Field bean 40 h deficient	H$_2$PO$_4^-$	46

Seeds were germinated for 2 or 3 days in moist filter paper, given 1 day in a full nutrient solution and treated with 10^{-3} M Ca (NO$_3$)$_2$ either with or with 10^{-4} M H$_3$BO$_3$. Absorption was measured from 10^{-5} M KH$_2$PO$_4$ or 10^{-3} M KCl at pH 6·0 at 25°C.

experiments is that plants raised in boron sufficient conditions still respond to the presence of boron during the period of ion absorption implying that the so-called boron sufficient plants operate at only two-thirds of their maximal potential unless boron is present during absorption. The absorption of charged molecules such as auxins and uncharged sugars is also stimulated by boron (Coke and Whittington 1968; Turnowska-Starck, 1960).

B. INTERACTIONS WITH AUXINS

Because auxins and phenols are known to affect membrane transport it is possible that the effects of B deficiency on ion absorption are mediated by the build-up of these compounds as are the long-term effects previously described. Whereas auxins can increase ion absorption by stem tissues, absorption by roots is usually inhibited by the presence of auxins or phenols. Bentrup *et al.* (1973) showed that IAA rapidly inhibits Cl^- absorption by cell cultures of *Petroselinium sativa* and phosphate absorption by field bean roots is significantly inhibited by treatment for only 1 h with 1 μM IAA or 2,4 dichlorophenoxyacetic acid. There is, however, little evidence that auxins and phenols build up to inhibitory concentrations during the short period necessary for the demonstration of impaired ion transport in deficient roots.

TABLE III

The effect of IAA on phosphate uptake by B deficient (4 day) and B sufficient maize roots

Status	Auxin Treatment	Phosphate Uptake (nmol $h^{-1}g^{-1}$)
+B	0	82
+B	10^{-4} M, 2 h	59
−B	0	62
−B	10^{-4} M, 2 h	61

Intact plants were treated with IAA and the root tips were excised for the determination of phosphate uptake capacity. Experimental details as in Table IV.

The effect of IAA on the capacity of excised root tips and roots of intact maize seedlings to absorb phosphate is shown in Table III and Table IV. Table III shows that root tips from B deficient roots are unaffected by 100 μM IAA. The effects of B deficiency and IAA treatment are similar and results of this kind are consistent with the view that B deficiency leads to the build up of IAA. If this occurs it would be expected that B deficient roots would not be inhibited by added IAA because they would already

TABLE IV

The effect of IAA on phosphate uptake by B sufficient maize roots

Excised root tips or intact roots	IAA concentration	Treatment time	Phosphate uptake (nmol hr^{-1} g^{-1})
Excised	0	—	74
Excised	5×10^{-6} M	3 h	40
Excised	5×10^{-5} M	3 h	34
Excised	0	—	39
Excised	10^{-4} M	2 h	31
Excised	10^{-4} M	15 h	20
Intact	0	—	92
Intact	10^{-5} M	3 h	65
Intact	10^{-4} M	3 h	73

Plants were grown in the presence of the required concentration of IAA and then either excised 2 cm root tips or intact plants used for the determination of phosphate uptake capacity. Absorption was from 10^{-5} M KH$_2^{32}$PO$_4$ + 10^{-3} M Ca(NO$_3$)$_2$ containing the appropriate concentration of IAA, and the absorption period was 20 min for excised root tips and 1 h for intact plants. In order to make the results of the two approaches comparable, results for intact plants were based on the uptake by the apical 2 cm section of the root, which was excised after the uptake period. Preliminary experiments revealed that IAA did not seriously alter any longitudinal transport of phosphate which occurred during the short uptake period.

TABLE V

The effect of B deficiency on the magnitude and sensitivity to exogenous auxin of the stimulation of phosphate uptake capacity which follows the incubation of excised maize root tips in fresh medium

Washing period	Auxin concn during washing	Phosphate uptake (nmol h^{-1} g^{-1})	
		+B	−B
Expt. 1.			
0		32	14
1·5 h	0	52	28
1·5 h	10^{-4} M IAA	28	12
Expt. 2.			
0		89	48
1·5 h	0	129	103
1·5 h	10^{-4} M IAA	73	62

Root tips were excised from B deficient and B sufficient plants, and either used immediately for the determination of phosphate uptake capacity or first incubated for 1·5 h in 10^{-3} M Ca(NO$_3$)$_2$ containing 0 or 10^{-4} M IAA. Absorption of phosphate was from 10^{-5} M KH$_2$PO$_4$ + 10^{-3} M Ca(NO$_3$)$_2$, with 10^{-4} M IAA also present for root tips which had received IAA during the "washing" phase.

contain increased endogenous levels. However, the increased capacity for absorption of phosphate that follows excision of maize root tips and incubation in fresh medium (the "washing" effect) is depressed by IAA (Leonard and Hanson, 1972) whereas Table V clearly shows that B deficiency has little effect on this phenomenon. The sensitivity of the "washing" increment to added IAA appears to be unrelated to the boron status of the roots.

C. INTERACTIONS WITH PHENOLS

Glass (1973) produced firm evidence that the presence of a wide variety of phenolic compounds including salicylic acid, p-coumaric acid, and sinapic acid reduced the absorption of phosphate by barley roots. That this is true for normal maize root tips is shown in Table VI but the inhibitory effect is

TABLE VI

The effect of salicylic acid on the uptake of phosphate by B deficient (4 day) and B sufficient maize roots

Concentration of phenol	Phosphate uptake (nmol h^{-1} g^{-1})	
	+B	−B
0	42 ± 1	22 ± 1 (mean $\pm$ S.E.)
5×10^{-6} M Salicylic acid	33 ± 2	22.5 ± 2.5
10^{-4} M Salicylic acid	28 ± 1	19 ± 0.5
10^{-4} M p-Coumaric acid	20 ± 0.5	
10^{-4} M Orcinol	34 ± 9	

Absorption was from 10^{-5} M KH_2PO_4 in 10^{-3} M $Ca(NO_3)_2$ containing the appropriate phenol as the K salt at pH 5·6.

much smaller in B deficient tissue. The effects of high levels of phenols on phosphate absorption are similar to those of boron deficiency. There is, however, no evidence that phenol levels alter during B deficiency to an extent capable of initiating the changes in membrane function. After 4 days deficiency the endogenous levels of phenolic compounds increase by only 5–15% (Table VII). No increased release of phenolic compounds occurs during boron deficiency and no browning of the tissue occurred. These results suggest that no change in compartmentation of phenols occurs during the early stages of boron deficiency.

It was reported by Pollard et al. (1977) that as boron deficiency developed, the capacity for Rb^+ absorption and the activity of the K^+

TABLE VII

The effect of B deficiency on the levels of phenolic compounds in maize roots

B status	Phenol content (μmol Gallic acid equivalents g^{-1})	
	Experiment 1	Experiment 2
+B	12·2 ± 0·3	12·9 ± 0·2 (mean ± S.E.)
−B	14·2 ± 0·1	13·7 ± 0·2

Excised B deficient (4 day) and B sufficient root tips were homogenised in 80% ethanol, centrifuged, and the supernatant then assayed for total phenols. Gallic acid was used as a standard.

stimulated ATPase decreased in parallel, and that both could be fully restored by the addition of boron 1 h before extraction of the enzyme or the measurement of uptake of Rb$^+$. Analysis of the membrane-bound enzyme preparations used in these experiments showed very low levels of phenolic compounds in both preparations. During the ATPase assay phenol levels would be diluted to approximately 0·4 μM and it is very unlikely that the reduced activity of the ATPase in B deficient plants is due to the presence of phenols. Further evidence against the phenol hypothesis is provided by experiments in which efflux of ions was studied. Although B deficiency reduces the absorption of Cl$^-$ by maize roots by 25% there is no significant difference in the rates of efflux suggesting that outward movement across the tonoplast and plasmalemma is unaffected (Fig. 3). Similar results were obtained with phosphate.

The methods of Cram and Laties (1971) and Cram (1973) for isolating cytoplasmic and vacuolar fluxes were used to examine influx of Cl$^-$ and H$_2$PO$_4^-$, and the plasmalemma transport systems for both ions and the tonoplast transport of Cl$^-$ were found to be reduced by B deficiency (Table VIII). It appears that B is required to maintain inward transport across

TABLE VIII

The effect of B deficiency (4 days) on the rate of influx of Cl$^-$ and H$_2$PO$_4^-$ ions into cytoplasm and vacuole of maize root cells

Ion	Compartment	Influx rate (nmol h^{-1} g^{-1})	
		+B	−B
Cl$^-$	Cytoplasm	2125 ± 65	1860 ± 37 (n=4)
Cl$^-$	Vacuole	814 ± 11	621 ± 5 (n=2)
H$_2$PO$_4^-$	Cytoplasm	61 ± 4	21 ± 1 (n=2)

Root tips were excised and influx into the cytoplasm measured using a 10 min absorption period followed by a 10 min wash. Influx into the vacuole was measured using a 60 min absorption period and a 120 min wash.

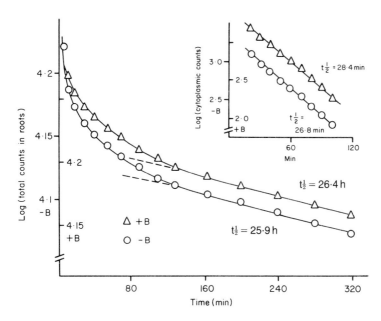

FIG. 3. The effect of B status on the efflux of chloride ions from maize roots. B sufficient and B deficient plants were grown in the presence of 10^{-3} M KCl for 2 days. Root tips were excised and loaded with ^{36}Cl by incubating for 40 min in 10^{-3} M $K^{36}Cl$ + 10^{-3} M $Ca(NO_3)_2$. The kinetics of chloride efflux were then determined by measuring the loss of ^{36}Cl to successive 5 ml portions of unlabelled 10^{-3} M KCl + 10^{-3} M $Ca(NO_3)_2$. pH 5·9, 25°C. ^{36}Cl content of +B tips = 119 000 cpm·g^{-1} ^{36}Cl content of −B tips = 92 000 cpm·g^{-1}.

both plasmalemma and tonoplast but has no effect on the processes of efflux. Glass and Dunlop (1974) have shown that phenols affect ion movement primarily *via* an enhancement of efflux brought about by a large nonspecific increase in membrane permeability and the addition of 0·1 mM salicylic acid to maize root tips increases efflux by about 20% (Fig. 4). As pointed out earlier no increased efflux is caused by B deficiency and it is unlikely therefore that the change in membrane function results from build up of phenols. A similar conclusion has been recently reached by Smyth and Dugger (1981) working with *Cylindrotheca fusiformis*.

The long term effects of B deficiency bring about large rises in auxins and phenolic compounds (Dear and Aronoff, 1965; Lewin and Chen, 1976) and this has led a number of authors to speculate that the central role of boron is concerned with control of the levels of these two groups of compounds (Lewis, 1980). In our view the primary effect of boron on membrane function is more direct than that of bringing about increased levels of auxins and phenols.

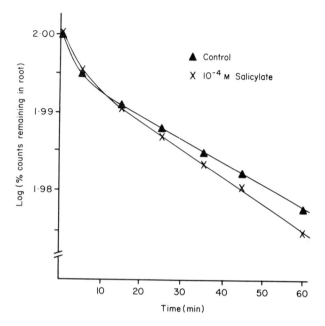

FIG. 4. The effect of salicylate on the efflux of phosphate from maize roots. Root tips were excised from B sufficient plants and loaded with $H_2^{32}PO_4^-$ for 40 min. The efflux of ^{32}P from the root tip was then followed either in the absence or presence of 10^{-4} M salicylic acid in the efflux medium. (pH adjusted to pH 5·0 in both cases.) Experimental details as in Fig. 3.

IV. Direct Interaction of Boron with Membranes

A. MEMBRANE CONFORMATION

Boron could interact directly with the membrane via glycoprotein or glycolipid components to maintain the most efficient conformation. A number of treatments can rapidly and drastically modify membrane conformation, e.g. low temperature or lack of calcium (Clarkson, 1974; Lyons and Raison, 1970). Although the interpretation of Arrhenius plots is difficult it is often assumed that discontinuities reflect a change in membrane conformation from "fluid" to "gel" as the temperature falls to a critical level (Raison, 1973). Such plots for membrane properties such as ion transport are biphasic with a break at the transition temperature corresponding to a change of phase in some component of the membrane lipid. The effect of temperature on phosphate absorption is shown in Fig. 5 and the effect of B deficiency is seen to be small although the transition temperature may rise by 1–2°C. More detailed examination of this type of relationship is clearly necessary before definite conclusions can be drawn

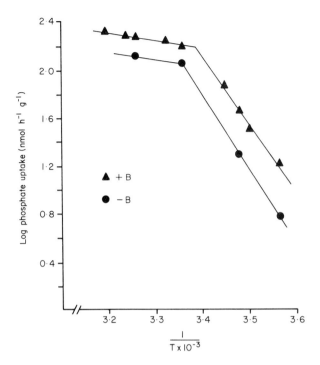

Fig. 5. The effect of temperature on the rate of phosphate uptake by B deficient (4 day) and B sufficient maize roots. Absorption of phosphate was for 60 minutes from 10^{-5} M $KH_2^{32}PO_4$ + 10^{-3} M $Ca(NO_3)_2$ at the appropriate temperature. A 10 minute wash in unlabelled 10^{-4} M KH_2PO_4 + 10^{-3} M $Ca(NO_3)_2$ at 20°C was then employed to remove $H_2^{32}PO_4^-$ in the root free space.

but the results suggest that deficiency had a greater inhibitory effect on phosphate absorption below the transition temperature.

B. ARTIFICIAL MEMBRANES

Because of the difficulties of correlating boron levels with efficiency of ion transport in intact roots and root tips, experiments were carried out with simplified membrane systems. The method of Bangham, Hill, and Miller (1974) was used to prepare artificial membrane bilayers or liposomes. Dilution of the medium in which liposomes are suspended results in uptake of water and swelling, which can be followed by measuring the light scattering of the suspension. Lipids were prepared from B deficient maize roots and incorporated into phosphatidyl choline liposomes. When compared with similar liposomes allowed to swell in the presence of $50\,\mu M$ borate the rate of swelling of the latter was significantly affected (Fig. 6).

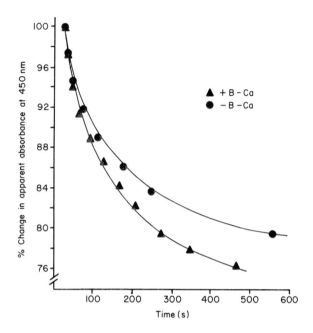

FIG. 6. The effect of boron on the osmotic swelling of liposomes containing 50% phosphatidyl choline and 50% Ca^{2+}-free maize root lipid. Liposomes were prepared in 10^{-2} M KCl + 0 or 5×10^{-5} M borate, then 200 μl suspension was rapidly injected into 3 ml B-free water (for the −B control liposomes) or 5×10^{-5} M borate (for the +B liposomes) at 20°C. The change in optical density of the diluted suspension as the liposomes swelled was followed continuously and expressed relative to a 25 s reading of 100.

Preliminary studies of this kind indicate that direct effects of boron on membrane function can be demonstrated.

The fact that the activity of the K^+ stimulated ATPase isolated from B deficient maize roots is considerably lower than that of the control plants again suggests a direct interaction between boron and enzyme activity. These results are consistent with the view that boron is directly associated with membranes and is concerned in maintaining their functional efficiency.

C. BORON CONTENT OF MEMBRANES

The levels of boron in membranes have not been examined in detail but differential centrifugation of maize root homogenates enables a crude estimate of the distribution of boron in cell fractions of B sufficient roots (Table IX). Whether the boron in the membrane represents a portion of that bound in the intact cell or that which has associated with the

TABLE IX

The boron content of subcellular fractions obtained by differential centrifugation of a B sufficient maize root homogenate

Fraction	B content (μg boron g^{-1})	% of boron in root
Wall fraction	71	36
Membrane fraction A	21	1·4
Membrane fraction B	24	1·6
"Soluble fraction"	—	61

Roots homogenized in 7×10^{-3} M MES/TRIS buffer at pH 6·5. Fraction A was sedimented at 10 000 g for 15 min and the supernatant centrifuged for 90 min at 80 000 g to give Fraction B.

membrane after disruption is difficult to establish but some degree of interaction is indicated by the results obtained.

V. NMR STUDIES

Earlier experiments showed that deficient root tips of field bean exhibited a much reduced capacity for incorporation of absorbed phosphate into esterified forms and that the distribution of ^{32}P between the esters differed from control plants (Robertson and Loughman, 1974a). Further evidence that the onset of boron deficiency causes changes in the distribution of phosphorylated intermediates is shown by ^{31}P-NMR studies. Samples of 100 freshly cut 4 mm root tips from 9-day field bean plants were placed in a spectrometer tube in MES buffer pH 6·0 for the accumulation of 4000 scans (about 17 min). Comparable samples from plants that had been grown without boron for the last 3 days were also examined and the spectra are shown in Fig. 7. The control root tips show the characteristic spectrum of active meristematic tissue described by Kime et al. (1982), whereas the deficient roots are more typical of those of the relatively inactive tissue of some storage organs. The fact that the distribution of inorganic phosphate between cytoplasm and vacuole is very different in the two samples suggests that B deficiency has in some way affected the transport capacity of the tonoplast. The B-deficient roots were growing more slowly and the ^{31}P-NMR spectrum reflects their reduced metabolic rate.

VI. CONCLUSIONS

The experimental information presented in this paper is consistent with our view that the early effects of withdrawal of boron from the environment of the root of many species manifest themselves at a membrane site. The

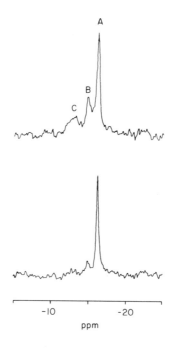

Fig. 7. ^{31}P NMR spectrum of root tips of 9-day field bean root tips. (a) Upper trace, control. (b) Lower trace, grown for last 3 days in absence of boron. Samples were run in 10 mM MES buffer at pH 6·0 containing 0·1 mM CaSO$_4$ at 121·49 MHz using a Bruker WM 300 spectrometer (8 mm tube). The 8 mm tube was arranged concentrically with a 10 mm tube containing D$_2$O for the field frequency lock and 1 mM methylene diphosphonic acid in 5 mM Tris Buffer pH 8·9 for the chemical shift reference at 0 ppm. Spectra were accumulated in the range 298 ± 2 K with the tube spinning. The accumulation conditions were 4000 transients with broad band proton decoupling and a 45° pulse angle, giving a total accumulation time of 17 min. Assignments: Peak A—Vacuolar free phosphate; Peak B—Cytoplasmic free phosphate; Peak C—Hexosemonophosphate and nucleoside monophosphate.

reduction in the capacity for absorption of ions appears to be a general response to deficiency and it is probable that many of the physiological, biochemical, and anatomical effects observed later, result from this interference with a number of processes dependent on close control of movement of metabolites between and within cells. During the development of boron deficiency, capacities of membranes of ion transport alter and as with the K$^+$ activated ATPase, can be restored by the addition of low concentrations of boron. The fact that control roots also respond to boron in that their ability to absorb ions increases, suggests that the continuous presence of the element in the environment is essential for the maximal absorptive capacity to be reached. The use of non-invasive techniques such as NMR to examine roots also indicates that changes in internal distribution of inorganic and organic phosphates accompany the onset of deficiency and further application of this technique with ^{31}P, ^{1}H, and ^{13}C should be helpful in following other metabolic changes.

The changes in ultrastructure observed during the early stages of deficiency further substantiate the view that membrane function is altered (Hirsch and Torrey, 1980). An increase in the thickness of the cell wall and a loss of membrane integrity are observed as early as 6 h after the removal of boron from the culture solution in which sunflower roots are growing and the authors conclude that the essential role of boron may involve maintenance of the integrity of membranes. They also conclude as a result of their experimental observations that the effects of boron deficiency are not duplicated by the addition of indoleacetic acid and it is unlikely that boron deficiency produces its effects via the production of supraoptimal levels of auxins in the root. We concur in this assessment of the role of boron in the metabolism of higher plants, preferring the view that some alteration in membrane function mediates the specific role of this element.

Our results taken together with those of others support the view that boron is essential for higher plants because it is a vital component of membranes and helps to control their functions. If boron is involved in linking glycoprotein and glycolipids the homogeneity of the membranes could be altered by aggregating these components into distinct domains exhibiting specific properties as suggested by Linden *et al.* (1973) or by changing the lipid packing and hence the fluidity (Wolfe, 1978). It is also possible that surface changes in the membrane as a result of complex formation between borate and sugars would result in alterations in transport capacity. There are well-known differences in chemical compositions between plant and animal membranes and the specific nature of the requirements for boron in plants is perhaps due to interaction with particular plant membrane components such as glycoglycerolipids and phytosterols. Phytosphingosine and its derivatives are other components specific to plants that have potential for boron binding, and it is clear that membranes of plants are sufficently different from those of animals to account for the fact that only higher plant membranes appear to have a major and perhaps absolute requirement for boron.

References

Bangham, A. D., Hill, M. W., and Miller, N. G. A. (1974). *Methods in Membrane Biology* **1**, 1–68.

Bentrup, F. W., Pfruner, H., and Wagner, G. (1973). *Planta* **110**, 369–372.

Birnbaum, E. H., Dugger, W. M., and Beasley, B. C. A. (1977). *Plant Physiol.* **59**, 1034–1038.

Bohnsack, C. W. and Albert, L. S. (1977). *Plant Physiol.* **59**, 1047–1050.

Chapman, K. S. R. and Jackson, J. F. (1974). *Phytochem.* **13**, 1311–1318.

Clarkson, D. T. (1974). "Ion Transport and Cell Structure in Plants", McGraw-Hill, London.

Coke, L. and Whittington, W. J. (1968). *J. exp. Bot.* **19**, 295–303.

Cram, W. J. (1973). *Aust. J. Biol. Sci.* **26**, 757–779.

Cram, W. J. and Laties, G. G. (1971). *Aust. J. Biol. Sci.* **24**, 633–646.

Dear, J. and Aronoff, S. (1965). *Plant Physiol.* **40**, 458–459.

Dugger, W. M. and Humphreys, T. E. (1960). *Plant Physiol.* **35**, 523–530.

Dugger, W. M. and Palmer, R. L. (1980). *Plant Physiol.* **65**, 266–273.

Dunitz, J. D., Hawley, D. M., Miklos, D., White, D. N. J., Berlin, Y., Marusic, R., and Prelog, V. (1971). *Helv. Chim. Acta* **54**, 1709–1713.

Gauch, H. G. and Dugger, W. M. (1953). *Plant Physiol.* **28**, 457–467.

Glass, A. D. M. (1973). *Plant Physiol.* **51**, 1037–1041.

Glass, A. D. M. and Dunlop, J. (1974). *Plant Physiol.* **54**, 855–858.

Hirsch, A. and Torrey, J. (1980). *Can. J. Bot.* **58**, 856–866.

Kibalenko, A. P. and Demchenko, T. T. (1971). *Dopov. Akad. Nauk, U.K. RSR* Ser. B. **33**, 844–867.

Kime, M. J., Loughman, B. C., Ratcliffe, R. G., and Williams, R. J. P. (1982). *J. exp. Bot.* **33**, 656–669.

Kouchi, H. and Kumazawa, K. (1976). *Soil Sci. Plant Nutr.* **22**, 53–71.

Lee, S. G. and Aronoff, S. (1967). *Science* **158**, 798–799.

Leonard, R. T. and Hanson, J. B. (1972). *Plant Physiol.* **49**, 430–435.

Lewin, J. (1966). *J. Phycol.* **2**, 160–163.

Lewin, J. and Chen, C. (1976). *J. exp. Bot.* **27**, 916–921.

Lewis, D. H. (1980). *New Phytol.* **84**, 209–229.

Linden, C. D., Wright, K. K., McConnell, H. M., and Fox, C. F. (1973). *Proc. Nat. Acad. Sci., U.S.A.* **70**, 2271–2275.

Loughman, B. C. (1961). *Nature* **191**, 1399.

Lyons, J. M. and Raison, J. K. (1970). *Plant Physiol.* **45**, 386–389.

McIlrath, W. J. and Skok, J. (1964). *Bot. Gaz.* **125**, 268–272.

Maevskaya, A. N. and Alekseeva, Kh.A. (1966). *Fisiologiya Rast.* **13**, 1054–1058.

Mengel, K. and Kirkby, E. A. (1978). "Principles of Plant Nutrition", International Potassium Institute, Berne.

Neales, T. F. (1960). *Aust. J. Biol. Sci.* **13**, 232–248.

Neales, T. F. (1967). *Aust. J. Biol. Sci.* **20**, 67–76.

Odhnoff, C. (1961). *Physiologia Pl.* **14**, 187–220.

Parish, R. W. (1969). *Z. Pflanzenphysiol.* **60**, 211–216.

Pollard, A. S., Parr, A. J., and Loughman, B. C. (1977). *J. exp. Bot.* **28**, 831–841.

Raison, J. K. (1973). *J. Bioenergetics* **4**, 285–309.

Rajaratnam, J. A. and Lowry, J. B. (1974). *Ann. Bot.* **38**, 193–200.

Robertson, G. A. and Loughman, B. C. (1974a). *New Phytol.* **73**, 291–298.

Robertson, G. A. and Loughman, B. C. (1974b). *New Phytol.* **73**, 821–832.

Shkol'nik, M. Ya. (1974). *Fiziologiya Rast.* **21**, 174–186.

Shkol'nik, M. Ya. and Kositsyn, A. V. (1962). *Doklady Biol. Sci.* **144**, 622–624.

Shkol'nik, M. Ya., Krupnikova, T. A., and Davydova, V. N. (1972). *Fisiologiya Rast.* **19**, 1240–1244.

Shkol'nik, M. A., Krupnikova, T. A., and Dmitrieva, N. N. (1964). *Fisiologiya Rast.* **11**, 188–194.

Slack, C. R. and Whittington, W. J. (1964). *J. exp. Bot.* **15**, 495–514.

Smyth, D. A. and Dugger, W. M. (1979). *Plant Physiol.* **63**, supplement, abstract 710.

Smyth, D. A. and Dugger, W. M. (1981). *Physiol., Plant.* **51**, 111–117.

Tanada, T. (1974). *Plant Physiol.* **53,** 775–776.
Timashov, N. D. (1969). *Fisiologiya Rast.* **16,** 518–523.
Turnowska-Starck, Z. (1960). *Acta Soc. Bot. Pol.* **29,** 219–247.
Vorob'ev., L. N. and Pelkhanov., S. E. (1973). *Chem. Abstr.* **83,** No. 53914.
Wilson, C. H. (1961). *Plant Physiol.* **36,** 336–341.
Wolfe, J. (1978). *Plant, Cell, and Environment* **1,** 241–247.

PART 2

So from the Root . . .

Aspects of the movement and storage
of iron and its incorporation
into prosthetic groups

So from the root
Springs lighter the green stalk, from thence the leaves
More airy, last the bright consummate flow'r
Spirits odorous breathes

(Milton, Paradise Lost)

CHAPTER 7

Phytoferritin and its Role in Iron Metabolism

H. F. BIENFAIT AND F. VAN DER MARK

Laboratory for Plant Physiology, Kruislaan, Amsterdam, The Netherlands

I. INTRODUCTION

Since Eusèbe Gris (1844) showed that chlorosis in plants could be relieved by treating roots or leaves with solutions of iron salts, it is known that iron is an indispensible element for the growth of plants. Iron apparently functions as a catalyst in redox processes, and it is also a cofactor in the synthesis of chlorophyll. The standard redox potential (E'_0) of the Fe(III)/Fe(II) couple, normally 0.77 V, can vary over a large range when incorporated into nature's ligands, as in heme and Fe-S centres. Thus practically the whole range between the NADP/NADPH and O_2/H_2O couples (E'_0 of -0.34 and $+0.81$ V, resp.) is covered by Fe-containing proteins that function in photosynthesis. It is widely held that most of the iron in the leaf is localized in the chloroplast (Liebich, 1941, Whatley *et al.*, 1951, Seckbach, 1972a) although there are reports to the contrary (Murphy and Maier, 1967).

The handling of iron presents problems to the living organism. Ferrous iron, the physiologically active form, e.g. as substrate for the enzyme ferrochelatase (Jones, this volume) and the form inserted into proteins such as pyrocatechase (Nakazawa *et al.*, 1969) and the storage protein ferritin (see below), is highly susceptible to oxidation by oxygen at pH values above 5. The superoxide ion that is formed in the reaction

$$Fe(II) + O_2 \rightarrow Fe(III) + O_2^-$$

is highly reactive and, with other radicals derived from it, can cause considerable damage to membranes and proteins (Halliwell, 1974, Trelstad et al., 1981). Ferric ions are practically insoluble at physiological pHs; the solubility constant for $Fe(OH)_3$ is 10^{-39}, which implies that the concentration of free ferric is $10^{-15}M$ at pH 6, $10^{-18}M$ at pH 7. Chelators such as EDTA or EDDHA, widely used in agriculture, or citrate, the natural ligand of iron in the xylem (Tiffin, 1966a; 1966b), are necessary to keep ferric in solution at pH values above 3. When bound to this kind of ligand, the E'_0 of the Fe(III)/Fe(II) couple is shifted to lower values, e.g. in EDTA to $+0\cdot13\,V$ (Grinstead, 1959). There are, however, several cellular compounds that are capable of reducing iron in this form, such as ascorbate, glutathione, or cysteine. Light is also able to reduce ferric ion when organic acids such as citrate or malate are present (Novak, 1965):

$$R{-}COO^- + Fe(III) + hv \rightarrow R + CO_2 + Fe(II).$$

This reaction may play a role in the transport of iron from the xylem vessels into the leaf cells (Brown et al., 1979).

Taken together, free iron ions, i.e. those which are not functionally bound to macromolecules, may turn into a kind of "redox mill" in the cell, with a net oxidation of cellular reductants and production of organic and oxygen radicals. This may eventually lead to the death of the cell, especially when there is not enough superoxide dismutase to keep the O_2^- level low (Halliwell, 1974, Abeliovich et al., 1974). It may also be the main cause of death for those plants that cannot restrict the entrance of Fe(II) in the roots and the subsequent transport to the leaves, when growing on waterlogged soils (Martin, 1968, Yoshida and Tadano, 1978).

It should now be clear that levels of free iron in the cell should be low, yet growing plants have a continuous need for iron. These two contrasting requirements are reconciled by a regulatory system that controls the uptake and the availability of iron.

II. Regulation of Iron Uptake

More is known about the mechanism of iron uptake in dicotyledonous than in monocotyledonous plants. The following is restricted to the situation in dicotyledons. Iron in aerobic soils, if not bound to more or less specific chelators (Whitehead, 1964; Powell et al., 1980), is mostly present as highly insoluble Fe(III) salts and complexes. Before it can be taken up, it must therefore be solubilized. In roots where the iron uptake system has been stimulated, uptake is preceded by the reduction of Fe(III) to Fe(II) (Chaney *et al.*, 1972). It is not certain whether this is also an obligatory step in those roots where uptake proceeds at a basic level (Wallace and Hale, 1961; Hill-Cottingham and Lloyd-Jones, 1965; Römheld and Marschner, 1981). In the roots of chlorotic plants the following activities are increased:

1. acidification of the medium around the root (van Egmond and Aktas, 1977);
2. reduction of soluble ferric chelates (Chaney *et al.*, 1972);
3. uptake of Fe(II) (Chaney *et al.*, 1972);
4. accumulation of organic acids, especially citrate (Landsberg, 1979; see also for a review Brown, 1977).

There is no certainty whether these four activities are obligatorily coupled; 1. and 2. may be subject to different regulation mechanisms (Bienfait, *et al.*, 1981). It is generally assumed that the activity of these processes is not just regulated by the root, but that the chlorotic condition of the leaves also plays a role in the regulation. This implies that the roots receive a signal from the leaves concerning their iron status. There is as yet no indication as to the identity of such a signal. Some suggestions are: that a precursor for an iron reductant is produced and exported by chlorotic leaves (Brown, pers. comm.); that a hormone (auxin) controls acidification (Landsberg, 1981); or that the iron level in the phloem controls at least one of the activities mentioned.

When a chlorotic plant receives a dose of iron, it takes up massive amounts and may take a week before returning to normal uptake levels. The massive uptake can result in extremely high levels of iron citrate in the xylem, e.g. $310\mu M$, compared with $2\mu M$ normally found in sunflower (Tiffin, 1966a). It is here that ferritin enters the picture as a storage protein to cope with this influx of iron in the leaf (Seckbach, 1968).

III. Ferritin

A. GENERAL

Many reviews have appeared on this iron storage protein; two recent and extensive ones being those of Munro and Linder (1978) and Bezkorovainy (1980). Most of the literature on ferritin is concerned with its biochemistry and its role in animals. It was first isolated by Laufberger from horse spleen (Laufberger, 1937). Hyde and his colleagues prepared ferritin from bean seeds and hypocotyls in 1963, after demonstrating its presence in plants by electron microscopy in the previous year (Hyde *et al.*, 1962; Hyde *et al.*,

1963). They called it phytoferritin. Ferritin is also found in moulds (David and Easterbrook, 1971) and bacteria (Stiefel and Watt, 1979). In this review, the structure and function of ferritin will be treated largely on the basis of what is known of the animal protein, for there are very little data on ferritin isolated from plant material. However, in those instances where plant and animal ferritins were compared, no pronounced differences were found (Crichton *et al.*, 1978a; Ponce-Ortiz, 1979; van der Mark *et al.*, 1981): therefore we feel justified in assuming that what holds for animal ferritin is essentially the same for the plant protein.

Phytoferritin from peas or lentils is built up of 24 subunits of MW 20 000 to 22 000, depending on the source (Ponce-Ortiz, 1979); the subunits are slightly larger than those (18 500) found in animal ferritins. The subunits aggregate to form the holoprotein, a hollow sphere, which in horse spleen has an outside diameter of 110 Å, an inner free space of 70 Å and six channels with a diameter of about 10 Å which penetrate the protein mantle (Banyard *et al.*, 1978). Through these pores iron can enter and leave. In the inner space an iron core may be present with maximally 4300 (horse spleen) or 5400–6200 (pea and lentil, Ponce-Ortiz, 1979) iron atoms in a hydrous ferric oxide phosphate micelle. It is this iron core which makes well-filled ferritin visible in the electron microscope. Ferritin protein contains little cysteine; thus in amino acid composition, as well as morphology, it is different from the storage protein, metallothionein which forms complexes with Zn, Cu, and Cd (Kojima and Kagi, 1978).

The knowledge of ferritin in plants is almost completely based on data from electron microscopy. Highly filled ferritin is generally observed in those cells where no active photosynthesis takes place, i.e. in roots and root nodules (Dart and Mercer, 1964; Bergersen, 1963; Newcomb, 1967; Amelunxen *et al.*, 1970), in etiolated leaves (Hyde *et al.*, 1963; Jacobson *et al.*, 1963). in cells yellowing by disease (Craig and Williamson, 1969) by mechanical damage (Wildman and Hunt, 1976) or by senescence (Barton, 1970), in reproductive cells (Sheffield and Bell, 1978), in seeds (Hyde, 1963) and in other tissues (see also Seckbach, 1972b). Ferritin is invariably found in the plastids, although Seckback (1972b) reports that in iron-treated *Xanthium* plants it may be associated with lipoid globules in the cytosol.

B. SYNTHESIS, IRON UPTAKE AND RELEASE

In animals, the translation of messenger RNA which codes for ferritin is controlled by the level of free iron in the cells. Zähringer *et al.* (1976) proposed a model for this regulation, in which translation of the messenger is prevented by adhering ferritin subunits. Iron should remove the inhibition by promoting aggregation of the subunits into the holoprotein. The existence of isoferritins has been shown; the question of whether these result from different messengers, or from post-translational modifications, or both, is not yet settled (Adelman et al., 1975; Treffry and Harrison, 1980a; Watanabe and Drysdale, 1981). The subunits spontaneously arrange themselves into the holoprotein.

Iron uptake is thought to occur by the oxidation of ferrous ions by oxygen at the inside of the ferritin molecule, for the empty holoprotein, or apoferritin, catalyses ferrous ion oxidation with concomitant formation of an iron core (Niederer, 1970; Harrison et al., 1974; Crichton and Roman, 1978). In the early stage of filling up, Zn^{2+} can inhibit this process (Niederer, 1970; Treffry and Harrison, 1980b), probably by binding to a catalytic site; but once a sizeable core has been built up, the influence of zinc on iron uptake is decreased (Treffry and Harrison, 1980b). According to Harrison this indicates that the surface of the iron core itself can catalyse ferrous ion oxidation. Iron uptake into horse apoferritin shows Michaelis-Menten kinetics with reported K_m values for Fe(II) of 2·8 mM (Bryce and Crichton, 1973), or 0·2–0·4 mM with different isoferritins (Russell and Harrison, 1978). Differences in iron affinity of isoferritins have also been reported by Wagstaff et al. (1978); according to Treffry and Harrison (1980c) the iron affinity of a ferritin depends on its past history in the cell before isolation.

How iron is released from ferritin remains a controversial subject. When incubated with chelators like EDTA, ferritin does not yield its iron readily. It is generally assumed therefore that breakdown of the protein is necessary, followed by reduction and chelation, or, alternatively, that reduction of the iron core occurs with reduced glutathione, ascorbate or reduced flavins, but without change in the protein.

Sirivech et al. (1974) showed that low concentrations of reduced flavins could mobilize ferritin iron within a few hours in vitro. Consequently many authors believe that flavins are the natural mobilizers of ferritin iron (Munro and Linder, 1978; Bezkorovainy, 1980; Ulvik and Romslo, 1979). However, as already pointed out by Sirivech et al. (1974) free reduced flavins are rapidly oxidized by oxygen (Gibson and Hastings, 1962). For example, ferritin iron at a concentration of 1 mM only competes effectively with oxygen for $FMNH_2$ when the oxygen concentration is as low as 1 μM; a condition which may be fulfilled in the human liver, but is not likely in plant leaves. Bienfait and van den Briel (1980) recently proposed that the ascorbate radical monodehydroascorbate is a physiologically active ferritin-iron reductant. The radical is formed when ascorbate is oxidized by oxygen in the presence of ascorbate oxidase or traces of free Cu^{2+}. In the latter case the oxidation is most rapid at higher pH values (≥ 7). Thus, the conditions in the stroma of actively photosynthesizing chloroplasts (pH 8, high ascorbate and oxygen levels) should be favourable for the production of ascorbate radicals and the release of iron from ferritin.

A different picture of ferritin iron mobilization is drawn by David (1974) working with *Phycomyces blakesleeanus*. Spores of this mould contain ferritin, and when they germinate in an iron-free medium, the ferritin molecules do not appear to yield their iron gradually and simultaneously, but one by one, by an all-or-none mechanism. In an iron-rich medium, mobilization occurs at a much lower rate. These results suggest a ferritin degradation mechanism which is active in the absence of iron, whereby a ferritin molecule, after attack, gives up its iron core to be dissolved rapidly

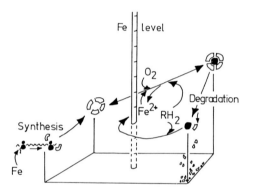

Fig. 1. Idealized scheme for ferritin synthesis, breakdown and iron content as a function of free iron level in the cell. The situation in the cell is thought to be a point moving in the tilted plane. x-axis: uptake and release of iron by oxidation or reduction; y-axis: level of free iron in the cell; z-axis: synthesis and breakdown of ferritin protein.

under the action of cellular chelators, reductants, or both. The way in which ferritin may function under the influence of the free iron level in the cell, is illustrated in Fig. 1.

C. FUNCTION OF FERRITIN IN THE PLANT

1. Ferritin Formation as a Response to High Iron Levels

Seckbach (1968) found that when chlorotic bean plants were given iron, they accumulated large deposits of ferritin within a few days. This was doubtless the consequence of the large uptake of iron by the activated roots, resulting in high iron concentration in the xylem. Using immuno-logical methods, van der Mark *et al.* (1982) found with *Phaseolus vulgaris* that chlorotic plants do not contain measurable quantities of ferritin protein. Thus the appearance of ferritin is a consequence of synthesis, not of loading of pre-existing apoferritin.

Figure 2 shows the course of ferritin synthesis and its iron content relative to total iron content in the leaf. Figure 3 shows what happens when a chlorotic plant is administered iron for one day only, and then returned to an iron-free medium. After a few days, during which the leaf is

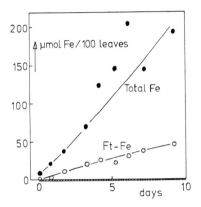

FIG. 2. Total and ferritin iron (Ft-Fe) in the first trifoliate leaf of a Fe-stressed plant of *Phaseolus vulgaris* var. Prélude when put on a Fe-containing medium. Total iron was determined after dry ashing and colorimetric estimation of ferric ions with rhodanide. Ferritin iron was determined immunologically (van der Mark *et al.*, 1982).

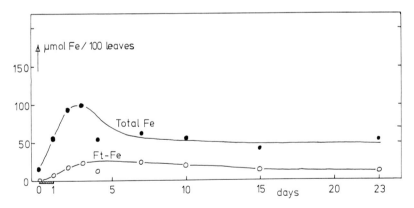

FIG. 3. Total and ferritin iron in a plant as described in the legend of Fig. 2, but iron given for one day only.

recovering from chlorosis, its iron content reaches its peak. Thereafter, the export of iron from the leaf predominates, although 50% of the peak iron content remains in the leaf. Ferritin protein and ferritin iron reach their peak later. Remarkably, the iron exported from the leaf does not appear to come from the ferritin pool. This suggests that it never entered the photosynthesizing cells, but simply was transferred from ferric deposits in the xylem to the phloem. After nine days, export from the leaf has come to an end and levels of ferritin protein and ferritin iron gradually decrease. This again indicates that there is no easy exchange between the ferritin pool in the stroma of the chloroplast and the phloem.

During the slow decrease in ferritin iron and ferritin protein levels, the mean ferritin iron content does not change. Thus, the mean life span of ferritin protein is not longer than that of its iron. This suggests a decay similar to that found by David (1974) in *Phycomyces*, but in view of the long half-time, it may very well reflect the turnover rate of the proteins in the chloroplast stroma rather than the action of a specific ferritin protease regulated by the cellular iron level.

2. Ferritin During De-etiolation

The presence of ferritin in etioplasts and its absence from chloroplasts suggests that ferritin iron is used during de-etiolation for the synthesis of iron-containing enzymes (Hyde, 1963; Robards and Humpherson, 1967). Sprey *et al.* (1978) studied the fate of ferritin during de-etiolation of tobacco cotyledons using both EM and immunological methods. They suggested that in the dark, iron is accumulated in the plastid in the form of large iron-containing precipitates. When put into the light, the iron in these clusters dissolved, and moved into newly formed ferritin. Thereafter the iron could be mobilized from the ferritin for the synthesis of chloroplast tissue. This implies that the storage iron should pass through ferritin before utilization. It is possible, however, that what was seen in the electron microscope as amorphous iron-rich clumps, are in reality dense clusters of ferritin, which upon development of the thylakoid apparatus are spread through the stroma so that the individual iron cores become discernible. Moreover, quantitative data on ferritin protein were not provided. The suggested pathway of iron: influx—precipitation—dissolution—incorporation into ferritin—mobilization—use, could therefore be simplified to: influx—incorporation into ferritin—mobilization—use.

Whatley (1977) studied the development of plastids in bean primary leaves and hypocotyls in the light and the dark, with the electron microscope. She described the appearance and disappearance of ferritin iron cores at specific stages of plastid development , which suggests a role for ferritin as a temporary store for iron.

Van der Mark *et al.* (1981) studied the role of ferritin in light grown, dark grown, and de-etiolating bean primary leaves, with immunological methods. Their findings are summarized in Figs 4a and b. In the dark as well as in the light there was a continuous influx of iron from the cotyledons (data not shown). In the light, the initiation of this flux was

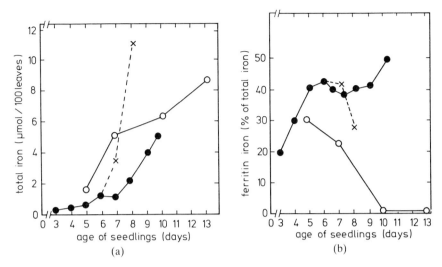

Fig. 4. Total iron (a) and ferritin iron (b) in primary leaves of darkgrown (●), lightgrown (○) and de-etiolating (×) bean plants.

about 3 days earlier than in the dark. On illumination after 6 days of growth in the dark, a strong increase in iron influx occurred. In all the stages of growth, ferritin protein as a fraction of the total protein remained rather constant, i.e. there was no significant influence of the incoming iron on the synthesis of ferritin protein (Table I). In the dark, however, there was a slow but continuous rise in the iron content of ferritin, from 180 to 560 irons per mol (at day 9). In the light, the higher iron content was already reached at day 5, after which it declined to nil at day 10. The decrease in ferritin iron content in the leaves growing in the light, between days 5 and 10, suggests a mechanism of iron release in which the protein has a longer lifetime than the iron, i.e. a reductive mechanism.

During de-etiolation, the iron influx and synthesis of protein and ferritin all increased in close harmony, so that ferritin still absorbed 40% of the incoming iron without any effect on the iron content per mol. Only after two days of de-etiolation did the ferritin iron as a percentage of total iron start to drop, thus moving in the direction of the situation in light-grown plants. If there were any inorganic iron precipitates, playing the role suggested by Sprey et al. (1978), ferritin iron as a fraction of total iron would be low before and high shortly after the start of de-etiolation. We should, however, be careful in comparing tobacco cotyledons and primary bean leaves.

From these data, ferritin emerges as a cellular buffer between the influx

TABLE I

Ferritin protein and ferritin iron in extracts of primary bean leaves under various light conditions

Age of seedlings in days	Ferritin μg 100 μg⁻¹ protein			Ferritin-Fe μmol Fe in 100 leaves			Ferritin Fe content atoms Fe mol⁻¹		
	dark grown	greening*	light grown	dark grown	greening*	light grown	dark grown	greening*	light grown
3	0·17	–	–	0·02	–	–	180	–	–
4	0·24	–	–	0·06	–	–	275	–	–
5	0·34	–	0·21	0·17	–	0·23	315	–	560
6	0·33	0·32	–	0·31	0·33	–	310	350	–
7	0·32	0·28	0·18	0·36	0·85	0·44	340	375	420
8	0·29	0·26	–	0·56	1·61	–	370	420	–
9	0·25	–	–	1·11	–	–	560	–	–
10	0·32	–	0·14	1·43	–	<0·1	540	–	<25
13	–	–	0·22	–	–	<0·1	–	–	<25

* Put in the light at day 6 (see Figs 4a and b).

of iron into a tissue from cotyledons or roots, and the local need for it. Thus, in a de-etiolating bean leaf, ferritin does not simply give up its iron because of the growing need for it, but instead, it takes iron up initially, because of the even greater influx from the cotyledons. In a later stage, when influx decreases relative to the efflux from the available pool, the iron is again released.

Ferritin can thus be considered an indicator of iron fluxes, rather than an immobile iron store unit, somnolent until the light or another stimulus wakes it up. We must, however, be careful in excluding other possible functions for ferritin, see, for example, the case of *Azotobacter* ferritin, reported to function as a cytochrome *b* (Stiefel and Watt, 1979).

REFERENCES

Abeliovich, A., Kellenberg, D., and Shilo, M. (1974). *Photochem. Photobiol.* **19,** 379–382.
Adelman, T. G., Arosio, P., and Drysdale, J. W. (1975). *Biochem. Biophys. Res. Comm.* **63,** 1056–1062.
Amelunxen, F., Thaler, I., and Hansdorff. (1970). *Z. Pflanzenphysiol.* **63,** 199–210.
Banyard, S. H., Stammers, D. K., and Harrison, P. M. (1978). *Nature* **271,** 282–284.
Barton, R. (1970). *Planta* **94,** 73–77.
Bergersen, F. J. (1963). *Austr. J. Biol. Sci.* **16,** 916–919.
Bezkorovainy, A. (1980). "Biochemistry of nonheme iron", pp. 207–269, Plenum Press, New York and London.
Bienfait, H. F. and van den Briel, M. L. (1980). *Biochim. Biophys. Acta* **631,** 507–510.
Bienfait, H. F., van der Bliek, A. M., and Bino, R. J. (1982). *J. Plant Nutr.* **5,** 447–450.
Brown, J. C. (1977). *In* Bioinorg. Chem. II, Adv. Chem ser. no 162, pp. 93–103, (K. N. Raymond, ed.).
Brown, J. C., Foy, C. D., Bennett, J. H., and Christiansen, M. N. (1979). *Plant Physiol.* **63,** 692–695.
Bryce, C. F. A. and Crichton, R. R. (1973). *Biochem. J.* **133,** 301–309.
Chaney, R. L., Brown, J. C., and Tiffin, L. O. (1972), *Plant Physiol.* **50,** 208–213.
Craig, A. S. and Williamson, U. J. (1969). *Virology* **39,** 616–617.
Crichton, R. R., Ponce-Ortiz, Y., Koch, M. J. H., Parfait, R., and Stuhrmann, H. B. (1978). *Biochem. J.* **171,** 349–356.
Crichton, R. R. and Roman, F. (1978). *J. Mol. Catal.* **4,** 75–82.
Dart, P. J. and Mercer, F. V. (1964). *Arch Mikrobiol.* **49,** 209–235.
David, C. N. and Easterbrook, K. (1971). *J. Cell. Biol.* **48,** 15–28.
David, C. N. (1974). *In* "Microbial iron metabolism", pp. 149–158, (J. B. Neilands, ed.), Academic Press, New York and London.
Egmond, F. van, and Aktas, M. (1977). *Plant and Soil* **48,** 685–703.
Gibson, Q. H. and Hastings, J. W. (1962). *Biochem. J.* **83,** 368–377.
Grinstead, R. R. (1959). *J. Am. Chem. Soc.* **82,** 3472–3476.

Gris, E. (1844). *C.R. Acad. Sci. (Paris)* **19**, 1118–1119.
Halliwell, B. (1974). *New Phytol.* **73**, 1075–1086.
Harrison, P. M., Hoy, T. G., Macara, I. G., and Hoare, R. J. (1974). *Biochem. J.* **143**, 445–451.
Hill-Cottingham, D. G. and Lloyd-Jones, C. P. (1965). *J. Exp. Bot* **16**, 233–242.
Hyde, B. B., Hodge, A. J., and Birnstiel, M. L. (1962). Proc. 5th Int. Conf. Electr. Microsc. Vol. 2, p. 1, (S. S. Breese, ed.), Academic Press, New York and London.
Hyde, B. B., Hodge, A. J., Kahn, A., and Birnstiel, M. L. (1963). *J. Utrastr. Res.* **9**, 248–258.
Jacobson, A. B., Swift, H., and Bogorad, L. (1963) *J. Cell. Biol.* **17**, 557–570.
Kojima, Y. and Kagi, J. H. R. (1978). *Trends Biochem. Sci.* **3**, 90–93.
Landsberg, E.–C. (1979). Ph.D. Thesis, Berlin.
Landsberg, E.–C. (1981). *J. Plant Nutr.* **3**, 579–591.
Laufberger, V. (1937). *Bull. Soc. Chim. Biol.* **19**, 1575–1582.
Liebich, H. (1941). *Z. Bot.* **37**, 129–157.
Mark, F. van der, de Lange, T., and Bienfait, H. F. (1981) *Planta,* **153**, 338–342.
Mark, F. van der, van der Briel, M. L., van Ders, J. W. A. M., and Bienfait, H. F. (1982). *Planta* (in press).
Martin, W. H. (1968). *J. Ecology* **56**, 777–793.
Munro, H. N. and Linder, M. C. (1978). *Physiol. Rev.,* **58**, 317–396.
Murphy, J. J. and Maier, R. H. (1967). *J. Agric. Food Chem.* **15**, 113–117.
Nakazawa, T., Nozaki, M., Hayashi, O., and Yamano, T. (1969). *J. Biol. Chem.* **244**, 119–125.
Newcomb, E. H. (1967). *J. Cell. Biol.* **33**, 143–163.
Niederer, W. (1970). *Experientia* **26**, 218–220.
Novak, J. (1965). *Talanta* **12**, 649.
Ponce-Ortiz, L. Y. (1979). Ph.D. Thesis, Univ. Catholique, Louvain-la-Neuve, Belgium.
Powell, P. E., Cline, G. R., Reid, C. P. P., and Szaniszlo, P. J. (1980). Nature **287**, 833–834.
Robards, A. W. and Humpherson, P. G. (1967) *Planta* **76**, 169–178.
Römheld, V. and Marschner, H. (1981). *J. Plant Nutr.* **3**, 551–560.
Russell, S. M. and Harrison, P. M. (1978). *Biochem. J.* **175**, 91–104.
Seckbach, J. (1968). *J. Ultrastr. Res.* **22**, 413–423.
Seckbach, J. (1972a). *Planta Medica* **21**, 267–273.
Seckbach, J. (1972b). *J. Ultrastr. Res.* **39**, 65–76.
Sheffield, E. and Bell, P. R. (1978). *Proc. R. Soc. Lond. B.* **202**, 297–306.
Sirivech, S., Frieden, E. and Osaki, S. (1974). *Biochem. J.* **143**, 311–315.
Sprey, B., Gliem, G., and Janossy, A. G. S. (1978). *Z. Pflanzenphysiol.* **88**, 69–82.
Stiefel, E. I. and Watt, G. D. (1979). *Nature* **279**, 81–83.
Tiffin, L. O. (1966a). *Plant Physiol.* **41**, 510–514.
Tiffin, L. O. (1966b). *Plant Physiol.* **41**, 515–518.
Treffry, A. and Harrison, P. M. (1980a). *Biochim. Biophys. Acta* **610**, 421–424.
Treffry, A. and Harrison, P. M. (1980b). *Biochem. Soc. Transac.* **8**, 655–656.
Treffry, A. and Harrison, P. M. (1980c). *Biochem. Soc. Transac.* **8**, 656–657.
Trelstadt, R. L., Lawley, K. R. and Holmes, L. B. (1981). *Nature* **289**, 310–312.
Ulvik, R. and Romslo, I. (1979). *Biochim. Biophys. Acta* **588**, 256–271.
Wagstaff, M., Worwood, M., and Jacobs, A. (1978). *Biochem. J.* **173**, 969–977.
Wallace, A. and Hale, V. Q. (1961). *Soil Sci.* **92**, 404–407.
Watanabe, N. and Drysdale, J. (1981). *Biochem. Biophys. Res. Comm.* **98**, 507–511.

Whatley, F. R., Ordin, L., and Arnon, D. I. (1951). *Plant Physiol.* **26,** 414–418.
Whatley, J. M. (1977). *New Phytol.* **78,** 407–420.
Whitehead, D. C. (1964). *Nature* **202,** 417–418.
Wildman, R. B. and Hunt, P. (1976). *Protoplasma* **87,** 121–134.
Yoshida, S. and Tadano, T. (1978). *In* "Crop tolerance to suboptimal land conditions", pp. 233–276, (G. A. Jung, ed.), ASA spec. publ. 32, Amer. Soc. Agron., Madison, Wisc.
Zähringer, J., Baliga, B. S., and Munro, H. N. (1976). *Proc. Natl. Acad. Sci. (USA)* **73,** 857–861.

CHAPTER 8

Ferrochelatase

O. T. G. JONES

Department of Biochemistry, University of Bristol, Medical School, Bristol, England

I. INTRODUCTION

The enzyme ferrochelatase (systematic name, protohaem ferro-lyase, E.C. 4.99.1.1.) catalyses the insertion of Fe^{2+} ions into porphyrins to make haems (see Fig. 1). Haems are found in animal, plant, and bacterial cells almost invariably as haem-protein complexes, the most abundant in vertebrates being haemoglobin and myoglobin. Another group of haem proteins, the cytochromes, are of almost universal distribution in aerobic organisms where they are components of respiratory and photosynthetic electron transport systems (see the articles in this volume by A. L. Moore and by M. F. Hipkins). Some haem proteins, the catalases and peroxidases, are concerned in the reactions involving peroxide breakdown and utilization and these too have very widespread distribution in aerobic

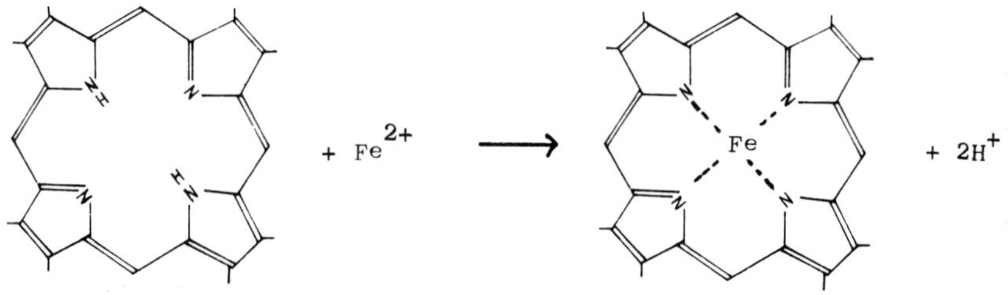

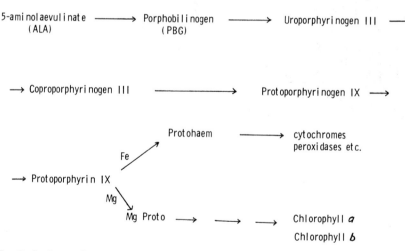

FIG. 1. The ferrochelatase reaction.

organisms. Consequently, haem synthesis is essential for the survival of nearly all species.

Photosynthetic organisms contain chlorophylls, which are magnesium complexes of tetrapyrroles and are almost certainly formed following the insertion of Mg^{2+} into protoporphyrin to make magnesium protoporphyrin (Granick, 1948; see also Jones, 1979) which is subsequently modified to form chlorophylls *a* and *b* or bacteriochlorophyll. A representation of these biosynthetic pathways is given in Fig. 2. This shows that the pathway

5-aminolaevulinate $\longrightarrow$ Porphobilinogen $\longrightarrow$ Uroporphyrinogen III $\longrightarrow$
(ALA)   (PBG)

$\longrightarrow$ Coproporphyrinogen III $\longrightarrow$ Protoporphyrinogen IX $\longrightarrow$

Protohaem $\longrightarrow$ cytochromes
peroxidases etc.

Fe

$\longrightarrow$ Protoporphyrin IX

Mg

Mg Proto $\longrightarrow$ $\longrightarrow$ $\longrightarrow$ Chlorophyll *a*
Chlorophyll *b*

FIG. 2. Outline pathway of haem and chlorophyll biosynthesis.

of biosynthesis of haem proteins and chlorophylls is believed to have common intermediates between 5-aminolaevulinate (ALA) and protoporphyrin IX. At protoporphyrin an important branch point in biosynthesis is located: iron may be inserted by ferrochelatase to make protohaem, or magnesium may be inserted to make magnesium protoporphyrin, a reaction about which surprisingly little is known. Although green plants may contain 50–250 times more chlorophyll than protohaem pigments (see, for example, Castelfranco and Jones, 1975) no measurements have convincingly demonstrated Mg^{2+} insertion into porphyrins by fully disrupted systems, although intact chloroplasts are reported to retain this activity (Fuesler, Wright, and Castelfranco, 1981).

The biosynthetic sequence given in Fig. 2 indicates that the tetrapyrrole intermediates between ALA and protoporphyrin IX are porphyrinogens, not porphyrins. In porphyrinogens the bridge carbon atoms linking the pyrroles are fully reduced, the macrocycle is flexible and does not coordinate iron (Porra and Jones, 1963b). The substrates for ferrochelatase in the tissues so far examined are porphyrins not porphyrinogens.

The porphyrin substrates of ferrochelatase are highly conjugated and have strong absorption bands in the visible region. It is therefore relatively easy to assay the enzyme by monitoring spectroscopically the disappearance of the porphyrin substrate (Nishida and Labbe, 1959; Jones and Jones, 1969); the formation of metalloporphyrin can also be measured spectroscopically. Other assays have involved the determination of newly formed haem by adding alkaline pyridine at the end of the incubation and measuring the reduced-minus-oxidized difference spectrum of the pyridine haemochrome (Porra and Jones, 1963a; Porra et al., 1967). A radiochemical assay follows the incorporation of $^{59}Fe^{2+}$ into the porphyrin (e.g. Jones, 1968). Because Fe^{2+} autoxidizes rapidly it has been found very convenient to use Co^{2+}, an alternative more stable substrate, in its place in spectroscopic assays (Jones and Jones, 1969).

II. LOCALIZATION OF FERROCHELATASE

A. ANIMAL TISSUES

Much of the fundamental work on the properties of ferrochelatase has been carried out using mammalian tissue as the enzyme source. It was found that the enzyme is an insoluble protein located in the mitochondrial fraction of liver (Nishida and Labbe, 1959). In further studies, mitochondria from rat liver were fractionated and the outer and inner membranes were separated from the soluble enzymes of the matrix. Ferrochelatase was found to be located on the inner mitochondrial membrane (Jones and Jones, 1969) and, further, it was found that the substrate binding sites are on the inner face of the inner mitochondrial membrane (Fig. 3). Rates of ferrochelatase were low when porphyrins and metal substrates were added under conditions such as to maintain mitochondrial integrity: disruption stimulated activity. This enzyme distribution imposes restrictions on the activity of ferrochelatase: its metal and porphyrin substrates must be transferred across two mitochondrial membranes into the matrix space.

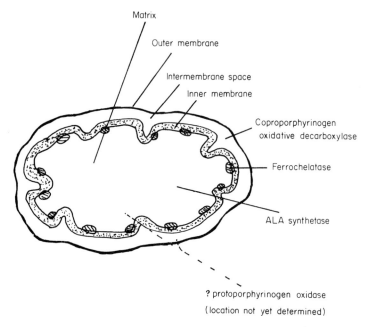

FIG. 3. Representation of a rat liver mitochondrian showing location of the mitachondrial enzymes of the haem biosynthetic pathway (ALA dehydratase, uroporphyrinogen III synthetase and cosynthetase, and uroporphyrinogen III decarboxylase are located in the cytosol.

The protein of many mitochondrial (and plastid) cytochromes is synthesised on cytoplasmic ribosomes and so in the final stages of haemprotein biosynthesis proteins or haems must be transported across membranes before the final active holoprotein is assembled. In the case of haemoglobin the rate of globin synthesis on the ribosome is regulated by haem which is exported from the mitochondria. This haem acts as an inhibitor of a kinase which phosphorylates the eukaryotic intiation factor' (elF-2) for globin synthesis. In the absence of haem the initiation factor is phosphorylated and is inactive (see Ochoa and De Haro, 1979, for a review). Thus, in mammalian systems haem has an important regulatory function in the synthesis of the apoprotein as well as forming part of the haemoglobin complex. The compartmentalization of the enzymes of haem biosynthesis is remarkable, and necessitates the transfer of ALA from inside the mitochondrial matrix to the cytosol for early reactions in the biosynthetic pathway and the transport of coproporphyrinogen to the intermembrane mitochondrial space and of protoporphyrin(ogen?) into the mitochondrial matrix. Haem, when formed by ferrochelatase may have to be transferred from the mitochondrial matrix to the cytosol for binding to different apoproteins.

The plastids are structures characteristic of higher plant cells, present in green tissue as chloroplasts which are the site of all chlorophyll and photosynthetic reactions. Other plastidic structures, called etioplasts, are found in etiolated, dark grown plants. These may be described as undeveloped chloroplasts: in place of chlorophyll they have small amounts of protochlorophyll and they lack the multilayered lamellar structure of the thylakoids of green plant chloroplasts. On illumination, the protochlorophyll is converted to chlorophyll and extensive membrane development commences, leading eventually to the transformation of an etioplast into what is recognizably a chloroplast. Other plastidic structures are found in stems and even in root tissue and it seems that these retain the capacity to develop into chloroplasts on illumination even though their function in the buried root is obscure. Early measurements of ferrochelatase activity in green plant tissues showed that partly purified spinach chloroplasts had the capacity to catalyse ^{59}Fe insertion into porphyrins (Jones, 1967; Jones, 1968). Etioplasts and mitochondria too, were found to contain ferrochelatase (Porra and Lascelles, 1968). Determination of the affinity of the chloroplast enzyme for iron (which was required in the Fe^{2+} state) gave a K_m of approximately 8 μM (Jones, 1968) although such determinations must be viewed with caution since Fe^{2+} is a difficult substrate, prone to autoxidation, precipitation and non-specific binding to proteins. The K_m for Fe^{2+} insertion into protoporphyrin by chloroplast ferrochelatase was, however, very close to that found for the same reaction in rat liver mitochondria by Labbe and Hubbard (1961). Chloroplast ferrochelatase is an insoluble enzyme, present in preparations stripped of the outer membrane (Jones, 1968) and so, presumably it is associated with the thylakoid membrane.

The reported presence of ferrochelatase in both plastid and mitochondrial factions (Porra and Lascelles, 1968) was interesting but not unexpected. Chloroplasts have frequently been reported to be self-sufficient in porphyrin biosynthesis, at least when supplied with ALA (Carell and Kahn, 1964; see also a review by Rebeiz and Castelfranco, 1973). Preparations of plant mitochondria are commonly contaminated with plastid fragments and in more recent work with barley and mung bean mitochondrial ferrochelatase, it has been found necessary to remove plastid fragments from mitochondria by density gradient centrifugation and check purity by monitoring marker enzymes or pigments characteristic of each organelle (Little and Jones, 1976; Hendry, Jones, and Wintle, 1981, unpublished work). Ferrochelatase was confirmed in both mitochondrial and plastid fractions; the enzymes are similar in affinity for the metal substrate Co^{2+}, and have the same pH optimum (Table I). The

TABLE I
Ferrochelatase of mung bean mitochondria and etioplasts (Hendry, Jones and Wintle, unpublished)

Organelle	Ferrochelatase (nmol min^{-1} mg^{-1} protein)	Cross contamination	Co^{2+} K_m (μM)	pH optimum
Mitochondria	0·21 ± 0·025	15% (by plastids)	8·3	7·4
Etioplasts	0·08 ± 0·02	2·3% (by mitos)	10·0	7·4

Organelles were prepared from briefly illuminated 3 day old dark grown seedlings by homogenisation and differential centrifugation. Cross contamination of fractions was assayed using reduced cytochrome c oxidase, succinate dehydrogenase and chlorophyll as markers.

enzyme from each organelle showed the same specificity to porphyrin substrates and sensitivity to heat inactivation (Little and Jones, 1976) and was equally sensitive to inhibition by -SH binding reagents, such as N-phenyl maleimide. N-methyl protoporphyrin IX, the tight binding inhibitor of ferrochelatase (De Matteis, Gibbs, and Smith, 1980), inhibited equally ferrochelatase from mitochondria or plastids. Currently, there is no evidence that ferrochelatase of plastids and ferrochelatase of mitochondria are different proteins although they have different locations.

The dual location of ferrochelatase in green plants is intriguing and may be important in regulating the balance of haem and chlorophyll pigments. Etiolated plants contain no chlorophyll but a near normal complement of haem proteins. On illumination the chlorophyll content rises dramatically whilst the haem protein content remains little altered. The pathway of protoporphyrin IX biosynthesis shown in Fig. 2 must be switched on fully but the protoporphyrin is directed almost exclusively to chlorophyll synthesis, not to haem. It is possible that in plants locally high concentrations of a regulator of ferrochelatase may accumulate in chloroplasts or mitochondria to control fluxes through the iron or magnesium branches of the biosynthetic path.

III. The Porphyrin Specificity of Ferrochelatase

A wide range of porphyrins has been tested as substrates for ferrochelatase, particularly in surveys of mammalian mitochondrial ferrochelatase (Porra and Jones, 1963b; Labbe, Hubbard, and Caughey, 1963; Honeybourne, Jackson, and Jones, 1979). The effect of changes in substitution on rings A and B of the tetrapyrrole whilst retaining the configuration of protoporphyrin IX on rings C and D are given in Fig. 4. It can be seen that protoporphyrin IX is a relatively poor substrate; reduction of its vinyl groups to ethyl yields mesoporphyrin, a better substrate. Removal of these

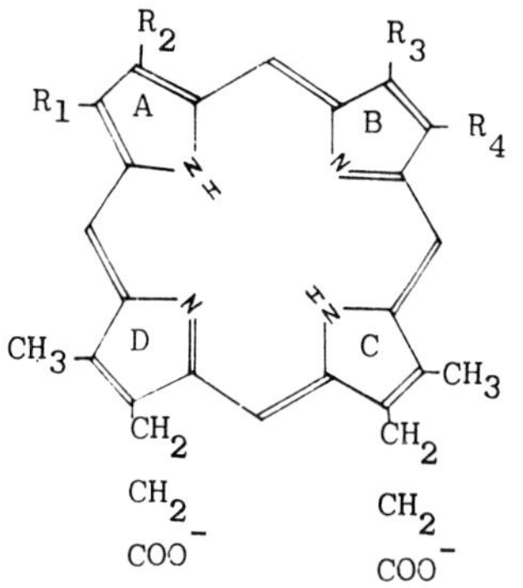

(a)

Trivial name	R_1	R_2	R_3	R_4	V_{max} (p mol min^{-1} mg^{-1} protein)
Protophorphyrin IX	Me	V	Me	V	42
Mesoporphyrin IX	Me	Et	Me	Et	280
Deuteroporphyrin IX	Me	H	Me	H	600
	H	H	H	H	1400
Coproporphyrin III	Me	P	Me	P	0

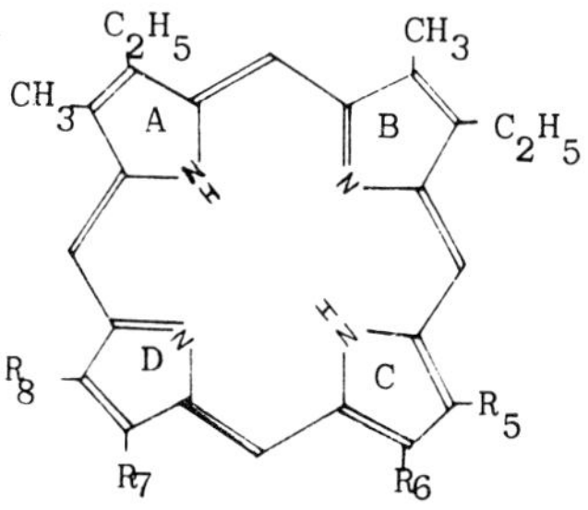

(b)

Trivial name	R_8	R_7	R_6	R_5	V_{max} (p mol min^{-1} mg^{-1} protein)
Mesoporphyrin IX	Me	P	P	Me	280
Mesoporphyrin I	P	Me	P	Me	262
Deuteroporphyrin III*	P	Me	Me	P	0
**	Me	P	Me	Et	30

Me is $-CH_3$; Et is $-C_2H_5$; P is $-CH_2CH_2COOH$; V is $-CH=CH_2$

* substitution on rings A and B as in deuterophophyrin IX (Fig. 4a)

** substitution on rings A and B as in mesoporphyrin IX

FIG. 4. Effect of substitution on rings A and B or on rings C and D on the effectiveness of porphyrins as substrates for ferrochelatase of rat liver mitochondria.

vinyl groups, as in deuteroporphyrin IX, makes a further improvement and complete removal of substituents in rings A and B yields the best substrate so far found. The natural substrate, protoporphyrin IX, is the most hydrophobic of this group of porphyrins and it is possible that it is a poor substrate because its product, protohaem IX, has a high affinity for the hydrophobic environment of the active centre of the enzyme. These structure-activity relationships have been found to hold good for plant mitochondrial and plastid ferrochelatase, as well as the mammalian enzyme. The presence of acidic substituents on rings A and B completely abolishes activity of the resulting porphyrins as substrates. These results are interesting since uroporphyrin and coproporphyrin have carboxylated side chains and can be formed by oxidation of the corresponding porphyrinogens from the biosynthetic pathway (Fig. 2).

Changes in the substitution of rings C and D of porphyrins also have considerable effect on ferrochelatase (Fig. 4). The best activities are found where propionate substituents are adjacent in positions 6 and 7, the natural isomer, or in positions 6 and 8. Separation of these propionates to 5 and 8 caused a complete loss of activity. When only one propionate substitutent was present at position 7 or together with a carboxylate at 6, then very low activity resulted. Neighbouring propionates appear to be necessary to attach the porphyrin to the appropriate site on ferrochelatase. Esterification of the propionates prevents the porphyrin ring acting as a substrate.

Protohaem is present in all aerobic cells as the prosthetic group of cytochromes b, haemoglobins, catalases or peroxidases. Other haem prosthetic groups are also found. Particularly important are haem a, a prosthetic group in the cytochrome oxidase complex, and haem c, part of the cytochrome c molecule (Fig. 5). The metal-free derivatives of these two haems have been prepared but are not substrates for ferrochelatase (Porra and Jones, 1963b; Jones and Jones, 1970). It is likely that haem a and cytochrome c are formed from protohaem by side chain modification and not by iron insertion into the modified porphyrin. Work with microorganisms supports this view. Sinclair, White, and Barrett (1967) were able to show that [14C]-labelled protohaem was taken up by Staphylococcus and incorporated fairly efficiently into haem a. Colleran and Jones (1973) carried out somewhat similar experiments with the slime mould *Physarum polycephalum*. This organism cannot grow in the absence of added protohaem. When protohaem was supplied, labelled with ^{59}Fe, it was incorporated very efficiently into cytochrome c. The cytochrome c, surprisingly, had a specific activity near half that of the added protohaem. It is possible that during the process of conjugation with protein the iron of the haem becomes susceptible to exchange with iron of the medium. It is known that in neutral aqueous solution iron is in the Fe^{3+} form in protohaem and is not readily exchanged with free iron in solution (Falk,

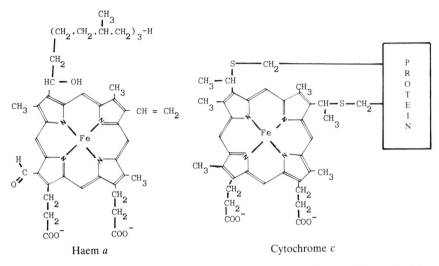

Haem *a*

Cytochrome *c*

Fig. 5. Structures of haem *a*, the haem prosthetic group of cytochrome oxidase and of the haem region of cytochrome *c*. In cytochrome *c* the haem is covalently attached to the protein.

1964, p. 39). During attachment to protein the Fe^{3+} of protohaem may be reduced to Fe^{2+} which is more exchangeable with soluble iron (Porra and Jones, 1963a).

The attachment of protohaem to apo-cytochrome *c* is likely to occur at or near the mitochondrial membrane. Apo-cytochrome *c*, not cytochrome *c*, is synthesized on the mammalian cytoplasmic ribosomes (Kadenbach, 1970) and in pulse labelling experiments apo-cytochrome *c* is shown to be a precursor of mitochondrial cytochrome *c* in *Neurospora crassa* (Korb and Neupert, 1978). A mitochondrial enzyme which catalyses this coupling of apo-cytochrome *c* with protohaem has now been reported (Basile, Di Bello, and Taniuchi, 1980).

In summary, ferrochelatase specificity and haem-feeding experiments support the view that protohaem is a common intermediate *en route* to all the respiratory cytochromes. Details of the reactions which modify its side chain are little known.

IV. The Metal Ion Specificity of Ferrochelatase
A. THE RANGE OF ION SUBSTRATES

Ferrochelatase from a variety of sources shows a wide specificity for metal ion substrates and there seems to be no difference in metal ion specificity between the enzyme from tissues that form chlorophyll, and so must have the capacity to insert Mg^{2+} into protoporphyrin, and those that form

TABLE II
Metal ion specificity of ferrochelatase of barley etioplasts

Metal ion	Concentration (μM)	Specific activity (nmol min^{-1} mg^{-1} protein)
Fe^{2+}	100	0·08
Mg^{2+}	400	0
Fe^{3+}	500	0
Ni^{2+}	400	0 or v. low
Co^{2+}	50	0·073
Mn^{2+}	500	0·033
Zn^{2+}	100	0·073[a]
Cu^{2+}	50	0·30[b]
Cd^{2+}	50	[c]

[a] Non enzymic activity was nearly equal to enzymic activity and was subtracted to obtain given value.
[b] Initial rate; rate declines after 50 s.
[c] Non-enzymic rate was so great (7nmol min^{-1}) that enzymic rate, if present, could not be assayed.
Assayed using a dual wavelength spectrophotometer at 498–510·5 nm. Reaction mixture (28°C) contained 100 mM Tris, pH 7·2, 0·1% Tween 80, 0·4 mg etioplast protein ml^{-1}, 25 μM deuteroporphyrin. Metal concentration was varied to determine maximum rate. Δ E_{nM} assumed to be 10.

haems only. Using a spectroscopic assay which measures porphyrin disappearance as metalloporphyrins are formed (Jones and Jones, 1969), it has been possible to investigate a range of metal ions as potential substrates. Results of such a survey for barley etioplast ferrochelatase are shown in Table II. The measurements of rates are not precise because some assumptions have been made about the absorption spectra of the metalloporphyrins but these are unlikely to introduce large errors. All the metal ions which are effective substrates are divalent cations of the transition series (but not all divalent metal cations are substrates!). Fe^{3+} was not incorporated; Fe^{2+} appears to be the natural substrate. This is an important observation because Fe^{2+} is relatively unstable at neutral or slightly alkaline pH and tends to autoxidise to Fe^{3+} and so some mechanism for maintaining iron in the ferrous state is necessary for ferrochelatase to function in synthesising haem. For the metal ions which were inserted enzymically, concentrations around 50 μM usually gave a measurable rate of metalloporphyrin synthesis.

Some metals, such as Cd^{2+}, Zn^{2+}, and Cu^{2+} are inserted into porphyrins non-enzymically as well as enzymically, and Zn and Cu-porphyrins are sometimes found in bacterial or yeast growth media. It is likely that at some stages of growth, small amounts of porphyrins are excreted by microorganisms and since Cu^{2+} and Zn^{2+} are present in the medium these metals are non-enzymically inserted into the porphyrins. Because Zn-protoporphyrin has spectroscopic properties resembling those of Mg-protoporphyrin, the former pigment is frequently confused with the latter, giving rise to false hopes that Mg^{2+} has been inserted during attempts at the enzymic synthesis of Mg protoporphyrin *in vitro*. There is usually sufficient contaminating Zn^{2+} in biological preparations and reagents to account for the metalloporphyrin produced.

Some metal ions, such as Pb^{2+}, Hg^{2+}, and Cu^{2+}, may inhibit ferrochelatase by binding to essential -SH groups of the enzyme. Hg^{2+} certainly inhibits normal ferrochelatase activity when Fe^{2+} is a substrate, and when Cu^{2+} is used there is an initial rate of metal incorporation, lasting about 50s, followed by a diminishing rate. This can be explained by Cu^{2+} acting both as substrate and inhibitor.

One of the most convenient cation substrates for assaying ferrochelatase is Co^{2+}. This is inserted, enzymically only, at about the same rate as Fe^{2+}, but unlike Fe^{2+} the divalent ion is the most stable form in aqueous solution. Indeed, Co^{3+} is a powerful oxidant and can oxidize water to oxygen. Competition studies show that the rate of porphyrin utilization is not increased when Fe^{2+} is added to a ferrochelatase assay mixture which is already working at its maximum rate with Co^{2+} as substrate, indicating that the same enzyme uses both metal ions. There is no difference in the porphyrin specificity of mammalian mitochondria or etioplasts with either Co^{2+} or Fe^{2+} as the metal ion substrates and Co^{2+} can be recommended as a metal substrate in investigations of ferrochelatase activity. It does have one drawback however: concentrations over 100 μM may be inhibitory, a phenomenon noted also by Labbe, Volland, and Chaix (1968) for yeast ferrochelatase.

In mammals, injection of solutions of some metal ions, particularly Co^{2+}, Sn^{2+}, Cd^{2+}, Cu^{2+}, Mn^{2+}, Zn^{2+}, and Ni^{2+}, causes the induction of the enzyme haem oxygenase in liver, spleen, or kidney microsomes (Maines and Kappas, 1976). This enzyme normally catalyses the break-down of protohaem from haemoglobin and cytochrome P-450 to form bile pigments. It is possible that "unnatural" metalloporphyrins formed as a result of the insertion of these ions into protoporphyrin by ferrochelatase may displace protohaem from haemoproteins, and the haem liberated may induce haem oxygenase. Although a process of protohaem breakdown is catalysed by plant microsomes (Hendry, Houghton, and Jones, 1981) the product has not been identified and nothing is known of the induction of the system.

B. THE FORMATION OF MAGNESIUM PROTOPORPHYRIN

Photosynthetic organisms contain a far higher concentration of magnesium

tetrapyrroles (chlorophylls) than haem pigments and early work of
Granick with algal mutants blocked in chlorophyll synthesis (Granick,
1948) strongly suggested that magnesium was inserted into protoporphyrin
to form magnesium protoporphyrin as an intermediate in chlorophyll
biosynthesis (see Fig. 2). Attempts to characterize the reaction *in vitro*
have been largely unsuccessful and most of the basic information was
obtained by Gorchein (1972, 1973) using whole cells of the photosynthetic
bacterium *Rhodopseudomonas sphaeroides*. In these bacteria, bacteriochlorophyll synthesis is regulated by both oxygen and by light. Oxygen
represses pigment production and the cells obtain energy by respiration
instead of photosynthesis. High light intensities lead to the growth of cells
containing low concentrations of bacteriochlorophyll. Gorchein found that
under semi-anaerobic conditions (low oxygen concentrations) in the light,
whole cells of *Rps. sphaeroides* incorporated magnesium ions into protoporphyrin which had been added to the suspension. The accumulated
product was magnesium protoporphyrin monomethyl ester, the next
intermediate in chlorophyll and bacteriochlorophyll synthesis and
Gorchein suggested that magnesium insertion was obligatorily coupled
with methylation. This latter suggestion cannot hold true for green plants
where some mutants which are blocked in chlorophyll synthesis and which
accumulate magnesium protoporphyrin, are known (see Granick, 1948).

Up to 50% of the added protoporphyrin was recovered as magnesium
protoporphyrin monomethyl ester and the reaction did not need concomitant protein synthesis. Partial pressures of oxygen greater than 15%
inhibited the reaction and cells grown under oxygen were completely
inactive. This fits a scheme for the control of pigment synthesis in the
photosynthetic bacteria proposed by Lascelles and Altshuler (1969) on the
basis of experiments with mutants. They suggested that magnesium
insertion was very sensitive to oxygen and so high aeration inhibits the
chlorophyll branch of tetrapyrrole pigment biosynthesis. The accumulated
protoporphyrin is then diverted to haem biosynthesis, in excess of the
available apo-protein of cytochromes. The excess haem acts as a feedback
inhibitor of ALA synthetase, the first enzyme in the biosynthetic pathway
leading to protoporphyrin. Thus a balance between the magnesium branch
and the iron branch of tetrapyrrole pigments is achieved (see Fig. 6).
Unfortunately, disruption of *Rps. sphaeroides* cells causes a loss of the
magnesium-insertion capacity and the Mg^{2+} insertion reaction has not
been further characterised in greater detail.

In higher plants it has been found (Castelfranco and Schwartz, 1978) that
crude plastid preparations from greening cucumber cotyledons catalyse the
production of magnesium protoporphyrin when supplied with L-glutamate
(a precursor of ALA), although the yields were small. The reaction
required ATP and oxygen and was stimulated by added $\alpha\alpha$ dipyridyl, an

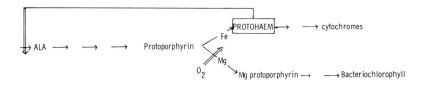

Oxygen is postulated to inhibit the insertion of Mg^{2+} into protoporphyrin. This leads
to a diversion of protoporphyrin to form protohaem in excess of apoprotein. The increased
free protohaem concentration inhibits ALA synthetase, the first enzyme in the pathway.

FIG. 6. Regulation of tetrapyrrole pigment synthesis in *Rhodopseudomonas sphaeroides* by
oxygen (after Lascelles and Altshuler, 1969).

iron-chelating agent known to stimulate production of magnesium protoporphyrin in whole leaves. The capacity of cucumber plastids to synthesize
magnesium protoporphyrin was increased when protoporphyrin was supplied (half saturation at 3·5 μM protoporphyrin) and was dependent upon
the presence of ATP (half maximum activity at 3·5 μM ATP). The
response of magnesium protoporphyrin formation to ATP, protoporphyrin
and Mg^{2+} was sigmoidal for each substrate (Fuesler, Wright, and Castelfranco, 1981), suggesting that the enzyme has allosteric properties, consistent with its position in the pathway of chlorophyll biosynthesis. A less well
defined system from the same source, which produces small amounts of
Mg-protoporphyrin monoester and Zn-protoprophyrin complexes has
been described by Rebeiz *et al.* (1975).

C. THE SUPPLY OF FE^{2+} FOR FERROCHELATASE

Iron is stored in mammalian spleen, liver, and bone marrow cells as iron
protein complex, ferritin; and in plants, within the plastids, as the closely
related phytoferritin (see article by Bienfait, this volume). Other transport
and storage complexes of iron are known in mammals, such as transferrin,
a protein which binds up to two atoms of iron per mole. Haemosiderin, a
less well defined aggregate of ferric hydroxide and protein, which may
contain up to 55% of its weight as iron, accumulates in iron overload. In
every case mentioned the iron is present in the ferric form and this is likely
to be true also in plants: in general, complexed iron has a lower
oxidation-reduction mid-point potential than free iron and so the oxidized
state is favoured in complexes. Utilization of iron by ferrochelatase is
therefore likely to involve the release of iron from complexed forms and its

Ferrochelatase was measured spectrophotometrically at 25^0 in 2.5ml of Tris buffer (I00mM, pH 7.2) containing I00µl of a post-climateric mitochondrial preparation (9 mg protein/ml), 0.1% Tween 80 and 40µM deuteroporphyrin. Co^{2+} (I0µl of I0mM $CoCl_2$), Fe^{3+} (I0µl of 50mM $FeCl_3$) and succinate (I0µl of IM) were added when shown.

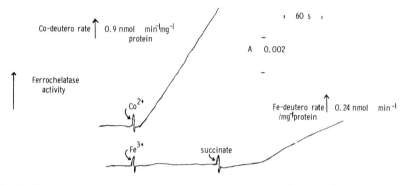

FIG. 7. Ferrochelatase of arum spadix mitochondria—use of Co^{2+} and Fe^{3+}.

reduction; these two processes may be almost simultaneous with reduction leading directly to release.

It has been found that rat liver mitochondria suspended in a ferrochelatase assay buffer which contains 0·1% Tween 80, added to solubilize porphyrins, will give a low rate of incorporation of iron into porphyrin using ferric chloride or ferric citrate as iron substrate if respiratory substrates such as NADH or succinate are also supplied (Barnes, Connelly, and Jones, 1972). When the mitochondrial suspension becomes anaerobic through respiratory activity then the rate of ferrochelatase greatly increases. A somewhat similar picture was observed with plant mitochondria, in this case prepared from the spadix of *Arum maculatum* (Fig. 7), and also with membrane particles from photosynthetic bacteria. The anaerobic rates of ferrochelatase are inhibited by rapid mixing of the cuvette contents with air, but are not inhibited by rotenone, when NADH is the electron donor, or by antimycin A, when succinate or NADH is electron donor. A likely explanation of these effects is that the respiratory substrates reduce a flavoprotein or quinone component of the electron transport system which can interact with Fe^{3+} and reduce it to Fe^{2+}. This Fe^{2+} can be used by ferrochelatase for haem synthesis, but in the presence of oxygen Fe^{2+} is rapidly oxidized back to Fe^{3+} and so ferrochelatase activity is relatively low. A representation of such a scheme for iron reduction is given in Fig. 8. Mitochondria can reduce oxygen directly to O_2^-, (superoxide), in the same region of the electron transport chain that we postulate for iron reduction (e.g. Boveris and Chance, 1973). Since we find that Fe^{3+} reduction is insensitive to superoxide dismutase and is increased in anaerobic conditions it is extremely unlikely that Fe^{3+} is

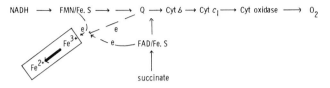

FIG. 8. A scheme to illustrate the likely sites of electron donation by the mitochondrial electron transport system to added Fe^{3+}, generating Fe^{2+} for ferrochelatase (FeS is iron sulphur protein, Q is ubiquinone). The Fe^{2+} produced can also react with oxygen, $Fe^{2+} + O_2 \rightarrow Fe^{3+} + O_2^-$.

reduced by superoxide: oxygen and Fe^{3+} may be alternative acceptors of electrons at this "leak" in the normal electron flow. The leaks to Fe^{3+} are a relatively minor diversion from normal respiration but may be sufficient to maintain a satisfactory rate of haem synthesis even under aerobic conditions.

An alternative method of studying the reduction of ferric complexes by mitochondria makes use of the reagent ferrozine (3-(2-pyridyl)-5,6-bis(4-phenyl-sulphonic acid)-1,2,4-triazine). This gives a strongly coloured complex with ferrous iron absorbing around 570 nm, and the progress of iron reduction can be monitored at this wavelength (Dailey and Lascelles, 1977). By using this reagent it has been shown that animal and plant mitochondria can reduce iron presented in a variety of complexed forms. Some examples of this are shown in Table III, where two microbial iron complexes were tested in addition to the more familiar iron complexes of eukaryotes. These two compounds, ferrichrome, and ferrioxamine B, are compounds produced by microorganisms to solubilize and transport inorganic iron for use in the cell (see Neilands, 1957, for a review). A large

TABLE III
Reduction of various iron complexes by isolated pea mitochondria

Iron complex	Conc. (μM)	Rate of Fe^{2+} formation (pmol min^{-1} mg^{-1})
Ferrichrome	5·4	6·3
Ferrioxamine B	6·8	6·3
Ferritin	? 10 (Fe^{3+})	21
Ovo-transferrin	12·5	3·4
Ferric citrate	10	2500
Ferredoxin	100	0
All rates abolished by stirring in air		

Release of Fe^{2+} was measured using ferrozine at 570–620 nm. The reaction mixture contained 9·6 mg mitochondrial protein, 300 μM ferrozine, 50 mM MOPS, 100 mM KCl, pH 7·0. It was preincubated with 4 mM succinate + 20 μM NADH at 25° before the iron complex was added.

group of such compounds was investigated as sources of Fe^{2+} in liver mitochondria by Barnes *et al.* (1972) using a less sensitive method depending on Fe^{2+} autoxidation and measuring the ensuing oxygen uptake. Even without searching for optimum conditions, iron was found to be reduced from a wide range of complexes of the sort that might be present in plant cells. The capacity for reduction of iron is not, apparently, likely to limit mitochondrial ferrochelatase. Access to iron may however, be important and this aspect will be discussed in the next section.

V. PERMEABILITY PROPERTIES OF MITOCHONDRIA AND THEIR EFFECTS ON FERROCHELATASE

In rat liver mitochondria it has been shown that the active site of ferrochelatase is on the inner membrane, facing inwards (Jones and Jones, 1969). Mitochondria are not freely permeable to cations and so movement of the iron substrate may present problems. Barnes *et al.* (1972) used intact and "inside out" mitochondrial preparations to demonstrate that the site of reduction of Fe^{3+} to Fe^{2+} was on the inside of the inner membrane, a location suitable for its use by ferrochelatase. Only one complex, ferrioxamine G, a microbicidal ferric iron hydroxamate, was able to transfer iron across the intact membrane in a form available for reduction. Ferrioxamine G, because of this property acted as an uncoupler of oxidative phosphorylation when supplied at 1 mM concentration, behaving like Ca^{2+} in this respect and suggesting that the iron complex is pulled across by the membrane potential and consequently dissipates the potential gradient. Solutions of $FeCl_3$ and ferric citrate did not uncouple in this way, confirming that the membrane is not freely permeable to them even when substrates are added to generate a membrane potential (negative on the inside of the mitochondria).

A different conclusion was arrived at by Romslo and Flatmark (1973a and b) who added radioactive $^{59}Fe(NO_3)_3$ complexed with sucrose to rat liver mitochondria, and used gradient centrifugation to recover the mitochondria. They then measured the radioactivity taken up into the mitochondrial pellet. They found two quantitatively similar uptake processes for iron, one energy-dependent (uncoupler and cyanide sensitive) and the other energy-independent. Since Ca^{2+} inhibited ^{59}Fe accumulation they suggest that Fe^{3+} may be taken up on a Ca^{2+} carrier. The maximum rate of iron uptake was obtained at concentrations of iron around 2–3 mM. Thus, Romslo and Flatmark believe that iron can be readily accumulated, probably on the Ca^{2+} carrier, whilst Barnes *et al.* (1972) speculate that a lipid soluble, or a specific complex may be necessary. These differences are puzzling and remain to be clarified. Different techniques involving different iron-complexes were used by the two groups and these differences may account for the lack of agreement.

The enzyme catalysing the oxidative decarboxylation of coproporphyrinogen III to protoporphyrinogen IX is located in the space between the inner and outer membrane of mitochondria (Elder and Evans, 1978; Grandchamp, Phung, and Nordmann, 1978) and the oxidation of protoporphyrinogen IX to protoporphyrin IX is catalysed by a mitochondrial enzyme (Poulson, 1976) although its precise location has not been determined. It is therefore necessary for either protoporphyrinogen or protoporphyrin to cross the mitochondrial membrane. The latter molecule is hydrophobic and so will readily enter the mitochondrial membrane. It has however been claimed that the uptake of protoporphyrin IX by isolated mitochondria can be driven by a transmembrane K^+ gradient, with high K^+ inside (Koller and Romslo, 1980), a process sensitive in part to respiratory uncouplers and to valinomycin. The metabolic significance of this process is unclear because an increase of the external concentration of K^+ to 100 mM *increases* the extent of uptake by protoporphyrin by the mitochondria although the uncoupler-insensitive component of K^+ uptake almost disappears at 100 mM external K^+.

VI. FERROCHELATASE OF *RHODOPSEUDOMONAS SPHAEROIDES*

Membranes from the photosynthetic bacterium *Rps. sphaeroides* contain an extremely active ferrochelatase. This system is attractive for study because *Rps. sphaeroides* can be grown with or without active chlorophyll synthesis (i.e. anaerobically or aerobically) and so oxygen must control the destination of protoporphyrin to either magnesium or iron pigments, just as in green plants. The high ferrochelatase activity and low light scattering of particles from a blue-green mutant of *Rps. sphaeroides* permitted an accurate kinetic study to be made of its ferrochelation using the convenient optical method (Jones and Jones, 1970). Ferrochelatase is a two substrate enzyme (porphyrin and metal) and the effects of changes in porphyrin concentration upon K_m of the metal substrate and V_{max} of the ferrochelatase and the effect of changes in metal concentration of the porphyrin K_m were investigated. Each substrate failed to affect the K_m of the other, suggesting that the reaction proceeds by a sequential mechanism, with both substrates binding before the product is released and that the binding of one does not affect the binding of the second substrate.

Very accurate figures for the K_m were obtained: 6·13 μM for Co^{2+} and 21·3 μM for deuteroporphyrin. It was also possible to investigate inhibitory effects of the products of metal insertion into protoporphyrin IX, i.e. protohaem and magnesium protoporphyrin. Both inhibited ferrochelatase with 50% inhibition obtained at around 5–10 μM. Protohaem inhibition was non-competitive with deuteroporphyrin (Ki, 15·1 μM) and with Co^{2+} (Ki, 26·9 μM). Magnesium protoporphyrin inhibition was kinetically less well defined ("mixed" and non-competitive inhibition) but the values of Ki were a little lower. A number of other metalloporphyrins were tested as inhibitors of ferrochelatase and it was found that cytochrome c, haemoglobin, magnesium protoporphyrin dimethyl ester, and protoporphyrin dimethyl ester were ineffective; Co-deuteroporphyrin and deuterohaem were less effective than protohaem or magnesium protoporphyrin.

These results present the possibility that ferrochelatase activity may be regulated by its product, protohaem. If it is produced in excess of requirements and not withdrawn by some specific binding protein, it may accumulate in the hydrophobic environment of the inner mitochondrial membrane and inhibit further haem synthesis. The inhibitory concentrations of protohaem involved are lower than those found necessary to inhibit ALA synthetase by feedback inhibition (Burnham and Lascelles, 1963), and the sensitivity to magnesium protoporphyrin and to protohaem offer scope for the regulation of haem and chlorophyll synthesis. In carrying out these kinetic studies it is necessary to maintain porphyrins and metalloporphyrins in solution by the addition of detergent (0·1% Tween 80). This may affect the relative sensitivity of the ferrochelatase to hydrophobic inhibitors.

VII. Factors Affecting the Activity of Ferrochelatase

Purification of ferrochelatase has proved difficult, with apparent rapid inactivation accompanying purification steps. An interesting attempt (Mailer *et al.*, 1980) using an affinity column with a porphyrin immobilized on Sepharose, gave a low recovery of ferrochelatase with a specific activity less than would be expected. The eluted protein ran as a single band in SDS gel electrophoresis with an apparent mol. wt. of 63 000. However, the enzyme has now been purified to apparent homogeneity from rat liver mitochondria (Taketani and Tokunaga, 1981). It has a minimum molecular weight of 42 000, a K_m for porphyrins near 30 μM and for iron near 35 μM. Its activity is stimulated by fatty acids.

The enzyme from mammalian mitochondria, plant mitochondria and plastids (Little and Jones, 1976) as well as bacteria, is sensitive to inhibition by the product, protohaem and 50% inhibition can be achieved at concentrations around 5 μM protohaem. If the product is not removed, the enzyme is inhibited.

Bacterial and mammalian ferrochelatases are stimulated by some organic solvents such as ether and acetone (Mazanowska, Neuberger, and Tait,

1966); increases of four or five-fold were obtained at appropriate concentrations. The stimulations took place without affecting the Km of ferrochelatase for porphyrin although the Km for metal ions was increased (Jones, O. T. G. and Randle J, unpublished). Changes in the fluidity of the lipid environment of the ferrochelatase may account for the stimulations, and similar effects may explain the reported stimulatory effects of phospholipids upon ferrochelatase (Yoshikawa and Yoneyama, 1964; Mazanowska et al., 1966; Labbe et al., 1968; Sawada et al., 1969; Simpson and Poulson, 1977). In general, depletion of vertebrate or yeast membranes of phospholipids causes a loss of ferrochelatase activity which can be restored by adding back phospholipids and some fatty acids such as palmitate. Reports of the optimum phospholipid addition vary and the significance of these effects is uncertain; ferrochelatase appears to have a requirement for certain lipids in its environment to stabilize its active conformation.

Thiol groups appear to be essential for ferrochelatase activity, since it is inhibited by thiol-binding reagents such as N-ethylmaleimide, p chloromercuribenzoate, and N-phenylmaleimide, as well as by Hg^{2+} and Pb^{2+}.

REFERENCES

Barnes, R., Connelly, J. L., and Jones, O. T. G. (1972). Biochem. J. **128**, 1043–55.

Basile, G., Di Bello, C., and Taniuchi, H. (1980). J. Biol. Chem. **255**, 7181–91.

Boveris, A. and Chance, B. (1973). Biochem. J. **134**, 707–716.

Burnham, B. F. and Lascelles, J. (1963). Biochem. J. **87**, 462–472.

Carell, E. F. and Kahn, J. S. (1964). Arch. Biochem. Biophys. **108**, 343–348.

Castelfranco, P. A. and Jones, O. T. G. (1975). Plant Physiol. **55**, 485–490.

Castelfranco P. A. and Schwartz, S. (1978). Arch. Biochem. Biophys. **186**, 365–375.

Colleran, E. M. and Jones, O. T. G. (1973). Biochem. J. **134**, 89–96.

Dailey, H. A. and Lascelles, J. (1977). J. Bact. **129**, 815–820.

De Matteis, F., Gibbs, A. F., and Smith, A. G. (1980). Biochem. J. **189**, 645–648.

Elder, G. H. and Evans, J. O. (1978). Biochem. J. **172**, 345–347.

Falk, J. E. (1964). "Porphyrins and Metalloporphyrins", Elsevier, Amsterdam.

Fuesler, T. P., Wright, L. A., and Castelfranco, P. A. (1981). Plant Physiol. **67**, 246–249.

Gorchein, A. (1972). Biochem. J. **127**, 97–106.

Gorchein, A. (1973). Biochem. J. **134**, 833–845.

Granick, S. (1948). J. Biol. Chem. **175**, 333–342.

Grandchamp, B., Phung, N., and Nordman, Y. (1978). Biochem. J. **176**, 98–102.

Hendry, G. A. F., Houghton, J. D., and Jones, O. T. G. (1981). Biochem. J. **194**, 743–751.

Honeybourne, C. L., Jackson, J. T., and Jones, O. T. G. (1979). FEBS Lett. **98**, 207–210.

Jones, O. T. G. (1967). Biochem. Biophys. Res. Commun. **28**, 671–674.

Jones, O. T. G. (1968). Biochem. J. **107**, 113–119.

Jones, O. T. G. (1979). In "The Porphyrins", Vol. 6. pp. 179–232, (D. Dolphin, ed.), Academic Press, London and New York.

Jones, M. S. and Jones, O. T. G. (1969). *Biochem. J.* **113,** 507–514.
Jones, M. S. and Jones, O. T. G. (1970). *Biochem. J.* **119,** 453–462.
Kadenbach, B. (1970). *Eur. J. Biochem.* **12,** 392–398.
Koller, M-E. and Romslo, I. (1980). *Biochem. J.* **188,** 329–335.
Korb, H. and Neupert, W. (1978). *Eur. J. Biochem.* **91,** 609–620.
Labbe, R. E. and Hubbard, N. (1961). *Biochim. Biophys. Acta* **52,** 130–135.
Labbe, R. F., Hubbard, N., and Caughey, W. S. (1963). *Biochem.* **2,** 372–374.
Labbe, P., Volland, C., and Chaix, P. (1968). *Biochem. Biophys. Acta* **159,**
    527–539.
Lascelles, J. and Altshuler, T. (1969). *J. Bact.* **98,** 721–727.
Little, H. N. and Jones, O. T. G. (1976). *Biochem. J.* **156,** 309–314.
Mailer, K., Poulson, R., Dolphin, D., and Hamilton, A. D. (1980). *Biochem.
    Biophys. Res. Commun.* **96,** 777–784.
Maines, M. D. and Kappas, A. (1976). *Biochem. J.* **154,** 125–131.
Mazanowska, A. M., Neuberger, A., and Tait, G. H. (1966). *Biochem. J.* **98,**
    117–127.
Neilands, J. B. (1957). *Bact. Review* **21,** 101–111.
Nishida, G. and Labbe, R. F. (1959). *Biochim. Biophys. Acta* **31,** 519–524.
Ochoa, S. and De Haro, C. (1979). *Ann. Rev. Biochem.* **48,** 549–80.
Porra, R. J. and Jones, O. T. G. (1963a). *Biochem. J.* **87,** 181–185.
Porra, R. J. and Jones, O. T. G. (1963b). *Biochem. J.* **87,** 186–192.
Porra, R. J. and Lascelles, J. (1968). *Biochem. J.* **108,** 343–348.
Porra, R. J., Vitols, K. S., Labbe, R. F., and Newton, N. A. (1967). *Biochem. J.*
    104, 321–327.
Poulson, R. (1976). *J. Biol. Chem.* **251,** 3730–3733.
Rebeiz, C. A. and Castelfranco, P. A. (1973). *Ann. Rev. Plant Physiol.* **24,**
    129–172.
Rebeiz, C. A., Mattheis, J. R., Smith, B. B., Rebeiz, C. C., and Dayton, D. E.
    (1975). *Arch. Biochem. Biophys.* **166,** 446–465.
Romslo, I. and Flatmark, T. (1973a). *Biochim. Biophys. Acta* **305,** 29–40.
Romslo, I. and Flatmark, T. (1973b). *Biochim. Biophys. Acta* **325,** 38–46.
Sawada, H., Takeshita, M., Sugita, Y., and Yoneyama, Y. (1969). *Biochim.
    Biophys, Acta* **178,** 145–155.
Simpson, D. M. and Poulson, R. (1977). *Biochim. Biophys. Acta* **482,** 461–469.
Sinclair, P., White, D. C., and Barrett, J. (1967). *Biochim. Biophys. Acta* **143,**
    427–428.
Taketani, S. and Tokunaga, R. (1981). *J. Biol. Chem.* **256,** 12748–12753.
Yoshikawa, H. and Yoneyama, Y. (1964). Int. Symp. on Iron Metabolism, p. 24,
    (F. Gross, ed.), Springer-Verlag, Berlin.

PART 3

Forms and Functions

Some ways in which metals are involved
in physiological and metabolic processes
in plants

. . . as I cast my eyes
I see what was, and is and will abide;
Still glides the stream, and shall for ever glide;
The Form remains, the Function never dies;

(Wordsworth, Valediction to the river Duddon)

CHAPTER 9

Metals and Photosynthesis

M. F. HIPKINS

Department of Botany, Glasgow University, Glasgow, Scotland

I. INTRODUCTION

The overall reaction of photosynthesis in higher plants involves the use of light energy to convert the "energy-poor" compounds, water and carbon dioxide, to "energy-rich" carbohydrate. This may be expressed in the approximate equation:

$$H_2O + CO_2 \xrightarrow{h\nu} \underset{\text{carbohydrate}}{(CH_2O)} + O_2$$

The overall reaction may be divided into two parts: the first part (the "light reactions") is the light-dependent oxidation of water and reduction of NADP, together with the phosphorylation of ADP to ATP; the second part (the "dark reactions") uses the ATP and reducing equivalents generated by the light reactions to fix carbon dioxide to the level of a sugar via the Calvin cycle. The division of photosynthesis into two parts in this manner is undoubtedly artificial, but it is nevertheless convenient, since each part may be discussed separately.

In this chapter I am going to make a selective review of the role of metals in photosynthesis, and in doing so will concentrate on the light reactions since the function of metals in these reactions is starting to be elucidated. The light reactions comprise a group of processes which include

 (i) the capture of light energy;
(ii) the transfer of light energy to specialized chlorophyll molecules called reaction centres;
(iii) the light-induced separation of charge at the reaction centres; followed by
(iv) the transfer of electrons from water to NADP and
 (v) the movement of protons through the thylakoid membrane which leads to the phosphorylation of ADP to ATP.

Metals are involved in each of these processes, from the presence of Mg in cholorophyll to the recently discovered, rather subtle mechanisms which control the flux of light energy into photosystem I and photosystem II.

When discussing the roles and effects of metals in photosynthesis it is important to know the ionic content of the chloroplast. Here it is necessary to distinguish between (a) intact chloroplasts which retain their bounding envelope containing the stroma, and which are capable of fixing carbon dioxide in a light-dependent reaction, and (b) broken chloroplasts (thylakoid membranes) whose bounding envelope and stroma have been lost during isolation, and which as a consequence can only reduce exogenous electron acceptors. Clearly, any complete description of the ionic content of chloroplasts must use intact chloroplasts (see Barber, 1976). The results of an experiment on the ionic analysis of intact pea chloroplasts using the technique of neutron activation are shown in Table I (Nakatani et al., 1979). While the major intrachloroplastic cation is K^+, the level of Mg^{2+} is only slightly less. The importance of using intact chloroplasts for such a

TABLE I
Ionic analysis of intact pea chloroplasts by neutron activation
(from Nakatani et al., 1979)

Element	μmol. $(mgChl)^{-1} \pm$ SEM of 17 sets
Na	0.22 ± 0.02
K	2.56 ± 0.06
Mg	1.72 ± 0.04
Ca	0.76 ± 0.03
Cl	0.21 ± 0.01
Mn	0.017 ± 0.0003

study is emphasized by the observation (Nakatani *et al.*, 1979) that isolated broken chloroplasts bind cations as a function of the ionic composition of the medium in which they are suspended, and that the major controlling factor is the electrostatic neutralization of surface negative charges on the surface of the thylakoid membrane.

II. METALS IN PHOTOSYNTHETIC ELECTRON TRANSPORT

The photosynthetic electron transport chain is illustrated in Fig. 1. It is an extension of the original "Z-scheme" proposed by Hill and Bendall (1960)

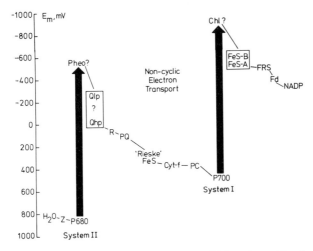

FIG. 1. The Z-scheme for non-cyclic electron transport in higher plants. The diagram is on a scale of oxidation-reduction potential, with the most reducing intermediates at the top. The components are fully described in the text: some have been omitted from the reducing side of photosystem I for clarity. Where there is doubt about the order or significance of components, they are enclosed in boxes.

and shows the pathway of light-driven electron movement from water to NADP. The pathway involves two photochemically active chlorophyll-protein complexes, P680 and P700, which serve to transform light energy into oxidation-reduction energy through light-induced charge separation. In photosystem II, P680 appears to have a pheophytin molecule as an electron acceptor (Klimov *et al.*, 1977), and electrons are then passed to the next member of the electron transport chain, *Q*. Up to a few years ago *Q* was thought to be a single entity, but recent experiments on the oxidation-reduction potential dependence of the chlorophyll fluorescence yield of photosystem II (see, for example, Horton and Croze, 1979), and of the flash-induced field-indicating absorbance change (Malkin, 1978;

Malkin and Barber, 1979; see also Junge, 1977) have led to the idea that there may be two pools of Q. The higher potential pool Q_{hp}, (mid-point redox potential at pH 7·6, $E_{m7·6} \sim 25$ mV, $n = 1$) is equivalent to the previously assigned electron transport component; the lower potential pool, Q_{lp}, ($E_{m7·6} \sim -270$ mV, $n = 1$) has not yet been assigned a function (Malkin and Barber, 1979).

In photosystem I, P700 may have monomer chlorophyll as primary electron acceptor with a mid-point redox potential in the region of -900 mV (Heathcote et al., 1979; Baltimore and Malkin, 1980), then an intermediate component X which is possibly an iron-sulphur centre with $E_m \sim -730$ mV (Bolton, 1977; Heathcote et al., 1979). Electrons then pass to two bound iron-sulphur centres, Centre B ($E_m \sim -590$ mV) and Centre A ($E_m \sim -540$ mV). All these electron transport components have been studied by use of electron paramagnetic resonance (epr): a relationship between them and the optical component P430 (Ke, 1973) has been suggested by Bolton (1977) who considers that P430 corresponds to Centre A or Centre B, or possibly both.

The chain of electron acceptors between photosystems II and I involves both electron carriers and hydrogen carriers, which is in agreement with Mitchell's chemiosmotic hypothesis for photophosphorylation (Mitchell, 1966). This region of the electron transport chain has been reviewed recently by Velthuys (1980). It begins with the component that accepts electrons from Q: since Q is a one-electron carrier, and plastoquinone (PQ) transports two hydrogen atoms, there is a component R (also called B) which accepts electrons singly from Q and donates them in pairs, together with a pair of protons, to PQ (Fowler, 1977a). Thus under flash illumination the degree of reduction of R exhibits cycles with period two in flash number (Bouges-Bocquet, 1973; Velthuys and Amesz, 1974). Plastoquinone is oxidized by a bound iron-sulphur centre ("Rieske" FeS), with an $E_m \sim 290$ mV which is itself oxidized by cytochrome-f. The cytochrome donates electrons to P700 via the copper-containing protein plastocyanin, (PC).

On the reducing side of photosystem I electron flow from the bound iron-sulphur centres proceeds through ferredoxin-reducing substance (FRS) (Yocum and San Pietro, 1969, 1970), and the soluble iron-sulphur protein ferredoxin (Fd) then through ferredoxin-NADP reductase to NADP, the terminal electron acceptor.

On the oxidising side of photosystem II the pathway from water to P680 is not entirely clear, but undoubtedly involves a manganese-protein complex.

In addition to the non-cyclic electron pathway just described there is a cyclic flow of electrons round photosystem I. Figure 2 shows that it involves the bound iron-sulphur centres, cytochrome-b_6 and several com-

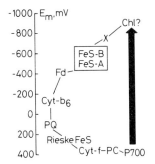

F<small>IG</small>. 2. A simplified scheme for cyclic electron flow round photosystem I. (For details, see text, and legend to Fig. 1).

ponents of the non-cyclic electron transport chain. Recent experiments on the slow phase of increase of the flash-induced field-indicating absorbance change suggest that this scheme is too simple, but more complex schemes which also deal with the oxidation of plastoquinone (see, for example, Crowther and Hind, 1980; Olsen et al., 1980; and Velthuys, 1980, for review) remain speculative at present.

It should also be pointed out that neither the strict independence of electron transport chains nor the one-to-one stoichiometry between photosystems I and II that is implied by the Z-scheme need be the case (Velthuys, 1980). There is evidence that up to ten electron transport chains may interact at the level of plastoquinone, which exists as a pool of several molecules per chain (Williams, 1977). Moreover, it has been estimated that as much as 25% of the total chlorophyll of photosystem I is functionally isolated from photosystem II (Haehnel, 1976). This probably reflects the finding that stromal lamellae appear to contain mostly photosystem I, and that in the granal stacks, photosystem I is confined to the edges (Arntzen and Briantais, 1975). The functionally isolated photosystem I centres may be involved in cyclic electron flow only.

A. COPPER

The copper-containing protein, plastocyanin, accounts for roughly half the copper found in the chloroplast. It has been found in a variety of higher plants and algae (Katoh, 1977). The protein has a molecular weight of 10 000 to 20 000 with some variation depending on the organism from which it is isolated (Ramshaw et al., 1973). Amino acid analyses of plastocyanins isolated from a number of sources show a large number of conserved residues (Aitken, 1975; Ramshaw et al., 1976). Recently the structure of poplar plastocyanin has been determined by X-ray crystallography (Colman *et al.*, 1978). The molecule is shaped like a flattened cylinder, with eight strands of the polypeptide chain roughly parallel to the axis of the cylinder. The copper is bound to two imidazole groups (His 37, His 87) a thiol group (Cys 84) and a thioether (Met 92) at a site that is embedded between the ends of two of the strands (Fig. 3).

Plastocyanin has a mid-point redox potential of about 370 mV in the pH range 5·4 to 9·9 (Katoh *et al.*, 1962), its oxidation and reduction may be detected spectrophotometrically, since oxidized plastocyanin absorbs at 597 nm, but this absorption is bleached on reduction (Katoh, 1977). Studies with antibodies (Hauska *et al.*, 1971) suggest that plastocyanin is located on the inside of the thylakoid membrane.

B. HAEM IRON

Three cytochromes have been identified in higher plant chloroplasts. Two are *b*-type, and one is *c*-type (see Jones, this volume). The *c*-type cytochrome is called cytochrome-*f*, and is involved in non-cyclic electron transport, where it donates electrons to plastocyanin. The mid-point redox potential is close to that of plastocyanin ($E_{m7} = 365$ mV). The molecular weight per haem has been found to be 65 000, but values half as large have also been found (see Cramer, 1977).

Both *b*-type cytochromes have now been isolated as multi-subunit lipoproteins. Cytochrome-b_6 has a protein part with a molecular weight about 40 000 (Stuart and Wasserman, 1975) whilst cyt *b*-559 has a proteinaceous part of molecular weight 46 000 composed of eight subunits each of roughly similar molecular weight (Garewal and Wasserman, 1974).

The roles of the two *b*-type cytochromes in electron transport have received considerable attention (Velthuys, 1980). Cytochrome b_6 with a mid-point potential in the region of 0 to −180 mV, is usually considered to be involved in cyclic electron flow mediated by photosystem I, but, as mentioned previously it has been suggested that cyt-b_6 also participates in non-cyclic electron flow on the oxidizing side of plastoquinone.

Cytochrome *b*-559 remains something of an enigma (Cramer and Whitmarsh, 1977). It can exist in two forms: high potential ($E_{m7} \geqslant 300$ mV) and low potential, ($E_{m7} \sim 75$ mV) and several roles have been discussed. It is well-established that at cryogenic temperatures where the oxidation of water is inhibited, cyt *b*-559 is the electron donor to photosystem II (Knaff and Arnon, 1969). In addition it has been associated with the oxidation of water (Horton *et al.*, 1975), in a cycle around photosystem II (Kok *et al.*, 1975), but it remains poorly understood (see Butler, 1978, for review).

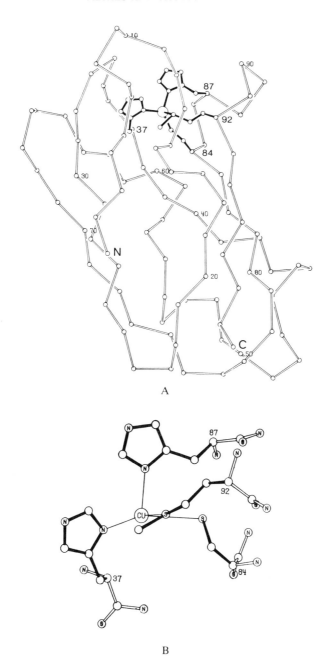

A

B

FIG. 3. A. The polypeptide chain of poplar plastocyanin (from Freeman 1981).
B. The copper binding site in plastocyanin, showing the four ligand residues (His 37, Cys 84,
His 87 and Met 92) (from Colman *et al.*, 1978).

C. NON-HAEM IRON

Both soluble and bound iron-sulphur centres participate in photosynthetic electron transport. Ferredoxin is perhaps the best known. It is a soluble iron-sulphur protein with a molecular weight of 10 500 to 11 000 and a low redox potential ($E_{m7} = -420\,mV$). There are two atoms of iron and labile sulphur per molecule. Ferredoxins from a number of higher plants and algae have been subject to amino acid sequence analysis, and show a striking homology (see Hall and Rao, 1977). The nature of the active centre has been studied by a variety of physical techniques including X-ray crystallography (Fukuyama et al., 1980). The iron atoms form a bimolecular cluster in which each iron atom is coordinated to two thiol groups of cysteine residues, and in which the labile sulphur atoms act as bridging ligands to the iron atoms (Fig. 4). It has been suggested that on reduction

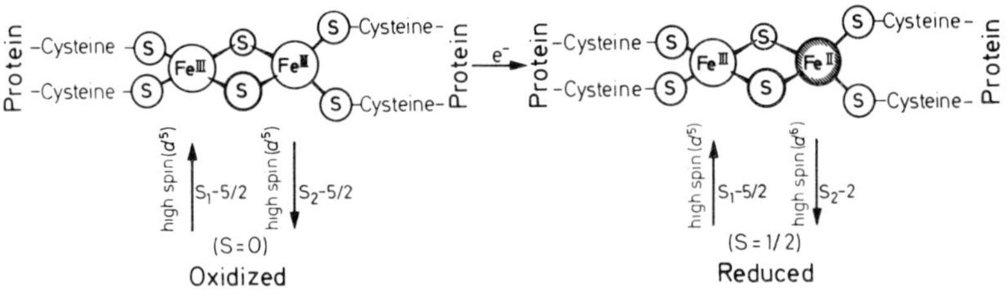

FIG. 4. The proposed structure of the prosthetic group of ferredoxin (from Hall and Rao, 1977).

only one of the two iron atoms is reduced from the Fe^{3+} form to Fe^{2+} (Rao et al., 1971). As shown in Fig. 1, ferredoxin accepts electrons from the bound iron-sulphur centres via FRS and donates them to the ferredoxin-NADP reductase. Because iron-sulphur centres show only small absorbance changes when oxidized and reduced, the technique of choice in their detection has been light-induced changes in the chloroplast epr spectrum. The technique, coupled with oxidation-reduction potentiometry, has been of great use in elucidating electron transport in photosystem I. The association of three further iron-sulphur centres, X, B, and A with photosystem I has already been described.

    Electron paramagnetic resonance experiments also revealed the high-potential iron-sulphur centre ($E_m = 290\,mV$ independent of pH in the range 6·5 to 8·0, $n = 1$, $g = 1·89$) that appears similar to that found in Centre III of mitochondria (Rieske et al., 1964). By using a mutant of Lemna perpusilla which lacks the epr signal of this so-called Rieske centre, and by oxidation-reduction potentionmetry, Malkin and co-workers (Malkin and Aparicio, 1975; Malkin and Posner, 1978) have assigned a position to the Rieske centre between plastoquinone and cytochrome-f in the electron transport chain.

D. MANGANESE

Manganese has been recognized as essential for the normal function of photosystem II for over 20 years (Kessler *et al.*, 1957). Since that time research into photosystem II has been intense, but the precise role that manganese plays in the evolution of oxygen from water is not yet clear.

The effects of Mn depletion on the reactions of photosystem II (Cheniae and Martin, 1966) suggest that manganese is involved on the oxidizing side of P680. The Mn content of isolated chloroplasts is variable but values of 27–29 Mn/400 Chl are typical where 400 Chl molecules are taken to correspond to one photosynthetic unit. The chloroplast Mn is not functionally homogeneous: the Mn abundance can be diminished by washing to 5–8 Mn/400 Chl without affecting photosynthetic reactions, suggesting that there is a large pool of photosynthetically non-functional Mn (Cheniae and Martin, 1970). In contrast, the 5–8 Mn 400 Chl^{-1} which are functional in photosystem II ("bound Mn") are not removed by washing, or washing with chelating agents. However, they can be released by

  (i) treatment with 0·8M Tris;
  (ii) treatment with hydroxylamine;
  (iii) high concentrations (>0·2M) Mg^{2+};
  (iv) acid treatment; or,
  (v) temperature shock.

There is a correlation of the loss of about two-thirds of the bound Mn with loss in photosynthetic activity (Cheniae and Martin, 1970). This suggests that the bound Mn is also functionally heterogeneous, with 3–6 Mn 400 Chl^{-1} associated with the evolution of oxygen, and a remaining pool of about 2 Mn 400 Chl^{-1} which is tightly bound, and not intimately involved in the oxidation of water.

The manganese detected by epr at room temperature is the aqueous Mn complex $(Mn(H_2O)_6)^{2+}$ which exhibits a characteristic hyperfine structure of six peaks. Other oxidation states of Mn are not detected, and complexing of Mn^{2+} with, for example, protein, so broadens the hyperfine structure that it usually becomes undetectable. Figure 5 shows the result of a typical experiment: sucrose-washed chloroplasts show only a weak Mn epr signal, but after Tris washing (which inhibits oxygen evolution) the aqueous Mn signal appears in the pellet, but not the supernatant. This experiment suggests that on washing with Tris, previously bound Mn is released into the intrathylakoid space as free Mn. Quantitative studies indicate that the released Mn is freely rotating, and that it represents some 60% of the total chloroplast-bound Mn (Blankenship and Sauer, 1974).

Research into the mechanism by which water is oxidized by photosystem II was greatly assisted by the development of a rapidly-responding oxygen electrode (Joliot, 1966). Use of the electrode makes it possible to

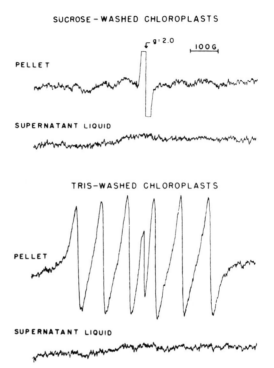

FIG 5. Electron paramagnetic resonance (epr) spectroscopy of isolated chloroplasts washed with sucrose or with 0·8 M Tris, pH 8·0. The signal at $g = 2·0$ is associated with the reaction centre of photosystem II. Sucrose-washed chloroplasts show weak Mn signals in both the pellet and supernatant, suggesting that the Mn in the chloroplast is bound. On washing with Tris the pellet, but not the supernatant, shows the Mn epr signal (from Blankenship and Sauer, 1974).

illuminate algae or chloroplasts with a train of flashes, and to measure the amount of oxygen elicited by each flash. Figure 6 shows a typical trace for chloroplasts that are thoroughly dark-adapted. The evolution of oxygen shows a cyclical pattern of period four, which damps out rather rapidly to yield an equal quantity of oxygen on each flash. The simplest model to account for these observations was presented by Kok *et al.* (1970) and is shown in Fig. 7. It proposes that each photosystem II centre is independent, and has a cyclic charge-storage mechanism consisting of four so-called S states. The subscript on each S state denotes the number of oxidizing equivalents stored on it. A quantum of light causes a transition $S_n \rightarrow S_{n+1}$; S_4 is transiently present, and oxidizes two molecules of water to yield molecular oxygen. The states S_0 and S_1 are stable in the dark; S_2 and S_3 decay in the dark, principally to S_1, so that in a dark-adapted preparation roughly 75% of centres are in state S_1 and 25% in state S_0. To account for the damping, it is assumed that on each flash a fraction of

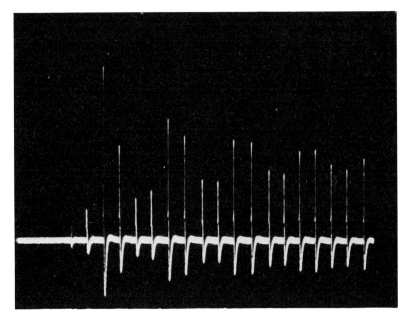

Fig. 6. A typical trace of oxygen output from a sample of isolated pea chloroplasts illuminated with a train of 5 μsec flashes separated by 500 msec in a Joliot-type oxygen electrode (Joliot, 1966). The output of the electrode is differentiated with respect to time for clarity; each of the spikes represents oxygen evolution elicited by a single flash. Note the absence of oxygen on the first flash, then maxima on flashes 3, 7 and 11 (M. F. Hipkins, unpublished).

centres suffers a "miss" $(S_n \rightarrow S_n)$ and a fraction undergoes "double-hits" $(S_n \rightarrow S_{n+2})$ which tends to destroy the synchrony between one centre and the next. The hypothesis is purely mechanistic, and it does not attempt a chemical identification of the charge storage entity. Nevertheless, it has often been suggested that Mn plays a part in charge storage, since it has such a variety of oxidation states.

The very characteristic period of four in flash-induced oxygen yield from dark-adapted photosynthetic systems provides a useful fingerprint for a

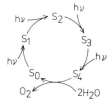

Fig. 7. The 'S-states' model for the evolution of oxygen in photosynthesis. The model proposes a cycle of five S-states which differ in their oxidizing power; the subscript n denotes the number of stored oxidizing equivalents. Only state S_4 can oxidize 2 molecules of water to yield molecular oxygen. For details, see text.

process which is closely linked to the oxygen-evolving process. A number of phenomena associated with photosystem II are known to exhibit period four oscillations (Diner and Joliot, 1977). But only recently has such behaviour been observed by the Mn epr signal (Wydrzynski and Sauer, 1980). The experiments examined the extent of the free Mn epr signal when a sample of isolated EDTA-washed spinach chloroplasts was pre-illuminated with a number of flashes and then subjected to mild heat treatment to release Mn. Wydrzynski and Sauer (1980) found that the quantity of heat-released Mn showed a cyclical oscillation with period four, and that these oscillations were essentially suppressed by an inhibitor of electron transport dichlorophenyldimethylurea (DCMU). These results provide evidence that Mn is closely associated with the oxidation of water in higher plants.

Govindjee and co-workers have made considerable use of another technique to study the state of manganese in photosynthetic systems. This is the measurement of the relaxation rates of water protons using nuclear magnetic resonance. The basis of the technique is that the proton relaxation rate (prr) is influenced by paramagnetic ions (see Govindjee et al., 1978). The results of the experiments show qualitatively that prr is also related to the flash number with a cycle of period four. But there is some doubt as to whether prr monitors functional Mn (Sharp and Yocum, 1980; Robinson et al., 1980).

Apart from these physical techniques, biochemical techniques have also been used to resolve the role of Mn in the oxidation of water. There have been several reports of Mn-protein complexes isolated from diatoms, green algae, and higher plants. Foyer and Hall (1979) have used a rapid technique to isolate the light-harvesting chlorophyll a/b protein complex (see Thornber et al., 1976), and found the complex to contain 50–70 ng Mn g^{-1} protein. The complex showed cyanide-insensitive superoxide dismutase activity. It is possible that the manganese-protein complex reported by Lagoutte and Duranton (1975) is also a light-harvesting protein complex, although a similar complex isolated by Henriques and Park (1976) was depleted in Mn. A large (850 000 molecular weight) Mn-containing complex isolated from the diatom *Phaeodactylum tricornutium* (Holdsworth and Juzu, 1977) appears to be a photosystem II complex. It catalyses the reduction of 2,6-dichlorophenol indophenol (DCIP) using diphenyl carbazide (DPC) as an artificial electron donor to photosystem II. The reaction is sensitive to DCMU. The complex also contains copper with a Cu : Mn ratio of about 8 : 1. von Kameke and Wegman (1978) have reported the isolation of two Mn-containing proteins from the unicellular green alga *Dunaliella tertiolecta*, of which one is probably superoxide dismutase. The other, of molecular weight 500 000, shows a Mn epr signal.

Another Mn-protein complex has been reported by Spector and Winget

(1980) who isolated it using the detergent cholate. The protein has a molecular weight of 65 000, and after extraction the remaining depleted chloroplasts may be incorporated into phospholipid vesicles yielding "depleted photosomes". The depleted photosomes are unable to oxidize water but show high rates of light-induced electron transport from DPC to DCIP. After reconstituting the depleted photosomes with the 65 000 M W protein the photosomes are reported to catalyse the oxidation of water when uncoupled with methylamine. The restoration of the capacity to evolve oxygen is sensitive to Tris. Confirmation of these properties is not yet available from other laboratories. Moreover, the identification of the protein as the Mn-protein directly involved in the evolution of oxygen has been challenged by Nakatani and Barber (1981) who, using the same isolation technique, have found that the protein contains only traces of Mn, and has a haem group. The protein still stimulates oxygen evolution in depleted photosomes, although it is argued that this may be due to a catalase-type activity.

Recently, a synthetic Mn-bipyridine complex has been described by Dismukes and Siderer (1980) which has an oxidised-minus-reduced epr difference spectrum at 10 K which resembles the epr difference spectrum (illuminated with one flash-minus-dark adapted) of isolated chloroplasts. The chloroplast spectrum has a complex hyperfine structure, which is abolished when chloroplasts are washed with Tris or extracted with cholate. The amplitude of the chloroplast epr signal oscillates with flash number, being enhanced on flashes 1 and 5, and diminished on flashes 2, 3, and 4.

Any model which attempts to describe how Mn is involved in the oxidation of water has to take account not only of the period four oscillations of oxygen yield with flash number, but also the oscillations of proton release with flash number. The sequence of proton release with flash number is not entirely clear (see Sauer, 1980) with two research groups finding the pattern 1, 0, 1, 2 for the steps $S_0 \to S_1$, $S_1 \to S_2$, $S_2 \to S_3$ and $S_3 \to (S_4) \to S_0$ respectively (Fowler, 1977b; Saphon and Crofts, 1977) and a third group finding the pattern 0, 1, 1, 2 for the same steps (Junge et al., 1977). Clearly further experimentation is needed.

Of the several models which have been proposed for the oxygen-evolving centre, that of Renger (1978) is attractive, particularly when modified to take into account recent epr data (Sauer, 1980).

E. ANIONS

Two anions, chloride and bicarbonate, have been identified as essential co-factors in photosynthetic electron transport. The data on chloride come from isolated chloroplasts which were washed in chloride-free media, and then the rate of light-induced oxygen evolution was measured. On addition of chloride the rate of oxygen evolution was increased four to tenfold (Izawa *et al.*, 1969). The site of inhibition caused by chloride depletion was located between the site of oxidation of water and the point at which NH_2OH donates electrons to photsystem II. Kelley and Izawa (1978) have recently confirmed this observation. In addition they showed that a number of electron donors that donate electrons between the site of the oxidation of water and photosystem II perform electron transport at rates which are unaffected by chloride depletion. It has also been proposed (Sauer, 1980) that a permeant anion, such as chloride, may be necessary for charge neutralization in the step in the oxygen-evolving cycle that does not involve proton release.

Bicarbonate has been found to be necessary for electron transport, particularly in the region between Q and plastoquinone (see Govindjee and Khanna, 1978). The effects of bicarbonate have been reviewed by Govindjee and van Rensen (1978). Whilst it is clear that the ultimate source of oxygen evolved in photosynthesis is water, there is still some discussion of whether bicarbonate is the immediate source (Metzner, 1978).

F. METALS AS INHIBITORS OF PHOTOSYNTHESIS

A number of heavy metals inhibit electron transport in isolated chloroplasts. In the case of mercury (Kimimura and Katoh, 1972), there is a specific antagonism towards plastocyanin, and in the reduced state the copper atom is replaced by mercury. Other heavy metals like cadmium (Li and Miles, 1975), copper (Cedeno-Maldonado *et al.*, 1972), lead (Miles *et al.*, 1972), and zinc (Tripathy and Mohanty, 1980), block electron transport on the oxidizing side of photosystem II. In addition it has been shown that copper (II) when applied to illuminated isolated chloroplasts, induces an increase in the permeability of the thylakoid membrane towards protons, thereby inhibiting ATP formation (Hipkins and Durant, unpublished observations). It is of course open to question whether the effect of these metals is physiological.

III. COUNTER-ION MOVEMENT AND CONTROL OF THE CALVIN CYCLE

The primary processes of photosynthesis produce both ATP and reduced NADP. NADP is reduced as the terminal electron acceptor of the non-cyclic electron transport chain (Fig. 1); ATP is phosphorylated from

ADP and orthophosphate using the energy that is released when electrons pass along the "downhill" sections of the Z-scheme. The precise mechanism by which ATP is formed is still unclear, but the chemiosmotic hypothesis (Mitchell, 1966) has proved to be remarkably successful in taking account of experimental observations.

Mitchell suggests that the high-energy intermediate between the redox energy of the electron transport chain and the phosphate bond energy of ATP is a transmembrane electrochemical gradient of protons, which is

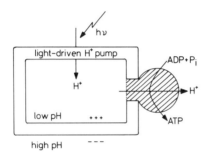

FIG. 8. A schematic diagram of Mitchell's chemiosmotic hypothesis of photophosphorylation, showing (a) the closed thylakoid vesicle bounded by a proton-impermeable membrane, (b) the vectorial, light-driven proton pump and (c) the membrane-bound vectorial ATP synthetase.

composed of a proton concentration gradient (Δ pH) and an electric field ($\Delta\Psi$). To complete the hypothesis, the following conditions are necessary:
  (i) that the chloroplast thylakoid is a closed vesicle, and that it is impermeable to protons;
 (ii) that there is a membrane-bound proton-pump that is linked to electron transport; and,
(iii) that there exists a vectorial membrane-bound, proton-conducting ATP synthetase. Figure 8 illustrates these essential features of the chemiosmotic hypothesis. The proton pump is the electron transport chain itself (Trebst, 1974), and the ATP synthetase has been well characterised (McCarty, 1980).

The terms Δ pH and $\Delta\Psi$ can be combined to give the proton motive force, pmf, which is measured in millivolts:

$$\text{pmf} = \Delta\Psi + 2{\cdot}303\,\frac{RT}{F}\,\Delta\,\text{pH}$$

Thermodynamically speaking, it is irrelevant whether the pmf is provided mainly by $\Delta\Psi$ or Δ pH, or equally by both. But in chloroplasts it is experimentally found that most of the pmf is provided by Δ pH, which can attain more than 3 pH units (Rottenburg *et al.*, 1972). That such a large pH gradient exists has important consequences. In particular it means that on illumination of dark-adapted chloroplasts, in order to prevent the build-up of a large transmembrane electric potential which would oppose the movement of protons into the intrathylakoid space there must be counter-ion flux to dissipate the potential. Clearly, either anion influx or cation efflux would be suitable: the experiments of Hind *et al.* (1974) using an array of ion-sensitive electrodes suggested both chloride influx and Mg^{2+} efflux served to compensate for the charge transferred by H^+ influx. Mg^{2+} movements from the intrathylakoid space to the stroma in intact chloroplasts have also been suggested by Portis and Heldt (1976), by Krause (1977) and by Barber *et al.* (1974). Magnesium ion movements induced by light, together with proton movements, may have a regulatory effect on the Calvin cycle. Since the chloroplast envelope is relatively impermeable to ions (Gimmler *et al.*, 1975) then both Mg^{2+} and H^+ fluxes will be apparent as concentration differences in the stroma. This might provide a mechanism to switch on the Calvin cycle in the light (Buchanan, 1980). Ribulose-bisphosphate carboxylase is certainly sensitive to Mg^{2+} which lowers the K_m for CO_2. Moreover, when divalent cations are released from the stroma into the incubation medium, CO_2 fixation is halted, but restored on readdition of Mg^{2+} (Portis and Heldt, 1976). In elegant experiments using permeant buffers and anions Heldt *et al.* (1973) were able to demonstrate an increase in pH of the stroma on illumination of intact chloroplasts. By examining levels of the intermediates of the Calvin cycle, Portis *et al.* (1977) and Purzeld *et al.* (1978) demonstrated that the enzymes fructose bisphosphatase and sedoheptulose bisphosphatase were both sensitive to Mg^{2+} levels and pH, but that ribulosebisphosphate carboxylase was unaffected. However, Buchanan (1980) has argued that light-induced changes in stromal Mg^{2+} levels may have little significance for enzyme activation.

IV. The Electrical Diffuse Double Layer

The literature on the primary processes of photosynthesis is rich in observations on the effect of cations on isolated chloroplasts (see Barber, 1976; Nakatani *et al.*, 1979). Among other things, cations have been found to influence

(i) the chlorophyll fluorescence yield of photosystem II (Homann, 1969) which is thought to reflect the distribution of light energy between the two photosystems (Murata, 1969);

(ii) the stacking of thylakoids into grana (Izawa and Good, 1972); and,

(iii) the degree of coupling between electron transport and ATP formation (Jagendorf and Smith, 1962).

Hitherto it has been difficult to see any pattern in cation-induced phenomena, but recently Barber and his colleagues have clearly indicated through a careful study of cation selectivity and valency effects, that an underlying unity emerges if the diffuse double layer near the negatively charged thylakoid membranes is taken into account.

That thylakoid membranes carry fixed negative charges has been demonstrated by particle electrophoresis (Nakatani et al., 1978; Nakatani and Barber, 1980). Thus, suspending thylakoid membranes in a medium which contains electrolyte would be expected to create a diffuse double layer (Barber, 1980). In a refinement of previous theoretical considerations, Rubin and Barber (1980) consider that the important parameter for control of photosynthetic processes is the integrated space charge density, which is a measure of (a) the degree to which an electrolyte medium can shield charge on a membrane surface, and (b) the repulsion interaction between charged surfaces. Integrated space charge density is strongly influenced by the cation composition of the medium and particularly by the valency of the cation. The diffuse-double layer theory accounts very well for cation-induced changes in the chlorophyll fluorescence yield of photosystem II (Barber and Searle, 1978) and the attachment of the ATP synthetase (coupling factor) to the thylakoid membrane (Telfer et al., 1980).

It might be thought that the double layer effects have limited physiological significance, since they have been studied primarily in isolated thylakoid membranes. But it is becoming clear that in at least one area, that of the distribution of light energy between the two photosystems ("spillover"), the diffuse layer may play a key role.

The regulation of the distribution of light energy between the two photosystems is important, since the two are in series, and if one were under-excited it would become the rate-limiting step. Spillover is usually measured as a decrease in the chlorophyll fluorescence yield of photosystem II, accompanied by an increase in the fluorescence yield of photosystem I, the latter usually measured at low temperature. It is assumed that changes in the rate of inter-photosystem energy transfer are brought about via the light-harvesting chlorophyll a/b protein (see Barber, 1980), since this protein is thought to capture incident light and channel it to both photosystems.

The problem in establishing the mechanism of regulation of spillover is that light absorbed by photosystem II or by photosystem I must have antagonistic effects. That is, it is not a question of the effect of light itself, but a question of the spectral quality of the light. This means that a hypothesis which involves, for example, the movement of Mg^{2+} ions from the intrathylakoid space to the stroma, is bound to be inadequate since it does not show photosystem II–photosystem I antagonism.

Recently Barber (1980) has proposed that changes in spillover are due to small movements within the thylakoid membrane of the pigment–protein complexes associated with photosystems I and II. The movements, it is suggested, will be controlled by the degree to which charge on the membrane is electrostatically screened, and which will in turn govern the distribution of pigment–protein complexes in the essentially fluid membrane. Barber (1980) recognizes that light-induced ion movements are unlikely to be a regulatory factor, and suggests that the introduction of electrical charge to the surface of one of the membrane proteins may be involved.

An experimental finding which is relevant to this hypothesis has now been made. It has been shown that the light-harvesting chlorophyll a/b protein is phosphorylated in the light in a DCMU-sensitive manner (Bennett et al., 1980). Moreover, the phosphorylation is paralleled by an enhanced excitation energy transfer to photosystem I. These results suggest that the phosphorylation of the chlorophyll a/b protein may be an integral part of the regulation of spillover, and that the diffuse double layer may be involved since phosphorylation will add an extra negative charge to each chlorophyll a/b protein, and hence modify the integrated space charge. Antagonism occurs because phosphorylation appears to depend on the redox potential of the plastoquinone pool, which itself depends on the distribution of light between photosystem II and I (Horton et al., 1981).

It should be mentioned that this hypothesis is not without problems, the major one being that the chlorophyll a/b protein does not appear to contribute to the surface charge (Nakatani et al., 1978), although it is not clear whether the measurements were made under conditions when the chlorophyll a/b protein would be expected to be phosphorylated. Nevertheless, the finding that the light-harvesting chlorophyll a/b protein is phosphorylated in a manner controlled by the redox poise of plastoquinone makes this hypothesis (Barber, 1980) very interesting.

V. CONCLUSIONS

In this review I have tried to demonstrate that metals are important at all stages of the photosynthetic processes, up to and including the fixation of carbon dioxide. Metals are not only involved as the prosthetic groups of metal–protein complexes, but it is becoming clear that metals may have an important role to play in the regulation both of the primary processes through spillover and of the Calvin cycle.

REFERENCES

Aitken, A. (1975). *Biochem. J.* **149**, 675–683.
Arntzen, C. J. and Briantais, J.-M. (1975). *In* "Bioenergetics of Photosynthesis",
  pp. 51–113, (Govindjee, ed.), Academic Press, London and New York.
Baltimore, B. G. and Malkin, R. (1980). *FEBS letters* **110**, 50–52.
Barber, J. (1976). *In* "The Intact Chloroplast. Topics in Photosynthesis Vol. 1",
  pp. 89–134, (J. Barber, ed.), Elsevier, Amsterdam.
Barber, J. (1980). *FEBS letters* **118**, 1–10.
Barber, J. and Searle, G. F. W. (1978). *FEBS letters* **92**, 5–8.
Barber, J., Mills, J., and Nicolson, J. (1974). *FEBS letters* **49**, 106–110.
Bennett, J., Steinbeck, K. E., and Arntzen, C. J. (1980). *Proc. Natl. Acad. Sci.
  USA* **77**, 5253–5257.
Blankenship, R. E. and Sauer, K. (1974). *Biochim. Biophys. Acta* **357**, 252–266.
Bolton, J. (1977). *In* "Primary Processes of Photosynthesis. Topics in Photosynth-
  esis Vol. 2", pp. 187–202, (J. Barber, ed.), Elsevier, Amsterdam.
Bouges-Bocquet, B. (1973). *Biochim. Biophys. Acta* **314**, 250–256.
Buchanan, B. B. (1980). *Ann. Rev. Plant Physiol.* **31**, 341–374.
Butler, W. L. (1978). *FEBS letters* **95**, 19–25.
Cedeno-Maldonado, A., Swader, J. A., and Heath, R. L. (1972). *Pl. Physiol.* **50**,
  698–701.
Cheniae, G. M. and Martin, I. F. (1966). *Brookhaven Symp. Biol.* **19**, 406–417.
Cheniae, G. M and Martin, I. F. (1970). *Biochim. Biophys. Acta* **197**, 219–239.
Colman, P. M., Freeman, H. C., Guss, J. M., Murata, M., Norris, V. A.,
  Ramshaw, J. A. M., and Venkatappa, M. P. (1978). *Nature (Lond.)* **272**,
  319–324.
Cramer, W. A. (1977). *In* "Encyclopaedia of Plant Physiology," Vol. V, pp.
  227–237, (A. Trebst and M. Avron, eds.), Springer-Verlag, Berlin.
Cramer, W. A. and Whitmarsh, J. (1977). *Ann. Rev. Plant Physiol.* **28**, 133–172.
Crowther, D. and Hind, G. (1980). *Arch. Biochem. Biophys.* **204**, 568–577.
Diner, B. A. and Joliot, P. (1977). *In* "Encyclopaedia of Plant Physiology", Vol.
  V, pp. 187–205, (A. Trebst and M. Avron, eds.), Springer-Verlag, Berlin.
Dismukes, G. C. and Siderer, Y. (1980). *FEBS letters* **121**, 78–80.
Evans, M. C. W. (1977). *In* "Primary Processes of Photosynthesis. Topics in
  Photosynthesis, Vol. 2", pp. 433–464, (J. Barber, ed.), Elsevier, Amsterdam.
Fowler, C. F. (1977a). *Biochim. Biophys. Acta* **459**, 351–363.
Fowler, C. F. (1977b) *Biochim. Biophys. Acta* **462**, 414–421.
Foyer, C. H. and Hall, D. O. (1979). *FEBS letters* **101**, 324–328.
Freeman, H. C. (1981). *Coordination Chem.* **21**, 29–51. (J. P. Laurent, ed.)
  Pergamon Press, Oxford.
Fukuyama, K., Hase, T., Matsumoto, S., Tsukihara, T., Katsube, Y., Tanaka, N.,
  Kakudo, M., Wada, K., and Matsubara, H. (1980) *Nature* **286**, 522–524.
Garewall, H. S. and Wasserman, A. R. (1974). *Biochem.* **13**, 4072–4079.
Gimmler, H., Schäfer, G. and Heber, U. (1975). *In* "Proc. 3rd Int. Cong. on
  Photosynthesis", Vol. II, pp. 1381–1392, (M. Avron, ed.), Elsevier, Amster-
  dam.
Govindjee and Khanna, R. (1978). *In* "Photosynthetic Oxygen Evolution", pp.
  269–282, (H. Metzner, ed.), Academic Press, London and New York.
Govindjee and van Rensen, J. J. S. (1978). *Biochim. Biophys. Acta* **505**, 183–213.
Govindjee, Wydrzynski, T., and Marks, S. B. (1978). *In* "Photosynthetic Oxygen
  Evolution", pp. 321–344, (H. Metzner, ed.), Academic Press, London and New
  York.

Haehnel, W. (1976). *Biochim. Biophys. Acta* **423**, 499–509.
Hall, D. O. and Rao, K. K. (1977). *In* "Encyclopaedia of Plant Physiology", Vol. V, pp. 206–216, (A. Trebst and M. Avron, eds.), Springer-Verlag, Berlin.
Hauska, G. A., McCarty, R. E., Berzborn, R. J., and Racker, E. (1971). *J. Biol. Chem.* **246**, 3524–3531.
Heathcote, P., Timofeev, K. N., and Evans, M. C. W. (1979). *FEBS letters* **101**, 105–109.
Heldt, H. W., Werdan, K., Milovancev, M., and Geller, G. (1973). *Biochim. Biophys. Acta* **314**, 224–241.
Henriques, F. and Park, R. B. (1976). *Biochim. Biophys. Acta* **430**, 312–320.
Hill, R. and Bendall, F. (1960). *Nature (Lond.)* **186**, 136–137.
Hind, G., Nakatani, H. Y., and Izawa, S. (1974). *Proc. Natl. Acad. Sci. USA* **71**, 1484–1488.
Holdsworth, E. S. and Juzu, H. A. (1977). *Arch. Biochem. Biophys.* **183**, 361–373.
Homann, P. H. (1969). *Pl. Physiol.* **44**, 932–936.
Horton, P. and Croze, E. (1979). *Biochim. Biophys. Acta* **545**, 188–201.
Horton, P., Allen, J. F., Black, M. T., and Bennett, J. (1981). *FEBS letters* **125**, 193–196.
Horton, P., Böhme, H., and Cramer, W. A. (1975). *In* "Proc. 3rd Int. Cong. on Photosynthesis", Vol. I, pp. 535–545, (M. Avron, ed.), Elsevier, Amsterdam.
Izawa, S. and Good, N. E. (1972). *Pl. Physiol.* **41**, 544–552.
Izawa, S., Heath, R. L., and Hind, G. (1969). *Biochim. Biophys. Acta* **180**, 388–398.
Jagendorf, A. T. and Smith, M. (1962). *Pl. Physiol.* **37**, 135–141.
Joliot, P. (1966). *Brookhaven Symp. Biol.* **19**, 418–433.
Junge, W. (1977). *Ann. Rev. Plant Physiol.* **28**, 503–536.
Junge, W., Renger, G., and Auslander, W. (1977). *FEBS letters* **79**, 155–159.
von Kameke, E. and Wegmann, K. (1978). *In* "Photosynthetic Oxygen Evolution", pp. 371–380, (H. Metzner, ed.), Academic Press, London and New York.
Katoh, S. (1962). *J. Biochem.* **51**, 32–40.
Katoh, S. (1977). *In* "Encyclopaedia of Plant Physiology", Vol. V, pp. 247–252 (A. Trebst and M. Avron, eds.), Springer-Verlag, Berlin.
Ke, B. (1973). *Biochim. Biophys. Acta* **301**, 1–33.
Kelley, P. M. and Izawa, S. (1978). *Biochim. Biophys. Acta* **502**, 198–210.
Kessler, E., Arthur, W., and Brugger, J. E. (1957). *Arch. Biochem. Biophys.* **71**, 326–335.
Kimimura, M. and Katoh, S. (1972). *Biochim. Biophys. Acta* **283**, 279–292.
Klimov, V. V., Klevanik, A. V., Shuvalov, V. A., and Krasnovsky, A. A. (1977). *FEBS letters* **82**, 183–186.
Knaff, D. B. and Arnon, D. I. (1969). *Proc. Natl. Acad. Sci. USA* **63**, 956–962.
Kok, B., Forbush, B. and McGloin, P. (1970). *Photochem. Photobiol.* **11**, 457–475.
Kok, B., Radmer, R., and Fowler, C. F. (1975). *In* "Proc. 3rd Int. Cong. on Photosynthesis", Vol. I, pp. 485–496, (M. Avron, ed.), Elsevier, Amsterdam.
Krause, G. H. (1977). *Biochim. Biophys. Acta* **460**, 500–510.
Lagoutte, B. and Duranton, J. (1975). *FEBS letters* **51**, 21–24.
Li, E. H. and Miles, C. D. (1975). *Plant Sci. Letts.* **5**, 33–40.
Lilley, R. McC., Holborrow, K., and Walker, D. A. (1974). *New Phytol.* **73**, 657–662.
Malkin, R. (1978). *FEBS letters* **87**, 329–333.
Malkin, R. and Aparicio, P. J. (1975). *Biochem. Biophys. Res. Comm.* **63**, 1157–1160.

Malkin, R. and Barber, J. (1979). *Arch. Biochem. Biophys.* **193**, 169–178.
Malkin, R. and Posner, H. B. (1978). Biochim. Biophys. *Acta* **501**, 552–554.
McCarty, R. E. (1980). *Ann. Rev. Plant Physiol.* **30**, 79–104.
Metzner, H. (1978). *In* "Photosynthetic Oxygen Evolution", pp. 59–76, (H. Metzner, ed.), Academic Press, London and New York.
Miles, C. D., Brandle, J. R., Daniel, D. J., Chu-Der, O., Schnare, P. D., and Uhlik, D. J. (1972). *Pl. Physiol.* **49**, 820–825.
Mitchell, P. (1966). *Biol. Rev.* **41**, 445–502.
Murata, N. (1969). *Biochim. Biophys. Acta* **189**, 171–181.
Nakatani, H. Y. and Barber, J. (1980). *Biochim. Biophys. Acta* **591**, 82–91.
Nakatani, H. Y. and Barber, J. (1981). *Photobiochem. Photobiophys.* **2**, 69–78.
Nakatani, H. Y., Barber, J., and Forrester, J. A. (1978). *Biochim. Biophys. Acta* **504**, 215–225.
Nakatani, H. Y., Barber, J., and Minski, M. J. (1979). *Biochim. Biophys. Acta* **545**, 24–35.
Olsen, L. F., Telfer, A., and Barber, J. (1980). *FEBS letters* **118**, 11–17.
Portis, A. R. and Heldt, H. W. (1976). *Biochim. Biophys. Acta* **449**, 434–446.
Portis, A. R., Chon, C. J., Mosbach, A., and Heldt, H. W. (1977). *Biochim. Biophys Acta* **461**, 313–325.
Purzeld, P., Chon, C. J., Portis, A. R., Heldt, H. W., and Heber, U. (1978). *Biochim. Biophys. Acta* **501**, 488–498.
Ramshaw, J. A. M., Brown, R. H., Scawen, M. D., and Boulter, D. (1973). *Biochim. Biophys. Acta* **303**, 269–273.
Ramshaw, J. A. M., Scawen, M. D., Jones, E. A., Brown, R. H., and Boulter, D. (1976). *Phytochem.* **15**, 1199–1202.
Rao, K. K., Cammack, R., Hall, D. O., and Johnson, C. E. (1971). *Biochem. J.* **122**, 257–265.
Renger, G. (1978). *In* "Photosynthetic Oxygen Evolution", pp. 229–248, (H. Metzner, ed.), Academic Press, London and New York.
Rieske, J. S., Hansen, R. E., and Zaugg, W. S. (1964). *J. Biol. Chem.* **239**, 3017–3022.
Robinson, H. H., Sharp, R. R., and Yocum, C. F. (1980). *Biochem. Biophys. Res. Comm.* **93**, 755–761.
Rottenburg, H., Grunwald, T., Schuldiner, S., and Avron, M. (1972). *In* "Proc. 2nd Int. Cong. on Photosynthesis", Vol. II, pp. 1035–1047, (G. Forti, M. Avron, and B. A. Melandri, eds.), Dr. Junk, The Hague.
Rubin, B. T. and Barber, J. (1980). *Biochim. Biophys. Acta* **592**, 87–102.
Saphon, S. and Crofts, A. R. (1977). *Z. Naturf.* **32C**, 617–626.
Sauer, K. (1980). *Acc. Chem. Res.* **13**, 249–256.
Sharp, R. R. and Yocum, C. F. (1980). *Biochim. Biophys. Acta* **592**, 185–195.
Spector, M. and Winget, G. D. (1980). *Proc. Natl. Acad. Sci. USA* **77**, 957–959.
Stuart, A. L. and Wasserman, A. R. (1975). *Biochim. Biophys. Acta* **376**, 561–572.
Telfer, A., Barber, J., and Jagendorf, A. T. (1980). *Biochim. Biophys. Acta* **591**, 331–345.
Thornber, J. P., Alberte, R. S., Hunter, F. A., Shiozawa, J. A., and Kan, K. S. (1976). *Brookhaven Symp. Biol.* **28**, 132–148.
Trebst, A. (1974). *Ann. Rev. Plant Physiol.* **25**, 423–458.
Tripathy, B. C. and Mohanty, P. (1980). *Pl. Physiol.* **66**, 1174–1178.
Velthuys, B. R. (1979). *Proc. Natl. Acad. Sci. USA* **76**, 2765–2769.
Velthuys, B. R. (1980). *Ann. Rev. Plant Physiol.* **31**, 545–567.
Velthuys, B. R. and Amesz, J. (1974). *Biochim. Biophys. Acta* **333**, 85–94.

Williams, W. P. (1977). *In* "Primary Processes of Photosynthesis. Topics in Photosynthesis Vol. 2", pp. 99–147, (J. Barber, ed.), Elsevier, Amsterdam.

Wydrzynski, T. and Sauer, K. (1980). *Biochim. Biophys. Acta* **589**, 56–70.

Yocum, C. F. and San Pietro, A. (1969). *Biochem. Biophys. Res. Comm.* **36**, 614–620.

Yocum, C. F. and San Pietro, A. (1970). *Arch. Biochem. Biophys.* **140**, 152–157.

CHAPTER 10

Characteristics of the Higher Plant Respiratory Chain

A. L. MOORE AND I. R. COTTINGHAM

Biochemistry Department, School of Biological Sciences, University of Sussex, Brighton, UK

Abbreviations and definitions

TMPD: N,N,N′,N′,-tetramethyl-p-phenylenediamine; TCA: Tricarboxylic acid cycle; DCPIP: 2.6-dichlorophenol-indophenol; PMS: phenazine methosulphate; FAD: Flavin adenine dinucleotide; FMN: Flavin mononucleotide; HiPIP: High potential iron-sulphur protein; EPR: Electron paramagnetic resonance; TTFA: 2-Thenoyltrifluoroacetone; UHDBT: 5-n-undecl-6-hydroxy-4,7-dioxybenzothiazole; State 3 and State 4-Respiratory activity in the presence and absence of ADP; E_{mpH} Midpoint potential at the indicated pH.

I. Introduction

Mitochondria are generally considered to be responsible for most ATP synthesis in non-photosynthetic eukaryotic cells which carry out aerobic metabolism. Organic substrates are oxidized by molecular oxygen via the electron transport chain situated on the inner membrane of the mitochondria. This chain comprises a series of multi-component enzyme complexes restrained within a two-dimensional membrane phase and the majority of its constituents are proteins possessing prosthetic groups. In a number of cases a metal is associated with the prosthetic group. Thus, iron occurs in the cytochromes and in the iron-sulphur centres and copper is present in the terminal complex, cytochrome oxidase.

Transfer of reducing equivalents, from the organic substrate to oxygen, along the electron transport chain causes the redox components to undergo alternate reduction and oxidation. Reducing equivalents are donated by carriers of low (negative) redox potential to carriers of high (positive) redox potential and since the redox components have closely spaced redox potentials, spanning the range −230 mv to +375 mv, the free energy from the overall reaction is released in relatively small amounts, suitable for the efficient synthesis of ATP. Three sites in the chain are generally recognized as furnishing sufficient energy for phosphorylation; Site I between NAD and ubiquinone, Site II between cytochromes b and c and Site III between a and oxygen. It is now generally accepted (Boyer et al., 1977) that electron transport is accompanied by the vectorial transport of hydrogen from the mitochondrial matrix to the cytoplasm. This results in the expulsion of protons and the concomitant generation of a membrane potential. The electron transport chain can therefore be regarded as an outwardly directed proton pump (Wikstrom et al., 1981). Its activity results in the generation of a proton-motive force which is considered to be the driving force for ATP synthesis (Mitchell, 1966). ATP synthesis occurs as a consequence of H^+ returning through the ATP synthetase which is located on the inner surface of the inner membrane. Thus electron transport and ATP synthesis are coupled via the proton gradient. The addition of ADP is required for maximal rates of electron transport (state 3) whereas in its absence (state 4) electron transport activity is very low.

Similarly to mammalian systems, the electron transport chain in plants can be fragmented into four multi-component enzyme complexes (NADH and succinate dehydrogenases, ubiquinol-cytochrome c reductase, and cytochrome oxidase) plus ubiquinone and cytochrome c by treatment with detergents and ammonium sulphate. Plant mitochondria are however, unique, since they possess routes of substrate oxidation and an alternative terminal oxidase not normally encountered. Thus, in this chapter we have attempted to compare and contrast the general properties of electron transport in plant mitochondria with animal systems but in addition emphasizing the peculiar features of plant mitochondria. In the final section the interaction between the dehydrogenases and the two oxidases has been considered in terms of integration at the level of ubiquinone. The reader is also referred to a number of relevant and recent reviews (Palmer, 1976, 1979, 1983; Storey, 1980; Hanson and Day, 1980; Day et al., 1980; Moore and Rich, 1980; Lance, 1981; Moore, 1983).

II. Components of the Respiratory Chain

A. THE CYTOCHROMES

All of the cytochromes are conjugated proteins containing a haem prosthetic group. The haem (porphyrin) is composed of 4 pyrrole rings joined by methene bridges and a central iron atom which carries a single electron and hence oscillates between the ferrous (Fe^{2+}) and ferric (Fe^{3+}) states. The individual cytochromes can be grouped into one of three classes, A, B, and C which differ in the nature of the substituents bound to the porphyrin ring (see Jones, this volume). These substituents are responsible for the spectral and redox potential properties of the individual cytochromes. The a-type cytochromes contain haem A as the prosthetic group, the b-type cytochromes contain protohaem whereas the c-type cytochromes contain haem C. The c-type cytochromes are the only ones in which haem is covalently bound to the protein; in all others the haem is held in position by a combination of co-ordination bonds to the iron atom and non-covalent interactions betwen the porphyrin ring and the protein. Plant mitochondrial cytochromes are similar in spectral and possibly functional properties to their mammalian counterparts being comprised of cytochromes a, a_3, at least 3 b-type cytochromes, and 2 c-type cytochromes.

1. The a-type Cytochromes

Cytochrome oxidase of plant mitochondria contains two a-type cytochromes. They are integral membrane proteins and can only be purified after solubilization with detergents. The purified enzyme contains 2 moles of haem A as a and a_3. Although a and a_3 are regarded as being separate entities they have never been physically separated from one another and do not appear to function independently. They can however be distinguished by their absorption spectra and redox potentials (see Table I). The α-absorption band of cytochrome a is located at 603 nm at room temperature, whereas at 77K there is a blue shift to 599 nm (Bonner, 1961). The spectral contribution attributed to cytochrome a_3 can only be

TABLE I
Characteristics of cytochromes in plant mitochondria

Cytochrome	α peak 25°C nm	α peak 77°K nm	$E_{m7.2}$ (mv)	Ref.
a	603	599	+190	(a, b)
a_3		445 (Soret)	+380	(a, b)
b-556	556	553–554	+75 to +100	(c)
		553	+75	(b)
b-560	560	557	+40 to +80	(c)
		557	+42	(b)
b-558	557–558	553–555	−70 to −105	(c)
b-566	566	561–563	−75	(c)
		562	−77	(b)
$S_2O_4^{2-}$ reducible	557–561	554, 560	−100	(c)
c (soluble)	550	547	+235	(b, d, e)
c_1 (membrane bound)	552	549	+235	(b, d, e)

(a) Bendall, D. S. and Bonner, W. D. Jr. (1966).
(b) Dutton, P. L. and Storey, B. T. (1971).
(c) Lambowitz, A. M. and Bonner, W. D. Jr. (1974).
(d) Lance, C. and Bonner, W. D. Jr. (1968).
(e) Ducet, G. and Diano, M. (1978).

measured in the Soret region between 430–450 nm. In this region cytochrome a can be resolved into 2 peaks at 77K, a major one at 438 nm and minor one at 445 nm. Sixty per cent of the absorption at 445 nm is due to cytochrome a, whereas the remainder is due to cytochrome a_3. The presence of a double peak in this region led to the suggestion that plant mitochondria possessed a second oxidase, in addition to cytochrome oxidase, which may be of the cytochrome a type. Bendall and Bonner (1971), however, concluded that the α peak at 600 nm and double Soret peak at 439 nm and 445 nm belong to the difference spectrum (reduced minus oxidized) of cytochrome a while the a_3 spectrum has little or no α absorption and a single Soret peak at 445 nm.

The optical properties of the two cytochromes can more easily be differentiated in the presence of azide or cyanide, both of which bind to cytochrome a_3. Binding occurs optimally with the oxidized form but may also take place with the reduced form under strictly anaerobic conditions. The reduced minus oxidized difference spectrum of cytochrome a can be obtained from the spectrum for substrate-reduced, aerobic mitochondria in the presence of cyanide minus that of the fully oxidized state. Under these conditions the contribution of a_3 to the spectrum cancels out due to the similar absorbance of the ferric a_3-cyanide compound in the sample cuvette and ferric a_3 in the reference cuvette. The difference spectrum of a_3 on the other hand can be obtained from the spectrum of substrate-reduced, anaerobic mitochondria minus substrate-reduced, aerobic mitochondria in the presence of cyanide. In this manner, the spectrum obtained is the difference ferrous a_3 minus ferric a_3-cyanide (Bendall and Bonner, 1971).

The two a cytochromes may also be readily resolved by their midpoint redox potentials. Dutton and Storey (1971) showed that cytochrome a has an $E_{m7.2}$ of $+190$ mV whilst the $E_{m7.2}$ for cytochrome a_3 is $+380$ mV.

In mammalian preparations, purified cytochrome oxidase has a molecular weight of approximately 250 000 (cf. Wikstrom and Krab, 1979). It contains 6–7 different sub-units and four one electron centres being comprised of two haems and two copper atoms. One of the copper atoms, Cu_A, has a lower E_m than the other, Cu_B, which interacts strongly with cytochrome a_3. The present evidence suggests that all electrons fed into the enzyme are primarily accepted by cytochrome a in association with Cu_A (see Fig. 1), and are then transferred to cytochrome a_3 which functions in

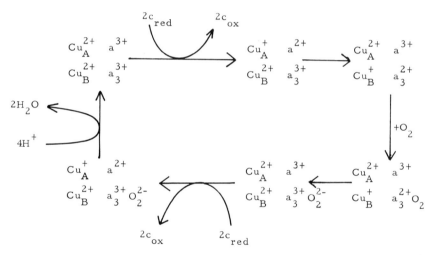

Fig. 1. Reaction mechanism proposed for cytochrome oxidase (adapted from Clore *et al.*, 1980). The fully oxidized enzyme accepts two electrons from cytochrome c before binding oxygen. Upon acceptance of two more electrons from cytochrome c water is formed and the oxidized enzyme regenerated.

combination with Cu_B as a closely coupled unit. Dioxygen is primarily bound to the haem iron of cytochrome a_3, after which electron transfer to the bound dioxygen ensues.

Recent evidence (Denis, 1981; Denis and Bonner, 1978; Denis and Clore, 1981) suggests that a similar reaction sequence operates in plant cytochrome oxidase. Interestingly, purified preparations of sweet potato cytochrome oxidase are only comprised of 5 subunits (Maeshima and Asahi, 1981). The reason for this discrepancy is unclear at present.

2. The b-type Cytochromes

The plant mitochondrial respiratory chain is complicated by the presence of multiple b-cytochromes. Bonner and his colleagues (Lance and Bonner, 1968; Lambowitz and Bonner, 1974) demonstrated that at 77K, the spectra of b-cytochromes in the presence of the respiratory inhibitor antimycin A, could be resolved into at least 3 bands found at 553, 557, and 562 nm. More purified plant mitochondria were subsequently shown to contain at least 4 b-components with α peaks at 556, 558, 560, and 566 nm and one b component which was reduced only by dithionite (α peak at 557 to 561 nm; Lambowitz and Bonner, 1974). This confirmed an earlier report by Lance and Bonner (1968) who observed that about 30% of the cytochrome b components could be reduced by dithionite but not by anaerobiosis in the presence of succinate. A single broad peak at 560 nm is observed with mung bean mitochondria while for potatoes two separate peaks at 557 and 561 are apparent (Lambowitz and Bonner, 1974). The dithionite reducible component is generally considered to be a unique species and not an artefact caused by incomplete reduction of known b-components, but its role in electron transport is obscure and it will not be considered further.

The b-cytochromes have been characterized also by potentiometric titrations and their mid-point potentials are summarized in Table I. It may be noted that cytochrome b-558 cannot be resolved potentiometrically from the dithionite reducible cytochrome b, and both components appear to have mid-point potentials of about −100 mV.

Two of the b-cytochromes, namely b-560 and b-566 have E_m values close to the succinate-fumarate couple and are thought to be analogous to the b-560 (b_k) and b-566 (b_T) of mammalian mitochondria. Cytochrome b-566 in animal systems has been implicated in energy transduction (Chance et al., 1970), since its redox potential increases substantially in the presence of ATP and the latter is required for the complete reduction of this cytochrome under anaerobic conditions. At present there is no evidence that in plants cytochrome b-566 shows an ATP-induced change in its E_m, although some energy-linked reduction of b-566 is observed (Lambowitz and Bonner, 1974). This has been attributed to reversed electron transport since rapid relaxation is obtained upon addition of the oxidation-reduction mediator, phenazine methosulphate (PMS). Lambowitz and Bonner (1974) have therefore suggested that in view of the overall similarity between the b-cytochromes in plant and mammalian mitochondria, it is conceivable that the change in Em observed in mammalian tissues may be due to incomplete mediation between cytochrome b-566 and PMS and that this cytochrome plays no special role in energy transduction.

Kinetic analysis of the plant b-cytochromes is complicated by the anomalous behaviour of b-566. Thus cytochromes b-560 and b-556 are readily reduced by most substrates under anaerobic conditions, although b-556 does appear to be more extensively reduced by succinate, and both are rapidly re-oxidized upon addition of oxygen, b-556 being somewhat the slower of the two. In contrast, b-566 is not extensively reduced by either NADH or succinate in uncoupled mitochondria upon anaerobiosis even in the presence of inhibitors of cytochrome oxidase (Storey, 1970b). In coupled mitochondria, cytochrome b-566 does become reduced in the presence of succinate and ATP upon anaerobiosis (Storey, 1972a, b). On the basis of this observation, Lambowitz and Bonner (1974) suggested that plant mitochondria contain a barrier which inhibits redox equilibration between NADH and the b-cytochromes, thereby preventing the full reducing power of NADH from being expressed in the b-region. Upon addition of the respiratory inhibitor, antimycin A, b-566 is readily reduced by substrate (Bonner, 1961; Bonner and Slater, 1970; Lambowitz and Bonner, 1974). Maximal reduction requires cytochrome c_1 to be fully oxidized which depends upon the simultaneous presence of oxygen (Lambowitz and Bonner, 1974). Under these conditions a reduction of b-558 is also observed. The mechanism whereby antimycin A causes reduction of b-566 is uncertain. From study of mammalian preparations, Von Jagow (1975) proposed that the inhibitor forms a complex with cytochrome b-566 and b-560, which blocks electron flow to cytochrome c_1. This proposal may possibly explain why antimycin A affects the absorbance spectrum of b-560 (Lambowitz and Bonner, 1974). Cytochrome b-566 also appears to be a sensitive indicator of the redox state of the respiratory chain since it undergoes cyclic oxidation and reduction upon addition of ADP (Moore and Bonner, 1977). Thus, in spite of its anomalous properties it can be considered a viable member of the respiratory chain, together with b-560. In comparison, the roles of b-556 and b-558 are presently unclear and must await further study of the parent complexes.

The b-cytochromes are integral membrane proteins being associated with cytochrome c_1 in ubiquinol-cytochrome c reductase, also referred to as the b–c_1 complex. Relatively little information is available for this enzyme in plants, although Ducet and Diano (1978) recently reported its isolation from potatoes.

In addition to the normal b-cytochromes indicated in Table I, aroid spadices contain an additional cytochrome of the b-type, b_7 (Bendall and Hill, 1956). This particular cytochrome is unusual in that it remains oxidized in the presence of substrate plus antimycin A and/or cyanide until anaerobiosis (Bendall, 1958). Interestingly, neither malate nor NADH can reduce cytochrome b_7 even on anaerobiosis and Rich and Bonner (1978a) have concluded that although the cytochrome is mitochondrial, it is probably not linked to the electron transport chain, and that the observations of Bendall and Hill (1956) can be explained by a slow equilibration between this cytochrome and succinate.

3. The c-type Cytochromes

Two c-type cytochromes can be detected in plant mitochondria (Lance and Bonner, 1968). One is readily extractable by salt solutions and has a single absorbance maximum at 550 nm which is comparable to the cytochrome c in animal mitochondria. Studies on its amino acid sequence in a variety of plants are in accord with the accepted structure for the prosthetic group of animal cytochrome c in which the haem is covalently linked to two cysteine residues and methionine and histidine act as axial ligands to the iron (Brown and Boulter, 1974). A second component can still be detected after salt extraction by its α absorption peak. This cytochrome, termed c_1 can be extracted with cholate as part of the ubiquinol-cytochrome c reductase from which it is obtained by treatment with β-mercaptoethanol (Ducet and Diano, 1978). Isolated c_1 has a room-temperature spectrum similar to that of cytochrome c with an α peak around 552 nm (Ducet and Diano, 1978), but with some differences in its low temperature spectrum compared to other preparations are observed, since only a single α peak is detected (cf. Yu et al., 1970).

The c-type cytochromes are directly reducible by ascorbate (plus TMPD), which differentiates them from the b-cytochromes. Both cytochromes c_1 and c have mid-point potentials of $+235$ mV (Dutton and Storey, 1971) and show rapid kinetics consistent with a role of reacting with cytochrome oxidase.

B. IRON-SULPHUR CENTRES

Iron-sulphur centres are relatively low molecular weight proteins which act as redox carriers and contain 2 or 4 atoms each of non-haem iron and acid-labile sulphur. They can be divided into two categories, namely the ferredoxin-type (2Fe-2S) containing 2 atoms of iron held in a lattice of four atoms of cysteine-sulphur and two atoms of labile sulphur (see Hipkins, this volume) and the high potential type (HiPIP, 4Fe-4S) in which the 4 iron atoms form a cube with four atoms of labile sulphur (Fig. 2). Although the prosthetic group contains more than one iron atom, all iron-sulphur centres function as single electron carriers by undergoing reversible Fe^{3+} and Fe^{2+} transitions. Reduction of the ferredoxin centres results in the charge at the redox centre decreasing from $(2+)$ to $(1+)$ whereas in the HiPIP-type, reduction is accompanied by a charge decrease from $(3+)$ to $(2+)$. Thus the ferredoxin-type centres show electron

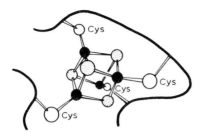

Fig. 2. The cuboid-like structure of the HiPIP (4Fe-4S) iron-sulphur centre (from Jones, 1981). Fe and labile S atoms are represented by filled and stippled circles respectively. The open circles represent the S atoms in the cysteine (Cys) side chains.

paramagnetic resonance in the reduced state (due to an unpaired electron) whereas the HiPIP-type are paramagnetic in the oxidized state.

The mitochondrial membrane contains at least twice as much non-haem iron and acid-labile sulphur as haem iron (Mackler *et al.*, 1954; Crane *et al.*, 1956). Thus in terms of species, iron-sulphur centres outnumber cytochromes as mitochondrial electron transfer components. Mitochondrial iron-sulphur centres have been mainly characterized in mammalian tissues and the reader is referred to a recent excellent review by Ohnishi (1979) in which a background and up-to-date account is summarized. Impetus into the study of iron-sulphur centres of plant mitochondria was brought about due to the extensive investigations of Rich and colleagues (Moore *et al.*, 1976; Rich *et al.*, 1977 and Rich and Bonner, 1978a, b, c, d) although some earlier reports are available (Cammack and Palmer, 1973; 1977).

The ferredoxin type centres of plant mitochondria based upon the data of Rich and Bonner (1978a) and Cammack and Palmer (1977) are summarized in Table II. Five dominant centres are recognized; S-1 and S-2 of succinate dehydrogenase and N-1, N-2 and N-3 of NADH dehydrogenase. Centres N-4 and N-5 are absent or present in undetectable amounts in plant mitochondria. Recently, Prince *et al.* (1981), reported the occurrence of the Rieske iron-sulphur centre in mung bean mitochondria. This high potential yet ferredoxin type centre which is an integral component of ubiquinol-cytochrome *c* reductase of animal mitochondria (Rieske, 1976) has previously not been found in plants, possibly due to its relatively broad signal. Upon addition of 5-n-undecyl-6-hydroxy 4, 7, dioxybenzothiazole (UHDBT), an inhibitor of the reductase (Trumpower and Haggerty, 1980), however, a prominent signal appeared at $g = 1.90$ with an E_m of $+300\,mV$, characteristic of the Rieske centre (Rieske *et al.*, 1964). Using this inhibitor, it is probable that the Rieske centre will also be found in other species of plant mitochondria, but there

TABLE II
Characteristics of iron-sulphur centres in plant mitochondria

Centre	g values	Temp °K	$E_{m7\cdot2}$ (mV)	Ref.
N–1	$g_{11} = 2\cdot025$, $g_\perp = 1\cdot935$	35	−260	(a)
		77	−240	(b)
N–2	$g_{11} = 2\cdot05$, $g_\perp = 1\cdot93$	12	+ 60	(a)
		18	−110	(b)
N–3	$g_z = 2\cdot11$, $g_y = 1\cdot93$, $g_x = 1\cdot87$	8	−275	(a)
		12	−275	(b)
S–1	$g_{11} = 2\cdot03$, $g_\perp = 1\cdot93$	25	+ 60	(a)
		77	− 7	(b)
S–2	$g_{11} = 2\cdot03$, $g_\perp = 1\cdot93$	12	−225	(a)
		18	−240	(b)
S–3	$g = 2\cdot02$	8	+ 65	(a)
		12	+ 85	(b)
Rieske	$g_{11} = 2\cdot03$, $g_\perp = 1\cdot90$	15	+300	(c)
Ubisemiquinone	(i) unsplit ($g = 2\cdot004$)	30	+ 56	(a)
		18	−100	(b)
	(ii) split (around $g = 2\cdot004$)	8	not estimated	
		12		

(a) Rich, P. R. and Bonner, W. D. Jr. (1978a).
(b) Cammack, R. and Palmer, J. M. (1977).
(c) Prince, R., Bonner, W. D. Jr. and Bershak, P. A. (1981).

may be some species variation such as occurs, for example, in *Arum* with respect to amounts of N centres (Cammack and Palmer, 1973).

Plant mitochondria, similarly to their mammalian counterparts, possess a HiPIP-type centre, S-3 of succinate dehydrogenase (Rich and Bonner, 1978a, b). A split signal is generally associated with S-3 (Moore *et al.*, 1976; Rich *et al.*, 1977) which is caused by an interaction between this centre and ubisemiquinone (Ruzicka *et al.*, 1975; Ingledew *et al.*, 1976; Rich *et al.*, 1977; Moore and Rupp, 1978). In addition to this spin coupled ubisemi-quinone pair, further paramagnetic species centred around $g = 2\cdot00$ may also be distinguished (Rich and Bonner, 1978a, b). These are the half-reduced species of either ubiquinone or flavin. Further unsplit $g = 2\cdot00$ signals can be observed as the temperature is raised with E_{m1} of +65 mV and E_{m2} of +180 mV suggesting they represent ubiquinone in its partially reduced form (Rich and Bonner, 1978a).

In addition to the ferredoxin and HiPIP-type centres, another group of signals can be detected in the low field end of the EPR spectrum at low temperature (Rich and Bonner, 1978c, d). The spectra consist of a $g = 4\cdot3$ high spin low symmetry iron signal together with a number of high spin haem species. One of the high spin haem species appears to be a catalase or peroxidase-type moiety (Rich and Bonner, 1978c). Whether these signals are representative of mitochondrial components is uncertain since purified mitochondria contained significantly less of these species than crude preparations. The $g = 4\cdot3$ iron component is attributed to a spin 3/2 iron-containing system. Apparently this component is normally EPR-silent, only detectable in the presence of nitric oxide (Rich et al., 1978a). Interestingly, the amount of iron present in the component is at least an order of magnitude greater than that present in the total iron-sulphur centres of the respiratory chain suggesting that it may be part of an iron-transporting or iron storage system (see Bienfait, this volume).

The interaction between iron-sulphur centres is shown by their power saturation behaviour. This powerful technique is not only a way of resolving overlapping signals but a method of detecting interacting species, since electron spin relaxation is characteristic for a particular iron-sulphur centre. For example, the overlapping signals attributed to the ubisemiquinone pair which are associated with centre S-3 at temperatures lower than 20 K suggest an interaction between ubiquinone and centre S-3 since free radicals (such as ubisemiquinone) tend to be saturated at temperatures less than 77 K (Moore and Rupp, 1978). Spin-spin interaction between centres S-1 and S-2 is also revealed by the power saturation technique as the electron spin relaxation of centre S-1 is enhanced upon reduction of centre S-2 (Rupp and Moore, 1979). For example, in the absence of signals from S-2, S-1 signals are half-saturated at $6\cdot0$ mW at 12 K while in their presence neither S-1 nor S-2 show saturation up to $16\cdot0$ mW. This clearly indicates a relaxation enhancement for centre S-1, and is most probably due to magnetic interaction with centre S-2, which, according to Salerno (1976) indicates that the distance between S-1 and S-2 is in the order of 10 Å. Similarly the saturation behaviour of centre N-2 also shows relaxation enhancement, possibly due to interaction with centre N-3 or the fast-relaxing ubisemiquinone species (Rupp and Moore, 1979). Saturation data thus indicate that magnetic interactions occur between the iron-sulphur centres, which would be anticipated if they were members of the electron transport chain.

C. FLAVOPROTEINS

Flavoproteins are enzymes consisting of an apoprotein plus either FMN or FAD with the flavin tightly (but not usually covalently) bound to the protein. The redox properties of the flavin isoalloxazine nucleus allow the prosthetic group to carry up to two electrons, although in vivo they probably oscillate between the oxidized and semiquinone form or the semiquinone and reduced forms. According to Storey (1980), flavoproteins can essentially be divided into two categories; those associated with the dehydrogenases and flavoproteins internal to the respiratory chain. Very little progress has been made on the isolation and purification of the plant flavoprotein containing dehydrogenases and a description of their role in electron transfer will be left until Section IIIA. Potentiometric titrations have however resulted in the identification of five flavoproteins, only one of which has any fluorescence (a characteristic property of flavins). This particular flavoprotein has an $E_{m7.2}$ of $-155\,mV$ and is probably not part of the electron transport chain but associated with lipoate dehydrogenase instead (Storey, 1980). The other four are non-fluorescent and their assignment as flavoproteins is based upon an absorption maximum at 464 nm and redox titration yielding slopes of n = 2. Since iron-sulphur centres absorb in this area (although they are 1 electron acceptors, i.e. $n = 1$) and they outnumber any other respiratory component, it is conceivable that they have been miscast as flavoproteins. Thus electron transport schemes assigning particular roles to the non-fluorescent components should be treated with caution.

D. UBIQUINONE

Ubiquinone is a substituted 1,4-benzoquinone with a long hydrophobic side-chain composed of sequential isoprenoid units. The length of this side-chain varies between different tissues but usually consists of ten isoprenoid units (hence the abbreviation Q-10) or fifty carbon atoms. This molecule is ideally suited to be a mobile carrier of protons and electrons within the electron transport chain because it is highly soluble in a lipid environment, it is capable of undergoing oxidation and reduction, and usually its concentration is in considerable molar excess over the other respiratory chain components. Electron transport in mammalian and plant mitochondria, bacterial, and photosynthetic systems has long been known to involve ubiquinone or other 1,4-benzoquinone analogues.

The presence of Q-10 has been demonstrated in pentane extracts of mitochondria from *Arum maculatum* (Huq and Palmer, 1978), *Phaseolus aureus* (Beyer *et al.*, 1968) and *Sauromatum guttatum* (Bonner and Rich, 1978). The mid-point potential of Q-10 in mung bean mitochondria was measured as $+70\,mV$ (Storey, 1973) which is close to the value of $+65\,mV$ obtained for Q-10 in beef heart mitochondria (Urban and Klingenberg, 1969).

The evidence to support the role of Q-10 as an obligatory carrier of reducing equivalents in the respiratory chain of plant mitochondria is not as convincing as that obtained for mammalian mitochondria. There is kinetic evidence from a study of the effects of substrate pulses on the individual rates of reduction of Q-10 and other respiratory chain components in mung bean mitochondria. Storey (1974a) interpreted these results in terms of an ordering theorem (Higgins, 1963) and concluded that Q-10 was the first redox component in the sequence to be reduced. The reduction of Q-10 under these conditions has two components; a fast phase in which 20% of the total substrate reducible Q-10 is reduced and a subsequent slower reduction of the remainder. This behaviour is predicted by the ordering theorem and does not indicate the presence of two kinetically distinct pools of Q-10 (Storey, 1974a; Storey, 1980).

The study of mammalian mitochondria has produced convincing evidence, other than kinetic studies, to show that Q-10 is an obligatory respiratory chain carrier. One approach involves so-called "extraction-reactivation" studies which show that solvent treatment of lyophyllized mitochondria interupts electron transport to oxygen. The solvent contains Q-10 which, when added back to the extracted lyophyllate, restores electron transport activity (Norling et al., 1974; Kroger and Klingenberg, 1973b). However, this protocol has drastic effects on the activities of the respiratory chain enzymes and the conclusions from these studies should be viewed with caution (Gutman, 1980). There is only one report of this type of study on plant mitochondria (Huq and Palmer, 1978a) where it was shown that NADH and succinate oxidase activities in lyophyllized Arum maculatum mitochondria were inhibited by extraction with pentane. The extract was shown to contain Q-10 and when this, or Q-10 obtained from a commercial source, was reincorporated, the oxidase activities were restored. Huq and Palmer (1978a) found that it was possible to reincorporate more Q-10 than was originally present in the membrane, which is consistent with a previous observation for mammalian mitochondria (Kroger and Klingenberg, 1973b). These observations contrast with those in another report where Q-10 could not be reincorporated beyond the original level (Ernster et al., 1969). This is quite an important point to resolve since the inability to reincorporate higher than original levels has been used to argue for the presence of "Q-binding protein" (Storey, 1980). Huq and Palmer (1978a) also reported that the reincorporation of higher than natural levels of Q-10 stimulated NADH oxidase activity beyond that of the control. This effect had already been observed in mammalian systems (Kroger and Klingenberg, 1973b) and probably represents an increase in the K_m of the dehydrogenase caused by freeze-drying. It seems likely that under physiological conditions the natural amount of Q-10 present in the membrane is not rate limiting for the overall respiratory process.

The best evidence that Q-10 is an obligatory carrier of reducing equivalents between the dehydrogenases and the cytochrome system in mammalian mitochondria, has come from the study of the electron acceptors or donors from or to the purified enzymes. Studies of this type on plant respiratory chain dehydrogenases have not been done for the obvious reason that none have been purified. However, the external NADH dehydrogenase (see Section IIIA-2) from *Candida utilis* has been isolated and is able to reduce Q-6 (Mackler *et al.*, 1980) and the alternative oxidase (see Section IIE) has been partially purified and shown to oxidize reduced quinone analogues. Further evidence that the alternative oxidase interacts at the level of Q-10 has come from EPR studies (Section IIE 1).

In summary, there is little published evidence that Q-10 is an obligatory carrier of reducing equivalents in the respiratory chain of plant mitochondria but all of the available evidence is consistent with the conclusions made about the role of Q-10 in mammalin mitochondria.

E. THE ALTERNATIVE OXIDASE

Perhaps one of the most characteristic features of plant mitochondria, is the possession, by some, of a cyanide and antimycin A resistant, alternative pathway to molecular oxygen. Cyanide-resistant respiration was first observed by Lamarck (1778) who noted that the spadices of Aroid plants were warm to the touch upon the onset of flowering. The rise in temperature of the spadix was associated with a foetid smell which has been described as "both foul and urinous" (Prime, 1960). The first investigation of the chemical mechanism of spadix respiration was made by Van Herk (1937) who discovered that the heating process of the spadix of *Sauromatum guttatum* was due to a high rate of respiration which displayed marked insensitivity to cyanide. These observations were extended to the Aroids by James and his colleagues (James and Beevers, 1950; James and Elliot, 1955), who showed that cyanide-resistant respiration is associated with mitochondrial preparations of Arum spadix. Ever since this date, more plant physiologists and biochemists have become interested in this phenomena and it is now apparent that not only is it widespread in the plant kingdom (see Henry and Nyns, 1975) but can also be demonstrated in a few animal species such as millipedes, pathogenic protozoa such as the African trypanosomes (Hill, 1978) and some species of bacteria and fungi such as *Candida lipolytica* (Henry *et al.*, 1974) and the poky strain of *Neurospora crassa* (Lambowitz *et al.*, 1972).

1. Characteristics

Electron transport by this pathway is characterized by a lack of sensitivity to respiratory inhibitors such as cyanide, azide, carbon monoxide, antimycin A and 2-heptyl-4-hydroxyquinoline N-oxide, all of which are potent inhibitors of mitochondrial respiration in animal tissues. The degree to which substrate oxidations are insensitive to inhibitors varies from as little as a few per cent, as in the case of fresh dormant white potatoes, to as much as 100% as in the spadices of *Arum maculatum*, *Sauromatum guttatum*, and the American skunk cabbage *Symplocarpus foetidus*.

Although absent from storage tissues such as the white potato and Jerusalem artichoke, it is possible to induce cyanide-resistant respiration upon slicing and washing these tissues in dilute salt solutions for 24 hours. That this adaption is dependent upon protein and lipid synthesis is demonstrated by the result that both actinomycin D and cerulenin inhibit the development of cyanide-resistant respiration (Waring and Laties 1977a, b). Indeed it has been suggested that a cyanide-resistant oxidase is present in the intact tissue but is lost upon slicing due to lipid breakdown. The re-establishment of cyanide-resistant respiration is consequently seen as a wound response dependent upon *de novo* phospholipid biosynthesis. For a fuller treatment of whole-tissue respiration the reader is referred to Day *et al.* (1980) and Laties (1982) in which this topic is reviewed in detail.

The work of Bendall and Bonner (1971) established that cyanide-resistant respiration in isolated plant mitochondria entails an additional oxidase. Bendall and Bonner (1971) working with skunk cabbage mitochondria confirmed the earlier findings of Wiskich and Bonner (1963) that the oxidation of ascorbate plus TMPD, a cytochrome oxidase substrate, was powerfully inhibited by cyanide. Oxygen uptake could be subsequently restored upon addition of succinate or NADH suggesting the existence of a second oxidase resistant to cyanide and azide. Direct evidence for the presence of two oxidases was also obtained by Ikuma *et al.*, (1964), from a study of oxygen affinity. They found that in the steady state, the apparent K_m was $0 \cdot 1$ μM whereas in the presence of cyanide it was $0 \cdot 5$ μM.

The stopped flow experiments of Storey and his colleagues (Storey, 1969; 1970a,b, 1972a,b, 1974b; Storey and Bahr, 1969; Erecinska and Storey, 1970), performed under conditions of slow electron input from substrate, have demonstrated that ubiquinone, the cytochromes *b* and part of the flavoprotein complement, can be oxidized via the alternative pathway when the cytochrome oxidase route is inhibited (Storey, 1976). Furthermore, these experiments suggest that antimycin A slows down the oxidation of the cytochromes *b*, but not of the ubiquinone or flavoprotein component by this pathway. It was initially suggested by Storey (1976) that one of the flavoprotein compounds acted as the control valve for the alternative oxidase (see also Day *et al.*, 1980), because of its relatively electronegative redox potential. However, in view of the difficulty in identifying separate flavoprotein components, this suggestion remains tentative (see Section IIIB).

On the basis of the above experiments and the observation that succinate or external NADH oxidation does not result in ATP synthesis in the presence of cyanide (Storey and Bahr, 1969; Passam and Palmer, 1972; Moore *et al.*, 1978; however, see Wilson, 1977; 1980) it is generally accepted that the branchpoint of the alternative oxidase from the main respiratory chain is on the substrate side of the b–c_1 complex at the level of ubiquinone.

EPR studies also support the notion that the interaction of the alternative oxidase with cytochrome pathway is at the level of ubiquinone. Rich *et al.* (1977) found that the split signal described by Moore *et al.* (1976) could be detected in the presence of cyanide suggesting a close association with the alternative oxidase. Furthermore an investigation of the redox behaviour of this signal indicated that it was not only in rapid equilibrium with the respiratory chain but also on the substrate side of the b–c_1 complex. Of further interest was the finding that primary hydroxamic acids, inhibitors of the alternative oxidase (Schonbaum *et al.*, 1971), removed the split signal, implying a binding at or close to one of the components involved in the interaction. Such data was in consensus with an earlier suggestion of Rich and Moore (1976) which assigned the location of the alternative oxidase to be closely associated with part of Mitchell's (1976) ubiquinone cycle. The interaction of the alternative oxidase with the main respiratory chain and the mechanism whereby it may be controlled will be discussed in detail in Section IIIB.

2. Nature of Alternative Oxidase

The nature of the alternative oxidase remains elusive although numerous possibilities have been eliminated. Several hypotheses have been proposed such as the incomplete inhibition of cytochrome oxidase (Chance and Hackett, 1957), a flavoprotein oxidase (James and Beevers, 1950) or an auto-oxidizable cytochrome b component (Yocum and Hackett (1957), but none have stood the test of time. In recent years iron-sulphur centres have been implicated as possible components of the alternative pathway (Bendall and Bonner, 1971) largely because of the nature of the inhibitors of this pathway. These are shown in Table III. The suggestion of an iron-sulphur centre as a component of the oxidase stems mainly from the observation that hydroxamic acids and some of the other inhibitors are metal chelators. However, this hypothesis is not wholly consistent with the finding that in the time scale of inhibitor experiments the iron of the centres is too tightly bound and inaccessible to chelators. Furthermore Rich *et al.* (1978b) have shown that hydroxamic acids also inhibit the catalytic activity of a number of redox enzymes which do not contain iron-sulphur centres such as tyrosinase, horse radish peroxidase, and α-glycerophosphate oxidase. Inhibition could not be related to either the

TABLE III
Inhibitors of the alternative pathway

Inhibitor	Approx. %	Inhibition	Conc^n.	Reference
	Cyt. pathway	ALT. pathway		
Thiocyanate	41	95	0·1 mM	(a)
8-hydroxyquinoline	5	78	1·5 mM	(b)
Piercidin A	36	83	0·5 mM	(c)
Thenoyltrifluoroacetone	50	81	0·3 mM	(c)
3'-Phenoxy-carboxin	65	85	0·075 mM	(d)
Disulfiram	6	87	0·025 mM	(e)
Hydroxamic acids	0	100	0·100 mM	(f)
n-Propyl gallate	0	100	0·005 mM	(g)

(a) Bonner *et al.* (1972).
(b) Lance (1972).
(c) Wilson (1971).
(d) Day *et al.* (1978).
(e) Grover and Laties (1978).
(f) Bahr and Bonner (1973a).
(g) Siedow and Bickett (1981).

metal site nor to the nature of the oxygen-containing substrate. Rather it appears that inhibition may involve either hydrogen bonding at the binding site of the reducing substrate or the formation of a charge transfer complex. Recently, other compounds have been found which inhibit the alternative pathway (Grover and Laties, 1978; Siedow and Girvin, 1980; Siedow and Bickett, 1981). Grover and Laties (1978) presented evidence that disulfiram is a more potent inhibitor than hydroxamic acids, being effective at μM concentrations. An investigation of the effectiveness of this inhibitor on cyanide-resistant oxidase in *Arum* mitochondria in our laboratory has confirmed that it is an inhibitor but in our hands is less potent than hydroxamic acids. More recently Siedow and Girvin (1980) found that n-propyl gallate inhibited the alternative pathway in a mutally exclusive fashion with respect to hydroxamic acid suggesting that the two inhibitors act at either the same site in a non-competitive manner or at spatially separate sites. The nature of the component(s) to which these inhibitors bind, however, is still uncertain. EPR studies (Rich *et al.*, 1977; Rich and Bonner, 1978a) indicate that it is not an iron-sulphur centre, *per se*, although it may be a quinone species. In this respect it is interesting to note that hydroxamic acids specifically interact with with the dipolar coupled semiquinone species (Moore and Rupp, 1978) again suggesting a close association of the alternative oxidase with quinone species. Initially it was suggested that the quinone species may be an autoxidizable pool which interacts with the main quinone pool associated with the cytochrome pathway (Rich and Bonner, 1978a). This was supported by the finding that pentane extraction of ubiquinone inhibited alternative oxidase activity (Huq and Palmer, 1978a; and see Section IIC). However, the lack of short chain quinones in mitochondria (Bonner and Rich, 1978; Vanderleyden *et al.*, 1980b) and studies with a solubilized alternative oxidase from *Arum maculatum* (Rich, 1978; Huq and Palmer, 1978b) do not support this notion. With respect to the isolated oxidase, in both cases a detergent solubilized fraction was obtained without apparent loss of activity when assayed using a hydroxamic acid-sensitive, antimycin-insensitive oxidation of either ubiquinol or menadiol. Both of these quinols appear to donate electrons efficiently at or close to the alternative oxidase as judged by their sensitivity to hydroxamic acids and insensitivity to antimycin or cyanide. However, it has been reported (Vanderleyden *et al.*, 1980a) that menadiol is autoxidizable and that this reaction is inhibited by hydroxamic acids. Nevertheless, in such partial purifications (Rich, 1978) there is a complete absence of cytochromes *a*, and a decrease in amounts of flavoprotein and cytochromes *b* and *c*. Only small amounts of NADH oxidase activity can be detected whilst succinate dehydrogenase activity is completely absent. EPR analysis revealed no centre S-3, a small signal at $g = 1.93$ due to non-haem iron and a copper signal at $g = 2.00$, all of which were not redox active. It may be deduced from these studies that the oxidase is associated with a protein species and not merely an autoxidizable quinone pool. Furthermore, it appears to be a quinol oxidase which catalyses the oxidation of quinols by molecular oxygen to produce the corresponding quinone and water as the end-products. The discovery that some quinols are capable of providing reducing equivalents close to the oxidase offers an invaluable tool for its further investigation and developments are awaited with eager anticipation.

The isolated oxidase does show some similarities to the particulate preparation of glycerol-3-phosphate oxidase of *Trypanosoma brucei* (Fairlamb and Bowman, 1977a,b). This preparation also contains copper and iron and is inhibited by hydroxamic acids. In contrast to the higher plant oxidase, however, it also contains flavin (however, see Huq and Palmer, 1978b). Fairlamb and Bowman (1977b) have suggested that the oxidase may be a copper protein similar to ascorbic acid oxidase. Interestingly the overall product of oxygen reduction is also water, as in the case of the alternative oxidase.

In contrast to these studies, experiments with micro-organisms indicate that alternative oxidase activity may be manipulated environmentally or genetically and its activity may be enhanced when the main cytochrome chain is damaged or impaired. For instance, the respiration of whole cells of the soil amoeba *Acanthamoeba castellanii* (Edwards and Lloyd, 1977, 1978) shows variations in sensitivity to cyanide at different stages of growth. This is somewhat reminiscent of the work on cultured sycamore

(*Acer pseudoplantanus*) cells (Wilson, 1971) in which cyanide sensitivity fluctuated during the growth cycle; particularly during cell division the addition of cyanide activated the alternative pathway (Blein, 1980). In some microorganisms induction of the alternative oxidase requires the presence of iron in the medium (Haddock and Garland, 1971; Henry *et al.*, 1977; Henry, 1978) implicating the involvement of iron-sulphur centres. Copper appears to be less important in yeast (Downie and Garland, 1973) since its preclusion results in the emergence of cyanide-resistant respiration. This does not appear to hold true with cultured sycamore cells, however, since deficiency does not induce cyanide-resistant respiration (Bligny and Douce, 1978).

In summary, although the nature of cyanide and antimycin resistant respiration has remained elusive for many years, it seems likely that its components will be identified in the very near future.

F. ORGANIZATION OF THE RESPIRATORY CHAIN

A currently acceptable form of the organization of the respiratory chain of plant mitochondria is illustrated in Fig. 3. This scheme is based upon oxidation-reduction kinetics, redox-potential studies, sites of action of respiratory inhibitors and inference from studies with mammalian mitochondria. Electron transport flavoproteins have not been included in this scheme due to the difficulty in their identification.

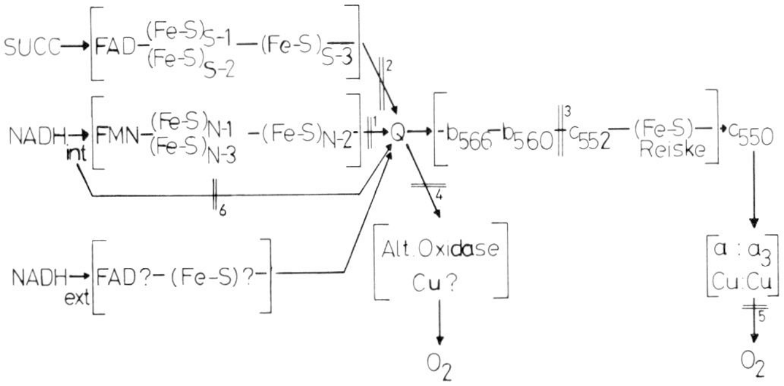

FIG. 3. The respiratory chain of higher plant mitochondria. FAD, FMN are flavoproteins associated with the dehydrogenases; Fe-S = iron-sulphur centre; Q = ubiquinone; b, c, and a are cytochromes. Energy conservation sites have been omitted for clarity but are discussed in text. Square brackets show the components of the various enzyme complexes and question marks denote uncertainties of composition. NADH produced either in the matrix or the cytoplasm and succinate are referred to as $NADH_{int}$, $NADH_{ext}$ and SUCC respectively. Sites of inhibitor action are numbered with numerals and refer to the following: 1 = rotenone, amytal and pieridicin A; 2 = TTFA; 3 = antimycin A, UHDBT; 4 = hydroxamic acids, n-propylgallate; 5 = cyanide, azide, carbon monoxide and 6 = dicoumarol.

Complexes I (NADH-ubiquinone reductase; Phosphorylation Site I) and II (succinate-ubiquinone reductase) contain the flavins and iron-sulphur centres of NADH and succinate dehydrogenases respectively. Complex III (ubiquinol-cytochrome c reductase or the $b-c_1$, complex; Site II) contains the b cytochromes, cytochrome c_1 and the Rieske iron-sulphur centre whilst Complex IV (cytochrome oxidase; Site III) contains cytochrome a, a_3 and the two copper centres. In addition to these components some plant mitochondria contain two other particulate components, namely the alternative oxidase (quinol : oxygen oxidoreductase) and an external NADH dehydrogenase. Inhibitors of the respiratory chain are also shown in Fig. 3. It may be noted that the site of action of inhibition is associated with the respiratory complexes. Thus rotenone inhibits electron transfer in Complex I, antimycin A in Complex III and cyanide in Complex IV. Primary hydroxamic acids and n-propyl gallate inhibit the alternative oxidase. To date there is no specific inhibitor of the external NADH dehydrogenase. Since the $b-c_1$ complex and cytochrome oxidase have been dealt with in Section IIA they will not be considered further. In the next section it is proposed to consider the peculiarities of the plant dehydrogenases and the manner in which they interact with the oxidases, with particular reference to the control of respiratory chain activity.

III. Special Features of the Respiratory Chain in Plants

A. MULTIPLE DEHYDROGENASES

Reducing equivalents, produced as a result of the activity of the TCA cycle enzymes, are transferred to the electron transport chain by way of the dehydrogenases. Plant mitochondrial dehydrogenases utilize only succinate and NADH as substrates. There is no flavoprotein dehydrogenase analogous to the ETF (Electron Transferring Flavoprotein) dehydrogenase that is involved in fatty acid oxidation in mammalian tissues, nor do they possess an α-glycerophosphate dehydrogenase. In addition to the TCA cycle enzymes, plant mitochondria also possess an NAD^+-malic enzyme which is located in the matrix (Day et al., 1979). It is generally considered that the presence of this enzyme, in addition to malate dehydrogenase, allows the continued oxidation of malate without the need to add an agent to remove oxaloacetate.

Plant mitochondria possess multiple pathways of NADH oxidation. In addition to the NADH dehydrogenase which is analogous to mammalian Complex I, they also possess three alternative systems capable of oxidizing NADH; two are capable of oxidizing external (i.e. cytoplasmic) NADH while the other can oxidize internal NADH.

1. Internal NADH Dehydrogenases

NADH generated within the mitochondrial matrix may be oxidized via two pathways, one of which is sensitive to NADH dehydrogenase inhibitors, and appears to involve Complex I, although little information on its structure and composition in plant mitochondria is available. According to mammalian studies (Ragan, 1980; Ohnishi, 1979) Complex I is a lipoprotein enzyme that catalyses the oxidation of intra-mitochondrial NADH with the concomitant reduction of ubiquinone. The purified form obtained by detergent solubilization and salt precipitation contains FMN, non-haem iron, acid-labile sulphide, ubiquinone, and phospholipid.

EPR spectroscopy of plant mitochondria reveals at least three (Cammack and Palmer, 1977; Rich and Bonner, 1978a) and possibly four (Cammack and Palmer, 1977; Rupp and Moore, 1979) iron-sulphur centres attributable to Complex I. Potentiometric titrations (Rich and Bonner, 1978a; Cammack and Plamer, 1977) and spin–spin interactions (Rupp and Moore, 1979) suggest that the sequence of iron-sulphur centres in this segment is N-1 $\rightarrow$ N-3 $\rightarrow$ N-2. This sequence is consistent with studies on mammalian systems where it has also been concluded that the site of inhibitor action is on the oxygen side of centre N-2 (Ohnishi, 1979). Oxidation of NADH by this dehydrogenase results in the synthesis of ATP (Site I) and is specifically inhibited by rotenone and piercidin A.

Plant mitochondria oxidizing internal NADH do, however, display a general lack of sensitivity to these inhibitors. For instance, both Brunton and Palmer (1973) and Day and Wiskich (1974a,b) reported that the oxidation of NADH produced from malate was only partially sensitive (up 60%) to piercidin A and rotenone respectively. It was also noted that NADH oxidation in the presence of these inhibitors was only coupled to ATP synthesis at Sites II and III. Brunton and Palmer (1973) interpreted their observations as being due to a second NADH dehydrogenase (see Fig. 3) that by-passes Site I and interacts with ubiquinone (Palmer, 1976). Their interpretation was further complicated by the suggestion that the NADH was compartmentalized. Thus NADH produced by the malic enzyme was more accessible to the inhibitor-insensitive, non-phosphorylating dehydrogenase, whereas NADH produced by malate dehydrogenase was associated with the inhibitor-sensitive phosphorylating pathway (Palmer, 1976; 1979) This view has been supported by the findings of Rustin and Moreau (1979) and Rustin et al. (1980). Furthermore, Palmer (1979) suggested that the relative activity of the two pathways is under the control of adenine nucleotides. A more plausible explanation recently offered by Palmer (1983) emphasizes the relative concentrations of NADH and NAD^+ and suggests that the redox state of the mitochondrial pyridine nucleotide pool dictates which electron transport sequence may be active, with the proviso that the inhibitor-sensitive pathway has a higher affinity for NADH than the inhibitor-insensitive dehydrogenase (see also Neuburger and Douce, 1980).

In contrast to these suggestions, Day and Wiskich (1974a, b, 1978) continue to explain inhibitor-insensitive respiration in terms of a trans-membrane transhydrogenase which is capable of transferring reducing equivalents from the matrix to external NAD^+ where it is re-oxidized by the external NADH dehydrogenase. This suggestion is based upon numerous observations on the ability of NAD^+ to stimulate malate oxidation in the presence of rotenone or piercidin A. More recently, Neuburger and Douce (1978; 1980) and Tobin et al. (1980) made the provocative suggestion that NAD^+ stimulation was caused by NAD^+ penetrating the membrane and interacting with the NAD^+-linked malic enzyme which presumably is associated with the inhibitor-insensitive pathway. They demonstrated that ^{14}C-NAD^+ uptake into potato mitochondria showed apparent first-order kinetics and saturation (Tobin et al. 1980) although initial rates were very low. On the basis of this work, Tobin et al., (1980) propose a specific transport system for NAD^+. However, it must be stressed that prior to postulating the existence of specific carriers, the following criteria suggested by Heber (1974), must be met; competition by suitable substrates, sensitivity to poisoning and saturation of transfer. Until these criteria are fulfilled it is difficult to discriminate between carrier-mediated transport and binding. Indeed, preliminary results from our laboratory (M. O. Proudlove and A. L. Moore, unpublished data) are more consistent with binding than suggestive of a specific NAD^+ carrier. Obviously considerably more data is required in order to resolve this question. Irrespective of the mechanism, the nature of the dehydrogenase is of some importance. Results have been obtained (A. L. Moore, unpublished data) that show that the internal NADH dehydrogenase can reduce DCPIP in a rotenone-insensitive dicoumarol-sensitive fashion. Furthermore, rotenone-insensitive internal NADH oxidation is antimycin and dicoumarol sensitive, in both mitochondria and sub-mitochondrial particles, suggesting that the enzyme is membrane bound. This enzyme bears close similarities to the DT diaphorase (NADH-lipoamide oxidoreductase) described by Ernster (1967), an enzyme that mediates electron transfer between NADH and menadione. However, in mammalian tissues no natural acceptor for this enzyme could be found. It is possible that in plant tissues napthoquinone, an abundant, naturally-occurring plant quinone (Brodie, 1965; Crane, 1965), may act as the electron acceptor and may function as an electron transport component between NADH and Complex III. This would account for continued respiration and the lack of ATP synthesis at Site I in the presence of inhibitors.

2. External NADH Dehydrogenases

One of the major characteristics of plant mitochondria is their ability to oxidize added NADH. Douce et al. (1973) has demonstrated that there are two external NADH dehydrogenases. One, which is located on the outer membrane, contains a flavoprotein and cytochrome b-555, and is insensitive to antimycin A. It is generally considered not to contribute reducing equivalents to the respiratory chain unless cytochrome c is added and consequently its physiological role is uncertain. Accordingly, it has been omitted from Fig. 3. The second dehydrogenase is located on the external face of the inner membrane and is linked to the respiratory chain (Douce et al., 1973; Day and Wiskich, 1974a, b). The oxidation of NADH by this system is insensitive to inhibitors of Complex I, such as rotenone and piercidin A, but sensitive to antimycin A, suggesting that it donates reducing equivalents into the respiratory chain at the level of ubiquinone, and is therefore coupled to two sites of ATP synthesis. This notion is confirmed by the finding that in intact mitochondria, NADH : ferricyanide oxidoreductase activity is antimycin A sensitive (Douce et al., 1973). The physiological significance and the mechanism whereby this dehydrogenase system donates reducing equivalents to ubiquinone remains unresolved. Its activity does appear to be controlled by divalent cations such as Ca^{2+} (Coleman and Palmer, 1971). Recently Møller et al. (1981) have suggested that stimulation by Ca^{2+} is due to both a general charge screening effect and a specific effect on the dehydrogenase. The mechanism of the specific effect has been left uncertain. One possibility is that calcium regulation may be mediated through the calcium binding protein, calmodulin. Calmodulin is generally considered to be the coupling link in many Ca^{2+}-mediated processes and is thought to regulate intracellular distribution of calcium (Jarrett, this volume; Charbonneau and Cormier, 1979). One criterion of calmodulin-regulated reactions is inhibition by trifluoperazine, an antipsychotic drug, and recent results in our laboratory suggest that this drug specifically inhibits the external NADH dehydrogenase (A. L. Moore and A. R. Slabas, unpublished results). Thus it is conceivable that external NADH dehydrogenase activity in vivo is controlled by calmodulin. This in turn may be regulated by phytochrome which has recently been implicated in the regulation of external NADH oxidation (Cedel and Roux, 1980) in oat mitochondria.

Recently Mackler et al. (1980) reported the isolation and purification of an external, rotenone-insensitive NADH dehydrogenase from Candida utilis. The enzyme has a molecular weight of approximately $1·5 \times 10^6$ and contains two moles of FMN per mole of enzyme. Iron and copper were found in low amounts and are considered contaminants. The enzyme oxidized NADH and NADPH and reduced quinone analogues. Although the identification of FMN awaits confirmation, the interesting observation that the enzyme can oxidize NADPH is consistent with recent reports on the capability of the plant external dehydrogenase to oxidize this substrate (Arron and Edwards, 1980).

3. Succinate Dehydrogenase

Succinate dehydrogenase has been extensively studied in mammalian tissues both in its membrane bound and solubilized form (Singer, 1976; Hatefi and Stiggal, 1977; Ohnishi, 1979). It has a molecular weight of approximately 100 000 and contains 1 mol. of FAD per mole and 8g-atoms of non-haem iron and acid-labile sulphide (Davis and Hatefi, 1971). The isolated enzyme is composed of two non-identical subunits, a ferro-flavoprotein and an iron-sulphur protein, having molecular weights of 70 000 and 27 000 respectively. The ferroflavoprotein contains FAD covalently bound to the polypeptide chain as 8α-[N(3)]-histidyl-FAD (Hemmerich et al., 1969) and two ferredoxin-type centres, whereas the smaller subunit contains one HiPIP-type centre.

Studies on the succinate dehydrogenase of plant mitochondria show its properties are very similar to the mammalian counterpart. For instance, the flavin is covalently bound (Singer et al., 1973) and EPR analysis shows the presence of a HiPIP-type and two ferrodoxin-type centres correspond-ing to centres S-3, S-1 and S-2 respectively (Cammack and Plamer, 1977; Rich and Bonner, 1978a, b; Rupp and Moore, 1797). Potentiometric titrations (Cammack and Palmer, 1977; Rich and Bonner, 1978a) and analysis of power saturation data (Rupp and Moore, 1979) suggest that electron transfer from succinate to ubiquinone probably occurs in the sequence (FAD, S-1) → S-3 → ubiquinone. This scheme is supported by the finding that centre S-3 is associated with a split signal caused by interaction with semiquinone species (see Section IID). Inhibitors of succinate dehydrogenase activity, thenoyl trifuoroacetone and carboxin, are presumed to inhibit between centre S-3 and the Q pool since their addition results in the loss of the semiquinone signals (Rich et al., 1977; Rich and Bonner, 1978a).

B. INTERACTION BETWEEN THE DEHYDROGENASE AND OXIDASE SYSTEMS

Of particular interest in the study of plant mitochondria is the mechanism by which the various membrane bound dehydrogenases and the two oxidase systems are inter-related and controlled. The evidence presented in the section on ubiquinone shows that the dehydrogenases and the cytochrome system in mammalian mitochondria are connected via ubiquinone and there is every reason to believe this is so with plant mitochondria. The best model to describe this interaction is the "Q-pool" model developed by Kröger and Klingenberg from the study of beef-heart sub-mitochondrial particles (Kröger and Klingenberg, 1973a, b). We intend to discuss this in detail as well as the extension made by De Troostembergh and Nyns, 1978 to include the simultaneous operation of two oxidase sytems in yeast. This is contrasted with the model proposed for the control of the alternative oxidase in plant mitochondria (Bahr and Bonner, 1973a, b). Finally, the Q-pool model is discussed when applied to the simultaneous oxidation of two respiratory substrates and to the differential effects of inhibitors on substrate oxidation.

1. Control of the Alternative Oxidase

The Q-pool model was developed from a very simple experimental observation; there is a linear relationship between the rate at which electrons flow from a single respiratory substrate to oxygen and the proportion of the ubiquinone present which is in the reduced form. The total amount of redox active Q-10 present, Q_T, is measured as the Q-10 which is reducible in the presence of excess substrate and cyanide. Under steady state conditions it is equal to the sum of the reduced and oxidized Q-10 present:

$$Q_T = Q_{RED} + Q_{OX} \qquad (1)$$

Kröger and Klingenberg (1973a) expressed the relationship between the overall respiratory rate, v, and the redox state of Q-10 thus:

$$v = V_{OX} \cdot Q_{RED}/Q_T \qquad (2)$$

V_{OX} is a first order rate constant and is defined as the respiratory rate when all of the Q-10 present is in the reduced form. Incidentally, this is not the same as saying that V_{OX} is the maximum possible activity of the oxidizing system under any circumstances since Q_T may not be at a saturating concentration. In the steady state the rate at which Q_{RED} is oxidized must equal that at which Q_{OX} is reduced. It follows that:

$$v = V_{RED} \cdot Q_{OX}/Q_T \qquad (3)$$

Where V_{RED} is the maximum rate at which the system donating electrons to Q-10 would operate if Q_T were completely oxidized. The rearrangement of equations (1), (2), and (3) to eliminate terms relating to Q-10 yields the expression:

$$v = \frac{V_{OX} \cdot V_{RED}}{V_{OX} + V_{RED}} \qquad (4)$$

which describes the respiratory rate in terms of the activities of the two systems which oxidize or reduce Q_T. Kröger and Klingenberg (1973a) confirmed the validity of using first order rate constants to describe the redox reactions of Q-10; they showed that the value of V_{OX} obtained in a steady state system was just as accurate in describing the rate at which Q_{RED} was oxidized when an oxygen pulse was added to anaerobic sub-mitochondrial particles.

The implications of the Q-pool model are important: Q-10 is an obligatory carrier of reducing equivalents between the dehydrogenases and the cytochrome system; Q-10 is freely mobile within the hydrophobic phase of the membrane; and reducing equivalents can be transferred between any dehydrogenase and any cytochrome system. The Q-pool model is particularly well supported by the study of reconstituted systems (Heron *et al.*, 1978; Cottingham and Ragan, 1980b). The behaviour of the NADH-cytochrome *c* oxido-reductase reconstituted system is accurately predicted by the model, even when the ratios of the constituent enzyme were varied well beyond those observed in physiological systems. However, Q-pool behaviour was only observed when the ratios of protein to phospholipid and phospholipid to Q-10 were approaching those found in mitochondria. This emphasizes the importance of the relative mobility betwen the respiratory chain enzymes and Q-10.

In the model discussed so far there is only one donor of electrons to Q-10 and one acceptor of electrons from it. De Troostembergh and Nyns (1978) demonstrated that the Q-10 pool model could be simply extended to describe the behaviour of a yeast system with one dehydrogenase and two oxidase systems. In their system the respiratory rates, v_T, v_{ALT} and v_{CYT}, are the rates of oxygen uptake in the presence of substrate, substrate and excess cyanide, and substrate and excess benzhydroxamic acid respectively; and V_{RED}, V_{ALT} and V_{CYT} are the maximum activities for the donor system and for the alternative oxidase and the cytochrome system respectively. A series of three Q-pool equations can now be written to describe the individual respiratory activities in terms of the donor and acceptor activities. In the equation relating to v_T the acceptor activity (V_{OX} in equation (4)) is replaced by the sum of the two acceptor activities, V_{ALT} and V_{CYT}, thus:

$$v_T = \frac{V_{RED} \cdot (V_{ALT} + V_{CYT})}{V_{RED} + V_{ALT} + V_{CYT}} \tag{5}$$

This equation is only valid if the alternative oxidase and the cytochrome system compete freely for reducing equivalents supplied by the dehydrogenase. There can be no mechanism for individually modulating the oxidase activities. The equations describing the respiratory activities with a single oxidase system are similar to equation (4),

$$v_{ALT} = \frac{V_{RED} \cdot V_{ALT}}{V_{RED} + V_{ALT}} \tag{6}$$

and

$$v_{CYT} = \frac{V_{RED} \cdot V_{CYT}}{V_{RED} + V_{CYT}} \tag{7}$$

De Troostembergh and Nyns (1978) demonstrated that their equations were valid under a variety of conditions which were designed to change activity of only one of the oxidase systems: the addition of ADP to respiring mitochondria increased V_{CYT} but did not affect V_{RED} or V_{ALT}; antimycin A decreased V_{CYT} only; and benzhydroxamic acid decreased V_{ALT} but did not affect V_{CYT} or V_{RED}.

A different model has been proposed by Bahr and Bonner (1973a) to explain the control of the alternative oxidase in plant mitochondria. They showed that there is a linear relationship between the total activity, v_T, and v_{ALT}, the activity in the presence of cyanide, when v_{ALT} is titrated with benzhydroxamic acid to inhibit the alternative oxidase progressively. (We have used the nomenclature of De Troostembergh and Nyns (1978). Bahr and Bonner (1973a) used V_T rather than v_T, $g(i)$ rather than v_{ALT} and V_{CYT} rather than v_{CYT}). The intercept of this plot (see Fig. 4) is the maximum rate of oxygen uptake in the presence of saturating amounts of benzhydroxamic acid, v_{CYT}, and the slope of the line is designated as ϱ. Their interpretation of this relationship is that the cytochrome system is always saturated with reducing equivalents and that ϱ represented the proportion of the alternative oxidase which is "switched on". Hence:

$$v_T = \varrho \, v_{ALT} + v_{CYT}$$

If the cytochrome system is stimulated by the addition of ADP then ϱ decreases to allow the cytochrome system to take more of the available reducing equivalents; the slope of the line decreases and the value of v_{CYT} increases. The proposed mechanism by which the cytochrome system takes priority is that there are two acceptors of reducing equivalents from any dehydrogenase, each of which can only supply electrons to one oxidase (Bahr and Bonner, 1973a). One of these acceptors, which exclusively provides reducing equivalents to the cytochrome system, has a higher mid-point potential than the other and, hence, being almost fully reduced, is capable of saturating the cytochrome system (if the donor activity is high enough). The other acceptor is only reduced if there is any surplus of reducing equivalents and these are then fed to the alternative oxidase. As a consequence of this mechanism the alternative oxidase would only be saturated at very high donor activities.

It is a remarkable coincidence that both of these theories explain the observed data adequately. However, the observations of Bahr and Bonner (1973a, b) are a special case of the more generally applicable Q-pool model. This can be demonstrated if values of v_{RED}, v_T and v_{ALT} are assumed and equations (5), (6), and (7) are used to predict the relationship between v_T and v_{ALT} when v_{ALT} is decreased incrementally as would occur during a titration with benzhydroxamic acid. This theoretical relationship is shown in Fig. 4, where it is obvious that the points do not lie on a straight line. However, the deviation from a straight line is so small that it would not be detected with experimental data because of the inherent error in such measurements. The model of Bahr and Bonner (1973a) is thus founded on the false premise that this is a linear relationship. Further theoretical considerations show that if the cytochrome system were stimulated by the addition of ADP then v_T would increase and the slope, ϱ, would decrease. This is shown by the top line in Fig. 4 where the value assumed for v_T was increased but those for V_{RED} and V_{ALT} were kept the same. The only differences are in the value of v_{CYT} (and V_{CYT}) and that the

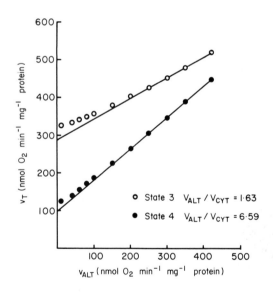

FIG. 4. In the upper line, which represents state 3, the points (○) are calculated from the Q-pool theory (see text) and do not lie on a straight line. The following values were assumed: $v_{RED} = 850$; $v_T = 520$; and $v_{ALT} = 410$. The calculated values of V_{CYT} and V_{ALT} were 509, and 830 respectively. The units of v_T and v_{ALT} are usually expressed as nmole oxygen consumed per min. per mg of mitochondrial protein. The units of V_{RED}, V_{CYT} and V_{ALT} are not important to the Q-pool calculations. In the lower line, which represents state 4 conditions, points (●) are recalculated with only v_T changed, from 520 to 450. This changes the calculated value of V_{CYT} to 126 but does not affect that for V_{ALT}.

line, if assumed to be straight, has a lower gradient. These are exactly the same changes as those observed by Bahr and Bonner (1973a) on addition of ADP. This shows that the Q-pool model theoretically predicts the observed behaviour of the system (compare Fig. 4 and Fig. 4B in the report by Bahr and Bonner, 1973a). Furthermore, this is indirect evidence to support the idea that the model of De Troostembergh and Nyns (1978) for two oxidase systems in yeast is equally valid for plant systems.

The Q-pool model can be used to explain some of the anomalies in the behaviour of plant mitochondria. For instance, it is frequently reported that the amount of cyanide-sensitive respiration in a single tissue is different for various substrates. One of the many examples is provided by cauliflower mitochondria (Rustin *et al.*, 1980) in which in state 3, malate oxidation is inhibited 70% by cyanide whilst under the same conditions succinate oxidation is inhibited 93%. This does not imply any special relationship between malate oxidation and the alternative oxidase but simply that the rate of succinate oxidation is faster than that of malate and that the ratio V_{ALT} to V_{CYT} in this system is low.

A practical advantage of the Q-pool model is that the simple experimental determination of v_T, v_{ALT} and v_{CYT} enables accurate calculations of V_{ALT}, V_{CYT} and V_{RED} which are representative of the absolute activities of the enzymes present. It would have been preferable to use these donor and acceptor activities in studying the development of cyanide-resistant respiration in potato slices (Theologis and Laties, 1978a, b), rather than quote the maximum oxygen uptake rates in the presence of cyanide or chlorobenzhydroxamic acid (CLAM), since v_{ALT} and v_{CYT} (using the De Troostembergh and Nyns (1978) notation) are dependent on V_{RED} and the ratio V_{ALT} to V_{CYT}. It is quite conceivable that during ageing in tissue slices the substrate and also the relative amounts of the two oxidase systems might change.

We believe that there are two major reasons for adopting the Q-pool model to explain the control of the alternative oxidase in plant mitochondria. Firstly, there is now considerable evidence for the involvement of Q-10 in respiratory chain dehydrogenases and oxidase systems and there have been numerous demonstrations of the general applicability and usefulness of the Q-pool equations (Rich *et al.*, 1981; and references therein). Also Q-pool behaviour has been demonstrated in plant mitochondria (Moreira *et al.*, 1980; Cottingham and Moore, 1981). Secondly, the model of Bahr and Bonner (1973a) requires another carrier of reducing equivalents between Q-10 and the alternative oxidase (a flavoprotein has been suggested for this role; Bahr and Bonner, 1973b) but recent investigations indicate that, like the cytochrome system, the alternative oxidase is capable of oxidizing ubiquinol yet it contains no chromophores (Rich, 1978 and Section IIE2).

2. Control of Substrate Oxidation

The foregoing discussion emphasizes that the alternative oxidase is not controlled at the level of ubiquinone but that the proportion of the reducing equivalents which go to either oxidase system is determined by the ratio of the activity of the cytochrome system to that of the alternative oxidase. This means that the control of the flow of reducing equivalents from different substrates must occur at the level of the dehydrogenases, an idea supported by studies with both animal and plant systems. For instance, in uncoupled beef heart sub-mitochondrial particles, the behaviour during the simultaneous oxidation of two substrates by the cytochrome system deviates from that expected from the Q-pool model (Gutman and Silman, 1972). The suggested explanation for this departure from the predicted behaviour is that the rate at which reducing equivalents flowed through the system exceeded that at which Q-10 could diffuse within the membrane. Thus domains are created in which one dehydrogenase predominates and the activity of that dehydrogenase determines the redox poise of Q-10 within its sphere of influence. The Q-10 in the domains of the more active dehydrogenase is more reduced than that in the domains of the other dehydrogenase and "spillover" of reduced Q-10 inhibits the less active oxidase simply by the law of mass action. Hence, the degree of departure from the Q-pool model is determined by the relative rate of this spillover and the more active dehydrogenase will inhibit the rate at which the less active one operates. This model is discussed in more detail in a recent review (Gutman, 1980). The simultaneous oxidation of two substrates has been investigated in plant mitochondria where a different control mechanism seems to be operating (Cowley and Palmer, 1980). The oxidation of exogenous NADH was inhibited by the presence of succinate even though succinate oxidation occurred at a considerably slower rate. This inhibition was only observed during state 3 or 4 conditions and not in the presence of uncoupler.

The departure from Q-pool theory during the oxidation of two substrates does not invalidate the idea of a mobile pool of Q-10 within the membrane but suggests that kinetic parameters are important in the regulation of substrate oxidation. However, observations have been made which conflict with the idea of a single Q-pool and even of kinetically determined multiple domains of Q-10. These observations are of two types. One is that the oxidase activities with different substrates vary greatly in their sensitivity to inhibitors. The oxidation of NADH by mitochondria from cassava (*Manihot esculenta*), for instance, is inhibited almost totally by antimycin but oxygen uptake can be restored by the addition of succinate and subsequently inhibited by the addition of a substituted benzhydroxamic acid (Huq and Palmer, 1978a). This data suggests that there is a specific association between dehydrogenases and oxidase systems and that reducing equivalents cannot be shared between the oxidase systems. This is perhaps best described as "Q-puddle" behaviour since it indicates the presence of completely isolated pools of Q-10. A similar affinity between dehydrogenases and oxidizing systems has been reported for bacterial systems (Porte and Vignais, 1980). Another departure from Q-pool theory is shown by the action of dibutylchloro-methyltin chloride on the oxidation of substrates by plant mitochondria. This suggests that the oxidation of succinate and malate occurs through a pool of Q-10 which is functionally distinct from that relating to the oxidation of exogenous NADH (Moore et al., 1980).

The "Q-puddle" theory seriously conflicts with the Q-pool theory since it is not consistent with freely diffusable Q-10 or a general movement of the respiratory enzymes within the membrane (Heron et al., 1978; Hacken-brock, 1981). Q-puddle behaviour may be caused by numerous unusual factors: the activation or inhibition of one dehydrogenase by the substrate or product of another; lower mobility of Q-10 within the membrane due to a lower ratio of protein to phospholipid or phospholipid to Q-10; a heterogeneous population of respiratory particles; or perhaps differences in substrate permeabilities. Finally, it is worth noting that, as considered earlier, an apparent preference between one dehydrogenase and one oxidase system may merely reflect the activities of the different dehydrogenase and is not an example of the more unusual "Q-puddle" behaviour.

In summary, the alternative oxidase does not appear to be controlled directly by its interaction with Q-10. Control occurs indirectly by the manipulation of the total rate of substrate oxidation and, perhaps in the long term, by the ratio of the activities of the two oxidase systems.

IV. CONCLUSIONS

It will be apparent to the reader that the sequence of electron transfer in plant mitochondria is very similar to that in mammalian tissue. In each case iron and copper complexes play a major role, undergoing reversible reduction and oxidation in the process. A particular feature is the repeated use of the two prosthetic groups of iron (Fe-S centres and haem) which have been adapted to form a series of complexes of differing redox potential.

Apart from the basic similarity it is also evident that plant mitochondria possess more complex pathways for oxidizing substrates and the considera-tions expressed in the final section show that the control of these pathways is very likely to occur at the level of the dehydrogenases, which implies that substrate availability may be an important regulator. This observation is consistent with plant mitochondria having a central role in biosynthetic reactions rather than in the production of energy since, in photosynthetic tissues, energy is readily available from the chloroplasts.

ACKNOWLEDGEMENTS

Experimental work described in this review was supported by the Science Research Council, the Agricultural Research Council and the Royal Society.

REFERENCES

Arron, G. P. and Edwards, G. E. (1980). *Plant Physiol.* **65,** 591–594.
Bahr, J. T. and Bonner, W. D. Jr. (1973a). *J. Biol. Chem.* **248,** 3441–3445.
Bahr, J. T. and Bonner, W. D. Jr. (1973b). *J. Biol. Chem.* **248,** 3446–3450.
Bendall, D. S. (1958). *Biochem. J.* **70,** 381–390.
Bendall, D. S. and Bonner, W. D. Jr. (1966). *In* "Hemes and Hemoproteins", pp. 485–488, (B. Chance, R. W. Estabrook and T. Yonetani, eds.), Academic Press, New York and London.
Bendall, D. S. and Bonner, W. D. Jr. (1971). *Plant Physiol.* **47,** 236–245.
Bendall, D. S. and Hill, R. (1956). *New Phytol.* **55,** 206–212.
Beyer, R. E., Peters, G. A., and Ikuma, H. (1968). *Plant Physiol.* **43,** 1395–1400.
Blein, J. P. (1980). *Plant Sci. Letts.* **19,** 65–71.
Bligny, R. and Douce, R. (1978). *In* "Plant Mitochondria", pp. 43–50, (G. Ducet and C. Lance, eds.), Elsevier, Amsterdam.
Bomhoff, G. H. and Spencer, M. (1977). *Can. J. Biochem.* **55,** 1114–1117.
Bonner, W. D. Jr. (1961). *In* "Haematin Enzymes", pp. 479–485, (J. E. Falk, R. Lemberg and R. K. Morton, eds.), Pergamon, Oxford.
Bonner, W. D. Jr., Christensen, E. L. and Bahr, J. T. (1972). *In* "Biochemistry and Biophysics of Mitochondrial Membranes", pp. 113–119, (G. F. Azzone, E. Carafoli, A. L. Lehninger, E. Quagliariello and N. Siliprandi), Academic Press, New York and London.
Bonner, W. D. Jr. and Rich, P. R. (1978). *In* "Plant Mitochondria", pp. 241–247, (G. Ducet and C. Lance, eds.), Elsevier, Amsterdam.
Bonner, W. D. Jr., and Slater, E. C. (1970). *Biochim. Biophys. Acta,* **223,** 349–353.
Boyer, P. D., Chance, B., Ernster, L., Mitchell, P., Racker, E., and Slater, E. C. (1977). *Ann. Rev. Biochem.* **46,** 955–1026.
Brodie, A. F. (1965). *In* "Biochemistry of Quinones", pp. 256–404, (R. A. Morton, ed.), Academic Press, New York.
Brown, R. H. and Boulter, D. (1974). *Biochem. J.* **137,** 93–100.
Brunton, C. J. and Palmer, J. M. (1973). *Eur. J. Biochem.* **39,** 283–291.
Cammack, R. and Palmer, J. M. (1973). *Ann. N.Y. Acad. Sci.* **222,** 816–822.
Cammack, R. and Palmer, J. M. (1977). *Biochem. J.* **166,** 347–355.
Cedel, T. E. and Roux, S. J. (1980). *Plant Physiol.* **66,** 704–709.
Chance, B. and Hackett, D. P. (1959). *Plant Physiol.* **34,** 33–49.

Chance, B., Wilson, D. F., Dutton, P. L., and Erecinska, M. (1970). *Proc. Nat. Acad. Sci. U.S.A.* **66**, 1175–1182.
Charbonneau, H. and Cormier, M. J. (1979). *Biochem. Biophys. Res. Commun.* **90**, 1039–1047.
Clore, G. M., Andréasson, L.-E., Karlsson, B., Aasa, R., and Malmström, B. G. (1980). *Biochem. J.* **185**, 139–154.
Coleman, J. O. D. and Palmer, J. M. (1971) *FEBS Lett.* **17**, 203–208.
Cottingham, I. R. and Ragan, C. I. (1980) *Biochem J* **192**, 19–32.
Cottingham, I. R. and Moore, A. L. (1981). *Biochem. Soc. Transac.* **9**, 429–431.
Cowley, R. C. and Palmer, J. M. (1980). *J. Exp. Biol.* **31**, 199–207.
Crane, F. L. (1965). *In* "Biochemistry of Quinones", pp. 193–206, (R. A. Morton, ed.), Academic Press, New York and London.
Crane, F. L. (1977). *Annu. Rev. Biochem.* **46**, 439–469.
Crane, F. L., Glenn, J. L. and Green, D. E. (1956). *Biochim. Biophys. Acta* **22**, 457–487.
Davis, K. A. and Hatefi, Y. (1971). *Biochemistry* **10**, 2509–2516.
Day, D. A., Arron, G. P., and Laties, G. G. (1980). *In* "The Biochemistry of Plants", Vol II, pp. 197–241, (E. Conn and P. K. Stumpf, eds.), Academic Press, New York and London.
Day, D. A., Arron, G. P., and Laties, G. G. (1979). *J. Exp. Bot.* **30**, 539–549.
Day, D. A. and Wiskich, J. T. (1974a). *Plant Physiol.* **53**, 104–109.
Day, D. A. and Wiskich, J. T. (1974b). *Plant Physiol.* **54**, 360–363.
Day, D. A. and Wiskich, J. T. (1978). *Biochim. Biophys. Acta* **501**, 396–404.
Denis, M. (1981). *Biochim. Biophys. Acta* **634**, 30–40.
Denis, M. and Bonner, W. D. Jr. (1978). *In* "Plant Mitochondria", pp. 35–42, (G. Ducet and C. Lance, eds.), Elsevier, Amsterdam and London.
Denis, M. and Clore, M. G. (1981). *Plant Physiol.* **68**, 229–235.
De Troostembergh, J-C. and Nyns, E-J. (1978). *Eur. J. Biochem.* **85**, 423–432.
Douce, R., Mannella, C. A., and Bonner, W. D. Jr. (1973). *Biochim. Biophys. Acta* **292**, 105–116.
Downie, J. A. and Garland, P. B. (1973). *Biochem. J.* **134**, 1045–1049.
Ducet, G. and Diano, M. (1978). *Plant Sci. Letts.* **11**, 217–226.
Dutton, P. L. and Storey, B. T. (1971). *Plant Physiol.* **47**, 282–288.
Edwards, S. W. and Lloyd, D. (1977). *J. Gen. Microbiol.* **103**, 207–213.
Edwards, S. W. and Lloyd, D. (1978). *Biochem. J.* **174**, 203–211.
Erecinska, M. and Storey, B. T. (1970). *Plant Physiol.* **46**, 618–625.
Ernster, L. (1967). *Methods in Enzymol.* **10**, 309–316.
Ernster, L., Lee, I. Y., Norling, B., and Persson, B. (1969). *Eur. J. Biochem.* **9**, 299–310.
Fairlamb, A. H. and Bowman, I. B. R. (1977a). *Int. J. Biochem.* **8**, 659–668.
Fairlamb, A. H. and Bowman, I. B. R. (1977b). *Int. J. Biochem.* **8**, 669–675.
Grover, S. D. and Laties, G. G. (1978). *In* "Plant Mitochondria", pp. 259–266, (G. Ducet and C. Lance, eds.), Elsevier, Amsterdam.
Gutman, M. (1980). *Biochim. Biophys. Acta* **594**, 53–84.
Gutman, M. and Silman, N. (1972). *FEBS Lett.* **26**, 207–210.
Hackenbrock, C. R. (1981). *Trends. Biochem. Sci.* **6**, 151–154.
Haddock, B. A. and Garland, P. B. (1971). *Biochem. J.* **124**, 155–170.
Hanson, J. B. and Day, D. A. (1980). *In* 'The Biochemistry of Plants", Vol. I, pp. 315–358, (E. Conn and P. K. Stumpft, eds.), Academic Press, New York and London.
Hatefi, Y. and Stiggall, D. L. (1977). *In* "The Enzymes", 3rd ed. pp. 175–295, (P. Boyer, ed.), Academic Press, New York and London.

Heber, U. (1974). *Annu. Rev. Plant Physiol.* **25**, 393–421.

Hemmerich, P., Ehrenberg, A., Walker, W. H., Erikson, L. E. G., Salach, J., Bader, P., and Singer, T. P. (1969). *FEBS Lett.* **3**, 37–40.

Henry, M-F. (1978). *In* "Plant Mitochondria", pp. 249–258, (G. Ducet and C. Lance, eds.), Elsevier, Amsterdam.

Henry, M-F., Bonner, W. D., Jr. and Nyns, E-J. (1977). *Biochim. Biophys. Acta* **460**, 94–100.

Henry, M-F., Hamaide-Deplus, M. C., and Nyns, E-J. (1974). *Antonie van Leeuwenhoek* **40**, 79–91.

Henry, M-F. and Nyns, E. J. (1975). *Sub. Cell Biochem.* **4**, 1–65.

Heron, C., Ragan, C. I., and Trumpower, B. L. (1978). *Biochem. J.* **174**, 791–800.

Higgins, J. (1963). *Ann. N.Y. Acad. Sci.* **108**, 305–321.

Hill, G. C. (1978). Proc. 11th FEBS Meet. 1977, Vol. 49. 149–158.

Huq, S. and Palmer, J. M. (1978a). *In* "Plant Mitochondria", pp. 225–232, (G. Ducet and C. Lance, eds.), Elsevier, Amsterdam.

Huq, S. and Palmer, J. M. (1978b). *FEBS Lett.* **92**, 317–320.

Ikuma, H., Schindler, F. J., and Bonner, W. D. Jr. (1964). *Plant Physiol.* **39**, 1x.

Ingledew, W. J., Salerno, J. C., and Ohnishi, T. (1976). *Arch. Biochem. Biophys.* **177**, 176–184.

James, W. O. and Beevers, H. (1950). *New Phytol.* **49**, 353–374.

James, W. O. and Elliott, D. C. (1955). *Nature* **175**, 89.

Jones, C. W. (1981). "Biological Energy Conservation: Oxidative Phosphorylation", 2nd Ed., Chapman and Hall, London.

Klingenberg, M. (1964). *Erbeg. Physiol., Biol. Chem., Exptl. Pharmakol.* **55**, 131–142.

Kröger, A., and Klingenberg, M. (1973a). *Eur. J. Biochem.* **34**, 358–368.

Kröger, A. and Klingenberg, M. (1973b). *Eur. J. Biochem.* **39**, 313–323.

Lamarck, J. B. (1778). *Flore Française,* **3**, 1150.

Lambowitz, A. M. and Bonner, W. D. Jr. (1974). *J. Biol. Chem.* **249**, 2428–2440.

Lambowitz, A. M., Slayman, C. W., Slayman, C. L., and Bonner, W. D. Jr. (1972). *J. Biol. Chem.* **247**, 1536–1545.

Lance, C. (1972). *Ann. Sc. Nat. Botanique,* 12e série 13, 477–495.

Lance, C. (1981). *In* "Recent Advances in the Biochemistry of Fruit and Vegetables", pp. 63–88, (J. Friend and M. J. C. Rhodes, eds.), Academic Press, London and New York.

Lance, C. and Bonner, W. D. Jr. (1968). *Plant Physiol.* **43**, 756–766.

Laties, G. G. (1982). *Ann. Rev. Plant Physiol.* **33**, 519–555.

Mackler, B., Haynes, B., Person, R., and Palmer, G. (1980). *Biochim. Biophys. Acta* **591**, 289–297.

Mackler, B., Repaske, R., Kohout, P. M., and Green, D. E. (1954). *Biochim. Biophys. Acta* **22**, 437–458.

Maeshima, M. and Asahi, T. (1981). *J. Biochem.* **90**, 391–397.

Mitchell, P. (1966). *Biol. Revs.* **41**, 445–502.

Mitchell, P. (1976). *J. Theor. Biol.* **62**, 327–367.

Møller, I. M., Johnston, S. P., and Palmer, J. M. (1981). *Biochem. J.* **194**, 487–495.

Moore, A. L. (1983). *In* "The Physiology and Biochemistry of Plant Respiration", (J. M. Palmer, ed.), Cambridge University Press. In press.

Moore, A. L. and Bonner, W. D. Jr. (1977). *Biochim. Biophys. Acta* **460**, 455–466.

Moore, A. L., Bonner, W. D. Jr., and Rich, P. R. (1978). *Arch. Biochem. Biophys.* **186**, 298–306.

Moore, A. L., Linnett, P. E., and Beechey, R. B. (1980). *J. Bioenerg. and Biomemb.* **12**, 309–323.

Moore, A. L., Rich, P. R., Bonner, W. D. Jr., and Ingledew, W. J. (1976). *Biochim. Biophys. Res. Commun.* **72,** 1099–1107.
Moore, A. L. and Rich, P. R. (1980). *Trends in Biochem. Sci.* **5,** 284–288.
Moore, A. L. and Rupp, H. (1978). *FEBS Lett.* **93,** 73–77.
Moreira, M. T. F., Rich, P. R., and Bendall, D. A. (1980). Short Reports of First European Bioenergetics Conference, pp. 61–62, Patrone Editore, Bologna, Italy.
Neuburger, M. and Douce, R. (1978). *In* "Plant Mitochondria", pp. 109–116, (G. Ducet and C. Lance, eds.), Elsevier, Amsterdam.
Neuburger, M. and Douce, R. (1980). *Biochim. Biophys. Acta* **589,** 176–189.
Norling, B., Glazek, E., Nelson, B. D., and Ernster, L. (1974). *Eur. J. Biochem.* **47,** 475–482.
Ohnishi, T. (1979). *In* "Membrane Proteins in Energy Transduction", pp. 1–87, (R. A. Capaldi, ed.), Marcel Dekker, Inc., New York and Basel.
Ohnishi, T., Ingledew, W. J., and Shiraishi, S. (1976). *Biochem. J.* **153,** 39–48.
Palmer, J. M. (1976). *Annu. Rev. Plant Physiol.* **27,** 133–157.
Palmer, J. M. (1979). *Biochem. Soc. Trans.* **7,** 246–252.
Palmer, J. M. (1983). *In* "The Physiology and Biochemistry of Plant Respiration", (J. M. Palmer, ed.), Cambridge University Press, in Press.
Passam, H. C. and Palmer, J. M. (1972). *J. Exp. Bot.* **23,** 366–374.
Porte, F. and Vignais, P. M. (1980). *Arch. Microbiol.* **127,** 1–10.
Prime, C. T. (1960). "Lords and Ladies", Collins, London.
Prince, R. C., Bonner, W. D. Jr., and Bershak, P. A. (1981). *Fed. Proc.,* **40,** 1667.
Ragan, C. I. (1980). *In* "Sub-cellular Biochemistry", Vol. 7, pp. 267–307, (D. B. Roodyn, ed.), Plenum Press, New York and London
Rich, P. R. (1978). *FEBS Lett.* **96,** 252–256.
Rich, P. R. and Bonner, W. D. Jr. (1978a). Proc. 11th FEBS Meet. 1977, Vol. 49, 149–158.
Rich, P. R. and Bonner, W. D. Jr. (1978b). *Biochim. Biophys. Acta* **501,** 381–395.
Rich, P. R. and Bonner, W. D. Jr. (1978c). *In* "Plant Mitochondria", pp. 61–68, (G. Ducet and C. Lance, eds.), Elsevier, Amsterdam.
Rich, P. R. and Bonner, W. D. Jr. (1978d). *Biochim. Biophys. Acta* **504,** 345–363.
Rich, P. R. and Moore, A. L. (1976). *FEBS Lett.* **65,** 339–344.
Rich, P. R., Moore, A. L., and Moreira, M. T. F. (1981). *Trends Biochem. Sci.,* Vol. 6 (March). pp. X.
Rich, P. R., Moore, A. L., Ingledew, J. W., and Bonner, W. D. Jr. (1977). *Biochim. Biophys. Acta* **462,** 501–514.
Rich, P. R., Salerno, J. C., Leigh, J. S., and Bonner, W. D. Jr (1978a). *FEBS Lett.* **93,** 323–326.
Rich, P. R., Weigand, N. K., Blum, H., Moore, A. L., and Bonner, W. D. Jr. (1978b). *Biochim. Biophys. Acta* **525,** 325–337.
Rieske, J. S. (1976). *Biochim. Biophys. Acta* **456,** 195–247.
Rieske, J. S., Hansen, R. E. and Zaugg, W. S. (1964). *J. Biol. Chem.* **239,** 3017–3022.
Rupp, H. and Moore, A. L. (1979). *Biochim. Biophys. Acta* **548,** 16–29.
Rustin, P. and Moreau, F. (1979). *Biochem. Biophys. Res. Commun.* **88,** 1125–1131.
Rustin, P., Moreau, F., and Lance, C. (1980). *Plant Physiol.* **66,** 457–462.
Ruzicka, F. J., Beinert, H., Schepler, K. L., Dunham, W. R., and Sands, R. H. (1975). *Proc. Nat. Acad. Sci. U.S.A.* **72,** 2886–2890.
Salerno, J. C. (1976). Ph.D. Thesis. The University of Pennsylvania.

Schonbaum, G. R., Bonner, W. D. Jr., Storey, B. T., and Bahr, J. T. (1971). *Plant Physiol.* **47**, 124–128.

Siedow, J. N. and Bickett, D. M. (1981). *Arch. Biochem. Biophys.* **207**, 32–39.

Siedow, J. N. and Girvin, M. E. (1980). *Plant Physiol.* **65**, 669–674.

Singer, T. P. (1976). "Flavins and Flavoproteins", Elsevier, Amsterdam.

Singer, T. P., Oestreicher, G., Hogue, P., Contreiras, J., and Brandao, I. (1973). *Plant Physiol.* **52**, 616–621.

Storey, B. T. (1969). *Plant Physiol.* **44**, 413–421.

Storey, B. T. (1970a). *Plant Physiol.* **45**, 447–454.

Storey, B. T. (1970b). *Plant Physiol.* **46**, 13–20.

Storey, B. T. (1971). *Plant Physiol.* **48**, 493–497.

Storey, B. T. (1972a). *Plant Physiol.* **49**, 314–322.

Storey, B. T. (1972b). *Biochim. Biophys. Acta* **267**, 48–64.

Storey, B. T. (1973). *Biochim. Biophys. Acta* **292**, 592–603.

Storey, B. T. (1974a). *In* "Dynamics of Energy Transducing Membranes", (L. Ernster, R. W. Estabrook, and E. C. Slater, eds.), pp. 93–102.

Storey, B. T. (1974b). *Plant Physiol.* **53**, 846–850.

Storey, B. T. (1976). *Plant Physiol.* **58**, 521–525.

Storey, B. T. (1980). *In* "The Biochemistry of Plants", Vol. II, pp. 125–195, (E. Conn and P. K. Stumpff, eds.), Academic Press, New York and London.

Storey, B. T. and Bahr, J. T. (1969). *Plant Physiol.* **44**, 115–125.

Theologis, A. and Laties, G. G. (1978a). *Plant Physiol.* **62**, 232–237.

Theologis, A. and Laties, G. G. (1978b). *Plant Physiol.* **62**, 238–242.

Tobin, A., Djerdjour, B., Journet, E., Neuburger, M., and Douce, R. (1980). *Plant Physiol.* **66**, 225–229.

Trumpower, B. L. and Haggerty, J. G. (1980). *J. Bioenerg. and Biomemb.* **12**, 151–164.

Urban, P. F. and Klingenberg, M. (1969). *Eur. J. Biochem.* **9**, 519–525.

Van Herk, A. W. H. (1937). *Proc. Acad. Sci.* **40**, 607–614.

Vanderleyden, J., Van Den Eynde, E., and Verachtert, H. (1980a). *Biochem. J.* **186**, 309–316.

Vanderleyden, J., Meyers, M., and Verachtert, H. (1980b). *Biochem. J.* **192**, 881–885.

von Jagow, G. (1975). *In* "Electron Transfer Chains and Oxidative Phosphorylation", pp. 23–30, (E. Quagliariello, S. Papa, F. Palmieri, E. C. Slater, and N. Siliprandi, eds.), Elsevier, Amsterdam.

Waring, A. J. and Laties, G. G. (1977a). *Plant Physiol.* **60**, 5–10.

Waring, A. J. and Laties, G. G. (1977b). *Plant Physiol.* **60**, 11–16.

Wikström, M. K. F. and Krab, K. (1979). *Biochim. Biophys. Acta* **549**, 177–222.

Wikström, M. K. F., Krab, K. and Saraste, M. (1981). *Ann. Rev. Biochem.* **50**, 623–655.

Wilson, S. B. (1971). *FEBS Lett.* **15**, 49–52.

Wilson, S. B. (1977). *Biochem. J.* **176**, 129–136.

Wilson, S. B. (1980). *Biochem. J.* **190**, 349–360.

Wiskich, J. T. and Bonner, W. D. Jr. (1963). *Plant Physiol.* **38**, 594–604.

Yocum, C. S. and Hackett, D. P. (1957). *Plant Physiol.* **32**, 186–191.

Yu, C. A., Yu, L. and King, T. E. (1970). *J. Biol. Chem.* **247**, 1012–1114.

Yu, C. A., Yu, L. and King, T. E. (1977). *Biochem. Biophys. Res. Commun.* **78**, 259–265.

CHAPTER 11

Calmodulin Activation of NAD Kinase and its Role in the Metabolic Regulation of Plants

H. W. JARRETT, T. DA SILVA, AND M. J. CORMIER

Department of Biochemistry, The University of Georgia, Georgia, USA

I. INTRODUCTION

Calmodulin is a small (16 700 MW) Ca^{2+}-dependent regulatory protein that appears to be ubiquitous to eukaryotic life. It has been most intensively studied in animal tissues where it has been shown that the calcium containing form activates at least eight distinct enzymes (Table I). These eight enzymes are crucial to various aspects of the cell's function

TABLE I
Mammalian enzymes activated by calmodulin

Enzyme	Location
Cyclic nucleotide phosphodiesterase	Brain
Adenyl cyclase	Brain
Tryptophan 5' monooxygenase	Brain
Phosphorylase kinase	Muscle
Myosin light chain kinase	Muscle
Glycogen synthase kinase	Muscle
Ca^{2+}-dependent ATPase	Erythrocyte membrane
Phospholipase A_2	Platelet membrane

including its saccharide metabolism, cyclic nucleotide metabolism, and membrane transport; additionally, calmodulin influences motility, secretion, and the mitotic cycle (for recent reviews, see Cheung, 1980; Means and Dedman, 1980; Klee *et al.*, 1980). Recently, our laboratory and others (Anderson and Cormier, 1978; Van Eldik *et al.*, 1980; Grand *et al.*, 1980) have reported that a calmodulin-like protein is present in extracts of various plant tissues and we have also shown that this protein, as well as mammalian calmodulin, will activate NAD kinase from pea seedlings (Anderson and Cormier, 1978). The NAD kinase activator was subsequently isolated from several species of higher plants and characterized as calmodulin (Charbonneau and Cormier, 1979; Anderson *et al.*, 1980). Since NAD kinase catalyses the formation of NADP from NAD and ATP and these two nicotinamide nucleotides serve different roles during the light and dark metabolism of plants, calmodulin activation of this enzyme is potentially important to the regulation of the metabolism of plant cells.

TABLE II
Amino acid composition of calmodulins from various sources

| Amino acid | Calmodulin from | | | |
	Plant[b] (*Arachis hypogea*)	Fungus[c] (*Agaricus bisporus*)	Coelenterate[b] (*Renilla reniformis*)	Animal[d] (bovine brain)
	mol 16 700 g^{-1}			
Lys	8	10	8	7
His	1	1	1	1
TML[a]	1	1	1	1
Arg	4	4	6	6
Asp	27	25	23	23
Thr	9	10	12	12
Ser	5	10	5	4
Glu	27	27	26	27
Pro	2	2	2	2
Gly	11	12	11	11
Ala	10	10	10	11
Cys	1–2	0	0	0
Val	6	5	7	7
Met	7	9	8	9
Ile	5	7	8	8
Leu	11	11	9	9
Tyr	1	2	1	2
Phe	8	8	9	8
Trp	0	0	0	0

[a] trimethyllysine.
[b] data from Anderson *et al.*, 1980.
[c] data from Charbonneau *et al.*, 1980.
[d] from the sequence, Watterson *et al.*, 1980.

Here, we have attempted to review calmodulin's role in plant cells and to report some preliminary characterization of purified NAD kinase.

II. Molecular Properties of Calmodulin

Calmodulin initially proved difficult to purify to homogeneity from plants. However, with the development of an affinity resin utilizing phenothiazine covalently attached to Sepharose (Charbonneau and Cormier, 1979), purification has been greatly facilitated. Briefly, it involves extracting the tissue into a Tris-EDTA buffer, absorbing protein onto DEAE-cellulose, eluting with 0·5 M salt and fractionating protein from the eluate with ammonium sulphate. Calmodulin is then specifically adsorbed from the preparation onto the affinity column, eluted by the chelating agent EGTA, and finally purified to homogeneity on DEAE-Sephacel.

Table II presents the amino acid composition of peanut seed calmodulin and calmodulin from other eucaryotes to illustrate how remarkably similar the proteins are. In fact, amino acid sequence studies of the proteins from bovine brain and the aquatic coelenterate, *Renilla reniformis*, have shown calmodulin to be one of the most highly conserved sequences known (Jamieson *et al.*, 1980). The similarity includes the presence of the unusual amino acid trimethyllysine and the absence of tryptophan. However, there are some differences also. The plant protein has one cysteine and only a single tyrosine while the animal and fungal proteins have none of the former and two of the latter. The sequence of plant calmodulin has not been reported but peptide maps of plant and animal calmodulins have demonstrated that some differences can be expected (Jarrett *et al.*, 1980). Table III compares the physical properties of a plant and an animal calmodulin. Again, the broad similarity of the proteins is obvious, anomalous behaviour on gel filtration and the low extinction coefficient at 276 nm being particularly noticeable, but subtle differences can be detected. Other studies have shown that the protein has a pI of 4, that it binds up to four moles of calcium per mole and that when complexed to Ca^{2+} it is remarkably resistant to denaturation by heat or urea. A recent study has shown that the calcium binding properties of the plant and animal calmodulins are indistinguishable (Anderson *et al.*, 1980).

III. Interaction of Calmodulin with NAD Kinase

It is only the Ca^{2+}-bound form of calmodulin which modifies enzymes. During the binding, calmodulin (CaM) undergoes a conformational change to the active conformer (CaM*-Ca^{2+}). This conformational change has been studied by several techniques (for reviews see Cheung, 1980; Klee *et*

TABLE III
Properties of plant and animal calmodulins

Properties	Source of calmodulin	
	Plant (*Arachis hypogea*)	Mammalian (bovine brain)
Stokes radius (Å)	22	23
$S_{20,w}$	1·9	1·9
Molecular weight		
from SDS-PAGE	17 300	18 000
from gel filtration	23 300	28 200
from sedimentation		
equilibrium	14 600	17 800
Extinction coefficient		
$E_{276}^{1\%}$	0·9	1·8
Phe/Tyr ratio	8	4
Absorption maxima (nm)	279, 268, 265, 259, 253	277, 269, 265, 259, 253
Ca^{2+}-dependent		
enhancement of		
Tyr fluorescence	+	+
Ca^{2+}-dependent inter-		
action with troponin 1	+	+

Data taken from Anderson *et al.*, 1980.

al., 1980; Means and Dedman, 1980). Thus, calmodulin activation of NAD kinase can be described as follows:

$$CaM + Ca^{2+} \rightleftharpoons CaM\text{-}Ca^{2+} \rightleftharpoons CaM^*\text{-}Ca^{2+}$$
$$CaM^*\text{-}Ca^{2+} + NAD\ Kinase \rightleftharpoons CaM^*\text{-}Ca^{2+}\text{-}NAD\ Kinase$$
$$\text{(inactive)} \qquad\qquad \text{(active)}$$

With the exception of the kinase which phosphorylates the light chain of myosin, animal enzymes that are activated by calmodulin are active to some extent in its absence. NAD kinase from plants, however, is completely inactive in the absence of Ca^{2+} and calmodulin as shown in Table IV. When either calmodulin or Ca^{2+} is absent, or when an inhibitor of calmodulin's function, trifluoperazine, is present, NAD kinase is totally inactive. The enzyme is active only when Ca^{2+}, calmodulin, and both of the nucleotide substrates are present. Whether this absolute dependency is a general property of calmodulin-regulated plant enzymes is unknown and must await the discovery of more of these enzymes.

That calmodulin is non-covalently bound by NAD kinase in a Ca^{2+}-dependent manner is shown in Fig. 1. A column of calmodulin covalently attached to Sepharose was prepared and partially purified NAD kinase was applied to the column in the presence of Ca^{2+}. Most of the applied protein but none of the enzyme activity passed through the column and

TABLE IV
Effects of calmodulin and trifluoperazine on plant
NAD kinase activity

Additions or deletions	NAD kinase activity [pmol min^{-1} mg^{-1}]
None	153·6 ± 8·1
−ATP	2·2 ± 1·5
−NAD$^+$	2·9 ± 2·2
−Ca^{2+}	1·6 ± 1·8
−Calmodulin	1·2 ± 2·7
+Trifluoperazine [50 μm, Final Concentration]	7·2 ± 6·7

The complete assay medium contained 50 mM Tris, 50 mM KCl, 10 mM MgCl$_2$, 2 mM NAD, 3 mM ATP, 0·2 mM CaCl$_2$, 20 μg ml^{-1} bovine brain calmodulin, and 0·2 ml of calmodulin-sepharose purified NAD kinase in 0·5 ml at pH 8·0. Additions and omissions from this standard assay mixture are as noted. When CaCl$_2$ was deleted it was replaced by 0·2 mM EGTA. After 30 min at 37°, the reaction was stopped and the NADP formed measured spectrophotometrically using glucose-6-phosphate and 2,6 dichlorophenol-indophenol as described by Jarret et al., 1980.

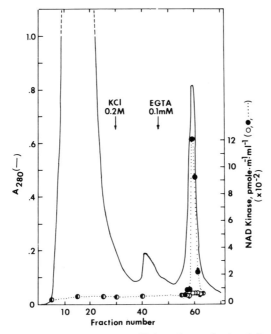

FIG. 1. Absorption of NAD kinase onto an affinity column of calmodulin-Sepharose. Data taken from Jarrett et al. (1980). NAD kinase from pea seedlings in 50 mM Tris, 50 mM KCl, 10 mM MgCl$_2$, 0·1 mM CaCl$_2$, pH 8·0 (Buffer A) was applied to a 2 × 8 cm calmodulin-Sepharose column equilbrated in the same buffer. The column was then washed with buffer A that contained 0·2 M KCl and finally eluted with Buffer A in which 0·1 mM EGTA replaced the 0·1 mM CaCl$_2$. (—) represents the A$_{280}$ of the eluate. (○) and (●) represent the activity without and with 20 μg ml^{-1} calmodulin added to the assay, respectively.

TABLE V
Purification of NAD kinase from pea seedlings

Stage of purification	Volume (ml)	Protein (mg ml⁻¹)	NAD kinase activity ($nmol\ min^{-1}\ mg^{-1}$)			Yield	Fold purification
			Minus Ca^{2+}	Minus added calmodulin	Complete		
1. Crude homogenate	1570	4·8	0·7	·57	·58	100%	1
2. Protamine precipitation	204	2·08	0·16	19·43	11·42	130%	23·1
3. Polyethylene glycol precipitation	64·4	2·59	0·12	21·72	20·25	86·4	39·0
4. DEAE chromatography	109	0·81	0·81	6·64	31·30	95·1	81·1
5. Calmodulin-Sepharose chromatography	15·1	0·041	3·29	4·88	2185	33·6	4130

NAD kinase was purified from 14-day-old pea seedlings grown on vermiculite. Plants (500 g) were homogenized in 1500 ml of 25 mM triethanolamine-HCl, 0·5% polyvinylpolypyrrolidone, 1 mM phenylmethylsulfonyl fluoride pH 7·5 and the enzyme purified through the polyethyleneglycol precipitation as described by Muto and Miyachi (1977). The precipitate was taken up in 50 mM Tris/HCl, 100 mM KCl, 3 mM $MgCl_2$, 0·1 mM EGTA, pH 7·5 and applied to a 2 × 8 cm DEAE-Sephacel column. The unabsorbed enzyme activity was collected, and sufficient 100 mM $CaCl_2$ was added to a final concentration of 0·3 mM. This was then applied to a 2 × 10 cm column of porcine brain calmodulin-sepharose. The column was prepared, washed, and eluted as described by Jarrett *et al.* (1980). Enzyme activity was measured with and without Ca^{2+} or calmodulin as in Table IV. Yield and purification factors were calculated from activity in the "complete" assay mixture.

activity was not eluted by washing with 0·2M KCl. Only when the Ca²⁺ chelating agent EGTA was added to the buffer and Ca²⁺ omitted was the enyme eluted from the column. This data provides evidence for the formation of a Ca²⁺-calmodulin-NAD kinase complex.

The formation of this complex can be utilized in the purification of NAD kinase as described in Table V. With one additional step (cation exchange chromatography), homogeneous NAD kinase can be obtained by this procedure. We will describe this purification in greater detail elsewhere and include this data to illustrate an important point, namely that calmodulin-activated NAD kinase is the major if not the only form of NAD kinase in pea seedlings. This is apparent from the data, provided one recognizes that in the absence of Ca²⁺, calmodulin is inactive. Thus the calmodulin-independent NAD kinase activity can be measured even when calmodulin is present in excess, by excluding Ca²⁺ from the assay mixture. The "minus Ca²⁺" activity at each stage of the purification represents this calmodulin-independent activity which in all cases is not significantly different from zero. The "minus added calmodulin" data shows that

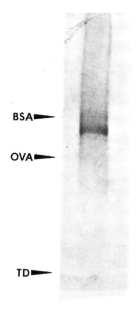

BSA◄

OVA◄

TD◄

Fig. 2. SDS-polyacrylamide gel electrophoresis of purified NAD kinase. NAD Kinase (~20 μg) was purified as described in the text and boiled for 3 min in a buffer containing 1% sodium dodecyl sulphate and 1% 2-mercaptoethanol. It was electrophoresed in a gel of 10% acrylamide and 0·25% bisacrylamide as described by Neville (1971). The position of bovine serum albumin (BSA, MW = 68 000), ovalbumin (OVA, MW = 43 000), and the track dye (TD) run in adjacent wells of the slab gel are indicated.

calmodulin is already present in the enzyme in sufficient quantity to activate the NAD kinase in the early stages of the purification. Only after endogenous calmodulin is removed by DEAE chromatography does the requirement for added calmodulin become apparent. That the calmodulin-dependent NAD kinase is the major form of this activity in pea seedlings is significant as will be discussed below.

Figure 2 presents an SDS-polyacrylamide gel of NAD kinase after the final cation exchange chromatography step. The protein has an apparent monomeric molecular weight of ~ 57 000 and at this stage is totally dependent on calmodulin. The enzyme is also quite unstable at this stage. Less than 1 h at room temperature will completely inactivate it, and at 4°,

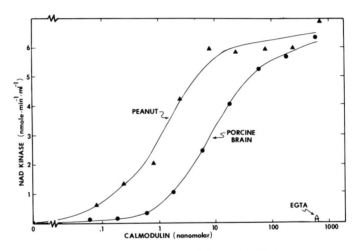

FIG. 3. Dose-response of NAD kinase for calmodulin. Serial dilutions of peanut seed (▲) and bovine brain (●) calmodulins were prepared in 0.1 mg ml^{-1} bovine serum albumin and 0.1 ml was added to the assay mixture containing $20 \mu l$ of the same NAD kinase preparation shown in Fig. 2 and assayed as described in Table IV. At one concentration shown, CaCl$_2$ in the assay was replaced with 0.2 mM EGTA, a Ca^{2+} chelator, to show the Ca^{2+}-dependence of calmodulin's effect.

complete inactivation occurs in less than 24 h. This instability makes it difficult to quantitate the degree of purification precisely, and the 4100-fold purification reported in Table V is probably a lower limit.

Figure 3 portrays the dose response curve for the activation of purified NAD kinase by calmodulins from peanut seeds and from porcine brain. As this figure clearly shows, the plant calmodulin is about seven-fold more active than its mammalian counterpart in the activation of NAD kinase. Thus, some of the subtle differences observed in the chemical and physical properties of plant and animal calmodulins are important to their biological properties.

IV. Possible Significance of Calmodulin in Plants

Here we have drawn together some previous results and some new data to make several crucial points:

1. Calmodulin from plants has properties very similar to those of other eukaryotes but not identical to them. These differences make the plant protein a more effective activator of the plant enzyme NAD kinase.

2. The calmodulin-activated NAD kinase has been purified and partially characterized and is the major or only form of NAD kinase in pea seedlings.

3. This NAD kinase has an absolute requirement for Ca^{2+} and calmodulin for its enzymic activity.

The significance of these results becomes apparent when one reviews the literature on Ca^{2+}, calmodulin, and NAD kinase.

Ca^{2+} exists in several forms within the cell and these different forms serve different roles. Insoluble salts of Ca^{2+} are found serving a structural role in many forms of life such as the structural Ca^{2+} in the cell walls of plants or in the bones of vertebrates. Besides these structural roles Ca^{2+} also serves as a "second-messenger" in the response to certain stimuli (Rubin, 1974; Rasmussen et al., 1975; Kretsinger, 1976). This is best characterized in animal cells where the concentration of free (uncomplexed) cytoplasmic Ca^{2+} is known to be $\leqslant 10^{-7} M$ in the resting state and rises upon electrical or hormonal stimulation of the cell to $10^{-5} M$. This transient rise in Ca^{2+} concentration in turn triggers various biochemical events within the cell. A particularly well-characterized example of this is the regulation of muscle contraction. In this case, the Ca^{2+} binds to a Ca^{2+}-receptor protein, troponin which in turn controls the interaction of actin and myosin and thus promotes contraction. The rise in Ca^{2+} concentration is only transient since in the muscle cell, the Ca^{2+} is rapidly resequestered by the sacroplasmic reticulum (Mannherz and Goody, 1976). In animal cells, it is clear that for many processes other than contraction, calmodulin serves as the Ca^{2+}-receptor protein and operates in a way somewhat analagous to troponin in muscle. Calmodulin binds Ca^{2+} with an apparent dissociation constant of about $1 \mu M$ and only in its Ca^{2+}-bound form does it activate the various enzymes it regulates. Thus in the resting cell, calmodulin is in its Ca^{2+}-free conformer and inactive; only during the transient rise in Ca^{2+} during stimulation is the active, Ca^{2+}-bound conformer available to activate the regulated enzymes.

In plants, there is less direct evidence about the concentration of free Ca^{2+} in the resting cell and the changes which may occur upon stimulation. Although these concentrations have not been directly measured, indirect evidence suggests that transient increases in cytoplasmic Ca^{2+} concentration occur over about the same range as they do in animal cells following stimulation. For example, actin has been reported to be involved in cytoplasmic streaming in the green algae *Nitella*, indicating that components of a contractile apparatus similar to those found in animal cells are involved in this streaming (Palevitz *et al.*, 1974). Perfusion of *Nitella* with various concentrations of Ca^{2+} has shown that concentrations greater than micromolar prevent streaming (Hayama *et al.*, 1979). Since the cytoplasm streams in living *Nitella*, the cytoplasmic concentration of free Ca^{2+} must be less than micromolar. Mechanical as well as electrical stimulation of these cells can cause an immediate cessation of streaming, suggesting that there is a rise in cytoplasmic Ca^{2+} concentration following stimulation. Recently, a Ca^{2+} concentration gradient has been shown in pollen tubes with Ca^{2+} concentrations being highest at the growing tip (Jaffe *et al.*, 1975; Reiss and Herth, 1978). Also it is known that directional growth of plant structures such as pollen tubes is reversibly stimulated by red light/far red light and is presumably mediated via phytochrome.

To utilize Ca^{2+} as a "second messenger", strict regulation of Ca^{2+} concentration is necessary, both by regulating the entry of Ca^{2+} into the cell and by regulating its exit or sequestration by Ca^{2+} transport enzymes. In some mammalian cells, Ca^{2+} transport is regulated by calmodulin (Larsen and Vincenzi, 1979; Jarrett and Penniston, 1978; Katz and Remtulla, 1978) and recently, Dieter and Marme (1980) have shown that Ca^{2+} uptake by microsomes isolated from zucchini squash (*Cucurbita pepo*) is also stimulated by calmodulin. Thus, in plants there is at least one of the components necessary for Ca^{2+} regulation and this Ca^{2+} transporter, as in animal cells, is responsive to calmodulin. We therefore hypothesize that, as occurs in animal cells, concentrations of free Ca^{2+} in the cytoplasm and organelles of plant cells change in response to various stimuli including light, and that Ca^{2+} functions as a second messenger in these cells.

NAD kinase and its regulation by calmodulin provide one example where changes in Ca^{2+} concentration following stimulation would alter the metabolism of the plant cell to elicit a useful response to the stimulus. NAD kinase catalyses the reaction between ATP and NAD to form NADP. We have shown that the major or sole enzyme which catalyses this reaction is absolutely dependent on both Ca^{2+} and calmodulin for activity and that calmodulin is normally present in crude extracts of plants in sufficient quantity to activate all of the NAD kinase present (see Table V). Assuming that NAD kinase activity is regulated by calmodulin *in vivo*, this most likely occurs by changes in the Ca^{2+} concentration in the immediate environment of the NAD kinase.

NAD kinase activity is indeed regulated *in vivo* and a stimulus involved appears to be light. Recently, Muto *et al.* (1981) have shown that the major form of NAD kinase in C_3 plants like pea plants is located in the chloroplast. These two results indicate that the calmodulin-activated NAD kinase is in the chloroplast. Furthermore, it was shown that upon illumination of dark-adapted tissue, the NAD concentration in the chloroplast decreased and the NADP concentration increased to the values characteristic of light-adapted tissue. These changes of NAD and NADP levels were likely catalyzed by NAD kinase and could be easily accounted for by the levels of NAD kinase activity found in the tissue. A useful model that explains these observations involves receptor-mediated, light-induced increases in the free Ca^{2+} concentration in the chloroplast. This Ca^{2+} would be available for binding to calmodulin which in turn binds to the inactive NAD kinase to yield an enzymatically active enzyme-calmodulin-Ca^{2+} complex. The active enzyme then converts the NAD to NADP in the chloroplast (see Fig. 4).

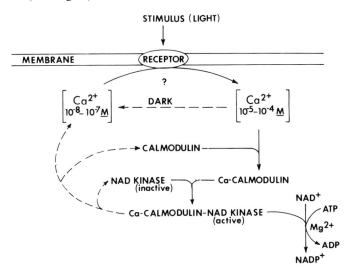

FIG. 4. Model for the possible role of Ca^{2+} as a "second messenger" in the activation of NAD kinase in plants.

During photosynthesis, the primary acceptor for electrons is NADP which must be available inside the chlorplast otherwise its concentration would be severely rate limiting. This newly formed NADP, along with photosynthetically produced ATP, would presumably act together to convert chloroplastic glyceraldehyde-3-phosphate dehydrogenase to the more active monomeric, NADP-specific form (Pupillo and Guiliani Piccari, 1973, 1975), thus facilitating the utilization of the photosynthetically reduced NADPH in the reductive pentose phosphate pathway in the chloroplast. This process would also facilitate the export of reducing power to the cytoplasm via the glyceraldehyde-3-phosphate/3-phosphoglycerate shunt (Kelly and Gibbs, 1973). In addition to these changes, reduced and oxidized NAD and NADP have been shown to be positive and negative modulators of a variety of metabolic enzymes (Turner and Turner, 1975). Thus, the light-induced conversion of NAD to NADP by NAD kinase would be expected to have profound influences on the cell's metabolism.

Whether as we suggest, light activation of NAD kinase is mediated via a change in the free Ca^{2+} concentration in the chloroplast is not known. Though this seems likely, there has been to our knowledge no report of such a change in free Ca^{2+} concentration and we are at present actively investigating whether such transient changes occur. It should be pointed out that NAD kinase could be regulated in several ways other than by changes in Ca^{2+} concentration. But since the *in vivo* conversion of NAD to NADP is nearly complete in ~4 min (Muto *et al.*, 1981) some possible modes of regulation are probably excluded. In so short a time, it seems unlikely that new calmodulin or NAD kinase could be synthesized, thus eliminating the possibility of regulation at the level of protein synthesis. Thus, the more likely mode of light induced regulation is via regulation of one of the substrates or effectors, i.e. ATP or Ca^{2+}. The ATP concentration in the chloroplast changes only four to six-fold, e.g. from 0·2 to 0·8 mM in *Elodea deusa* during the dark-light transition (Santarius and Heber, 1965), and since NAD kinase follows simple Michaelis-Menten kinetics towards this substrate ($K_{ATP} = 1·1$ mM; Yamamoto, 1966), the changes in ATP concentration would cause at most a three-fold increase in NAD kinase activity. However, the response of NAD kinase in the presence of calmodulin to Ca^{2+} is an all or nothing phenomenon and would be capable of causing very large changes in NAD kinase activity over a very narrow range of Ca^{2+} concentration. These considerations lead us to propose the model shown in Fig. 4. This model supposes that Ca^{2+}-transients do occur and are responsible for the *in vivo* light-induced conversion of NAD to NADP.

In view of this model, it is interesting to consider what may be the light receptor responsible for such Ca^{2+}-transients. Recently, evidence has been presented concerning the ability of phytochrome to modulate the level of Ca^{2+} in various plant tissues and organelles (for a discussion of this see Hale and Roux, 1980). It will be of interest to examine the question of whether the *in vivo* conversion of NAD to NADP is mediated via a change in free Ca^{2+} concentration as has been suggested here and whether phytochrome is involved.

ACKNOWLEDGEMENTS

We would like to thank Dr Harry C. Charbonneau for his kind gift of peanut seed calmodulin. We would also like to thank Alice Harmon and

Harry Charbonneau for reviewing this manuscript and Rita Hice and Richard McCann for excellent technical assistance. Dr Jarrett is a recipient of an USPHS individual postdoctoral fellowship award (NIH HL–05698). This work was supported by the National Science Foundation (PCM 79–05043).

REFERENCES

Anderson, J. M. and Cormier, M. J. (1978). *Biochem. Biophys. Res. Comm.* **84,** 505–602.

Anderson, J. M., Charbonneau, H. C., Jones, H. P., McCann, R. O., and Cormier, M. J. (1980). *Biochemistry* **19,** 3113–3120.

Charbonneau, H. C. and Cormier, M. J. (1979) *Biochem. Biophys. Res. Comm.* **90,** 1039–1047.

Charbonneau, H. C., Jarrett, H. W., McCann, R. O., and Cormier, M. J. (1980). *In* "Calcium-Binding Proteins: Structure and Function", pp. 155–164, (F. L. Siegel *et al.*, eds.), Elsevier North Holland, New York.

Cheung, W. Y. (1980). *Science* **207,** 19–27.

Dieter, P. and Marme, D. (1980). *Proc. Nat. Acad. Sci* (USA), **77,** 7311–7314.

Grand, R. J. A., Nairn, A. C., and Perry, S. V. (1980). *Biochem. J.* **185,** 755–760.

Hale, C. C. and Roux, S. J. (1980). *Plant Physiol.* **65,** 658–662.

Hayama, T., Shimmen, T., and Tazawa, M. (1979) *Protoplasma* **99,** 305–321.

Jaffe, L. A., Weisenseel, M. H., and Jaffe, L. F. (1975), *J. Cell. Biol.* **67,** 488–492.

Jarrett, H. W. and Penniston, J. T. (1978). *J. Biol. Chem.* **253,** 4676–4682.

Jarrett, H. W., Charbonneau, H. C., Anderson, J. M., McCann, R. O., and Cormier, M. J. (1980). *Annals. N.Y. Acad. Sci.* **356,** 119–129.

Jamieson, G. A., Hayes, A., Blum, J. J., and Vanaman, T. C. (1980). *In* "Calcium Binding Proteins: Structure and Function", pp. 165–172, (F. L. Siegel *et al.*, eds.), Elsevier North Holland, New York.

Katz, S. and Remtulla, M. A. (1978). *Biochem. Biophys. Res. Comm.* **83,** 1373–1379.

Kelly, G. J. and Gibbs, M. (1973). *Plant Physiol.* **52,** 674–676.

Klee, C. B., Crouch, T. H., and Richman, P. G. (1980). *Ann. Rev. Biochem.* **49,** 489–515.

Kretsinger, R. H. (1976). *Ann. Rev. Biochem.* **45,** 239–266.

Larsen, F. L. and Vincenzi, F. F. (1979). *Science* **204,** 306–309.

Mannherz, H. G. and Goody, R. S. (1976). *Ann. Rev. Biochem.* **45,** 427–465.

Means, A. R. and Dedman, J. R. (1980). *Nature* **285,** 73–77.

Muto, S. and Miyachi, S. (1977). *Plant Physiol.* **59,** 55–60.

Muto, S., Miyachi, S., Usuda, H., Edwards, G. E., and Bassham, J. A. (1981), *Plant Physiol.* **68,** 324–328.

Neville, D. M. (1971). *J. Biol. Chem.* **246,** 6328–6334.

Palevitz, B. A., Ash, J. F., and Hepler, P. K. (1974). *Proc. Nat. Acad. Sci.* (USA) **71,** 363.

Pupillo, P. and Giuliani Piccari, G. (1973) *Arch. Biochem. Biophys.* **154,** 324–331.

Pupillo, P. and Giuliani Piccari, G. (1975). *Eur. J. Biochem.* **51,** 475–482.

Rasmussen, H., Jensen, P., Lake, W., Friedman, N., and Goodman, D. B. P. (1975) *Adv. Cycl. Nucleo. Res.* **5,** 375–394.

Reiss, H. -D. and Herth, W. (1978) *Protoplasma* **97,** 373–377.

Rubin, R. P. (1974). "Calcium and the Secretory Process", Plenum Press, New York.
Santarius, K. A. and Heber, U. (1965). *Biochim. Biophys. Acta* **102,** 39–54.
Turner, J. F. and Turner, D. H. (1975). *Ann. Rev. Plant Physiol.* **26,** 159–186.
Van Eldik, L. J., Grossman, A. R., Iverson, D. B., and Watterson, D. M. (1980) *Proc. Nat. Acad. Sci.* **77,** 1912–1916.
Watterson, D. M., Sharief, F., and Vanaman, T. C. (1980). *J. Biol. Chem.* **255,** 962–975.
Yamamoto, Y. (1966). *Plant Physiol.* **41,** 523–528.

CHAPTER 12

Micronutrients and Nitrate Reductase

B. A. NOTTON

University of Bristol, Long Ashton Research Station, Bristol

I. INTRODUCTION

Under normal circumstances of adequate nitrogen and mineral supply, inorganic nitrogen metabolism in higher plants is controlled by the concentration of active NADH-nitrate reductase (EC 1.6.6.1.). This enzyme occurs in the cytoplasm and catalyses the rate limiting reaction, (1) below, in the conversion of nitrate to ammonium. The pathway also comprises nitrite reductase, an enzyme present in chloroplasts, which catalyses the ferredoxin-dependent reduction of nitrite shown in reaction (2).

$$NO_3^- + NADH + H^+ \rightarrow NO_2^- + NAD^+ + H_2O \qquad (1)$$
$$NO_2^- + 6Fd_{red} + 6H^+ \rightarrow NH_4^+ + 6Fd_{ox} + 2H_2O \qquad (2)$$

As discussed by Hewitt (this volume), nitrate reductase is the principal molybdenum protein of vegetative tissues and usually occurs in plants grown in the presence of nitrate and Mo. The enzyme shows a preference for NADH over NADPH as electron donor but various dehydrogenase acceptors such as cytochrome c, dichlorophenolindophenol, and ferricyanide can replace nitrate *in vitro* as electron acceptors in a reaction which does not require the molybdenum prosthetic group to be active or even present in the enzyme.

This contribution describes some features of the nitrate reductase of higher plants (particularly spinach) before considering the effects of molybdenum, tungsten, and manganese on nitrate reductase structure and activity.

II. Purification and Composition of Nitrate Reductase

A. PURIFICATION

Spinach nitrate reductase has been purified to homogeneity using a multistage procedure (Notton *et al.*, 1977). This involved removal of nucleic acids with streptomycin sulphate, ammonium sulphate fractionation, hydroxylapatite absorption chromatography, molecular sieve chromatography, and Blue-Sepharose affinity chromatography to give a 2600-fold increase in specific activity (Table I). The inclusion of $10\,\mu$M FAD in buffers used in the preparation and use of Blue-Sepharose has now enabled us to elute the enzyme with $0\cdot1$ mM NADH and this has improved the yield of active enzyme without however increasing the specific activity above $24\,\mu$mol NO_2^- produced min^{-1} mg protein^{-1}.

TABLE I

Purification of spinach nitrate reductase

Stage	Total enzyme activity[a]	Specific enzyme activity[b]	Purification	Recovery
Crude extract[c]	143·8	0·009	1	100
Streptomycin sulphate	130·7	0·009	1	91
Ammonium sulphate	172·9	0·034	3·7	120
Hydroxylapatite	114·1	0·292	31·8	79
Biogel chromatography	53·8	1·108	120·0	37
Blue-Sepharose	9·6	24·120	2600·0	7

[a] Total enzyme activity expressed as μmoles NO_2^- produced, min^{-1}, total volume^{-1}.
[b] Specific enzyme activity expressed as μmoles NO_2^- produced, min^{-1}, mg protein^{-1}.
[c] Crude extract obtained from approximately 1 kg of chilled leaves.
For experimental details see Notton *et al.* (1977).

B. SUBUNIT COMPOSITION

When purified nitrate reductase was incubated with sodium dodecyl sulphate (SDS) and mercaptoethanol and then submitted to acrylamide gel electrophoresis in the presence of SDS, a number of protein subunits were revealed. These had molecular weights of 37 000, 74 000, and 120 000 while that of the intact enzyme was 160 000 (Notton and Hewitt, 1979). This suggested a 1 : 2 : 3 : 4 ratio and a tetrameric structure for the enzyme (Table II).

TABLE II
Subunit composition of spinach nitrate reductase

Methods		Molecular weight ($\times 10^{-3}$)			
SDS gel electrophoresis	major peaks	37	74	120	
(protein stain)	minor peaks	34	50	102	160
Sucrose density centrifugation	healthy plants	46			152
(Cytochrome c reductases)	Mo-deficient plants grown on ammonium nitrate	45	85	120	152

For experimental details see Notton *et al.* (1976) and Notton and Hewitt (1979).

Further supporting evidence for this structure was obtained by examination of cytochrome c reductase activities obtained after sucrose density centrifugation of crude extracts of healthy and molybdenum-deficient spinach plants (Notton *et al.*, 1976). Healthy plants produced two peaks of activity (Fig. 1) corresponding to sedimentation coefficients of 8·1S, which was coincident with nitrate reductase activity, and approximately 3·6S, which was a group of cytochrome c reductases, only one of which is thought to be a subunit of the nitrate reductase (Wallace and Johnson, 1978). Molybdenum-deficient plants grown with ammonium nitrate produced a complex pattern of four peaks (Fig. 2) which when submitted to a statistical curve analysis (P. Brain, B. A. Notton, and E. J. Hewittt, unpublished work) revealed significant peaks at 8·1S, 6·9S, 5·5S, and 3·6S; the 5·5S being the dominant species. The sedimentation coefficients corresponded to molecular weights of 152 000, 120 000, 85 000 and 45 000 which is again an approximately 1 : 2 : 3 : 4 ratio.

These results with healthy and molybdenum-deficient plants imply that breakdown of the apo(demolybdo) nitrate reductase, produced in the absence of molybdenum, gives rise to the formation *in vivo* of cytochrome c reductases made up of various combinations (mono, di, and trimeric) of

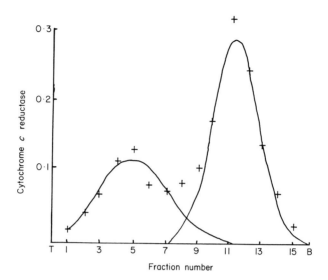

FIG. 1. Cytochrome *c* reductase activities of fractions obtained after sucrose density centrifugation of extracts from leaves of healthy spinach plants grown with nitrate. Observed activity (+); statistically analysed underlying curves (−). T, top; B, bottom of centrifuge tube.

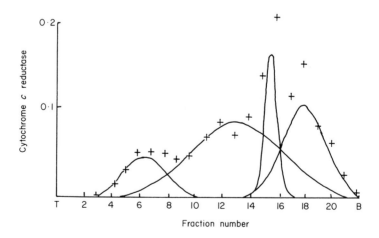

FIG. 2. Cytochrome *c* reductase activities of fractions obtained after sucrose density centrifugation of extracts from leaves of molybdenum-deficient spinach plants grown with ammonium nitrate. Observed activity (+). Statistically analysed underlying curves (−). T, top; B, bottom of centrifuge tube.

subunits of nitrate reductase, and that the holoenzyme is composed of four subunits plus a molybdenum-containing cofactor. If the subunits are arranged in the holoenzyme in pairs, only one of the pairs needs to have constitutive cytochrome c reductase activity to satisfy this assumption. Also, one subunit having cytochrome c reductase activity can be associated with two subunits without any activity to produce an active trimer.

C. THE HAEM COMPONENT

1. Light Absorption Spectrum

The absorption spectrum of the pure enzyme (Notton et al., 1977) revealed a haem component with a Soret band at 412 nm and a 280 nm/412 nm ratio of 3·5 : 1. On reduction with NADH the Soret band shifted to 424 nm and α and β bands were displayed at 557 nm and 528 nm respectively (Fig. 3).

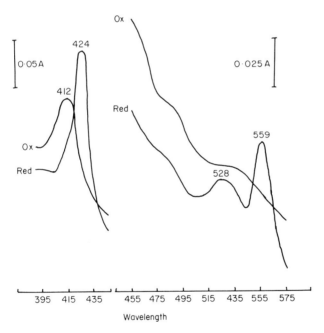

FIG. 3. Visible light absorption spectra of oxidized (native) and NADH-reduced spinach nitrate reductase measured at room temperature (Notton et al., 1977).

Involvement of the haem (designated cytochrome b_{557}) in overall electron transport in the enzyme was shown by addition of excess nitrate, which reoxidized the haem and produced nitrite. When the NADH-reduced haem was titrated to slight excess with dichlorophenolindophenol, the haem was again converted to the oxidized form. This indicates either direct involvement of the haem in the dehydrogenase function of the enzyme or, more likely, some "back electron-flow" from the reduced haem to the dehydrogenase site on the enzyme.

The visible light absorption spectrum of the NADH-reduced enzyme was measured at liquid nitrogen temperature (77°K) (Fido *et al.*, 1979). A shift of the α and β peaks towards shorter wavelengths relative to those obtained at 20°C was observed. The α peak split into two, located at 557 nm and 553 mn; the β peak occurred at 526 nm with shoulders at 533 nm and 516 nm; the position of the Soret band was not affected by the decrease in temperature but was intensified (Fig. 4). Such shifts in the spectrum towards shorter wavelengths and band splitting are commonly obtained in this region of the spectrum with cytochrome at low temperature.

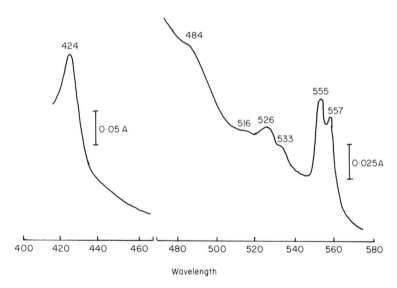

FIG. 4. Visible light absorption spectrum of NADH-reduced spinach nitrate reductase measured at liquid nitrogen temperature (77 K) (Fido *et al.*, 1979).

2. Nature of the Haem Component

The haem was characterized as protoporphyrin IX (Jones, this volume) through the spectrum of its pyridine haemochromogen derivative (Fido *et al.*, 1979). Peaks occurred at 555 nm, 524 nm, and 418 nm and the spectrum was identical to that obtained when a sample of hemin chloride was treated in the same way.

3 Mid-point Potential

The mid-point potential of the haem was determined by reductive titration with NADH in the presence of mediators together with simultaneous measurement of redox potential of the mixture (Fido *et al.*, 1979), and was estimated to be −60 mv. This potential would allow electron flow from an NADH-reduced haem to dichlorophenolindophenol (mid-point potential + 210 mv) after excess NADH (mid-point potential −315 mv) had been oxidized. Involvement of the haem in the dehydrogenase function of the enzyme is therefore not resolved by knowledge of its redox potential.

D. THE FLAVIN COMPONENT

A difference spectrum of NADH-reduced versus oxidized enzyme (Fig. 5)

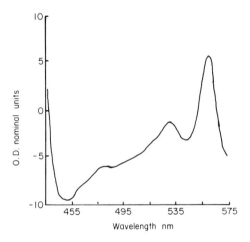

FIG. 5. Difference spectrum of NADH-reduced versus oxidized spinach nitrate reductase (Notton and Hewitt, 1979).

was indicative of the presence of a flavin having a trough at 450 nm and a broad shoulder at 475 nm. This was confirmed by spectrofluorimetry of the enzyme after heat treatment to release the flavin. The flavin was identified as FAD by its ability to cause reactivation of the apo(deflavo) form of D-amino acid oxidase (Notton *et al.*, 1979).

Higher plant nitrate reductase therefore resembles the enzyme isolated from other eukaryotic sources in having FAD, cytochrome b_{557} and molybdenum as prosthetic groups.

E. RATIO OF IRON TO MOLYBDENUM

The small yield of pure, active nitrate reductase from the Blue-Sepharose affinity chromatography stage of purification (Notton *et al.*, 1977), led to a study of the metal content of various eluted fractions and breakdown products of the enzyme (Table III) which may be produced during the

TABLE III

Analysis of eluates from Blue-Sepharose affinity chromatography of semipurified spinach nitrate reductase (Notton *et al.*, 1978; 1979b)

Solution	Metal[a]			Cytochrome *c* reductase[b]		
	Mo	Fe	Fe/Mo	3·7S	5·5–6·9S	8·1S
Applied protein	11·4	44·0	3·85	23	25	52
0·08M phosphate wash	2·5	35·2	14·08	42	0	58
0·4M phosphate wash	4·0	1·1	0·28	12	35	53
Enzyme eluate	2·9	6·9	2·38	0	0	100
(% recoveries	82·5	96·2)	2·0[c]			

[a] Metal analysis expressed as nmoles, total vol^{-1}.
[b] Cytochrome *c* reductase expressed as percent size distribution measured as μmoles cytochrome *c* reduced, min^{-1}.
[c] Ratio of metals corrected for recoveries.

various chromatographic stages of purification in a similar way to those observed using AMP-Sepharose (see later). A sample of semi-purified enzyme containing 44 nmoles of iron and 11·4 nmol of molybdenum (a Fe/Mo ratio of 3·85), was applied to the Blue-Sepharose. The solution passing through the column and that eluted with 0·08M phosphate buffer contained an iron-rich fraction while the 0·4M phosphate buffer wash contained a molybdenum-rich fraction. The eluted enzyme contained iron and molybdenum in a ratio of 2·38 which, after corrections for the recovery of the two metals, resulted in a final ratio of 2·0 (Notton *et al.*, 1978).

Sucrose density centrifugation of the four solutions and analysis of fractions for cytochrome *c* reductases revealed that the solution applied to the affinity column contained all the previously mentioned polymers of cytochrome *c* reductase (3·7S to 8·1S), only 52% was present as the 8·1S molecule associated with nitrate reductase. The 0·08M phosphate eluate contained only 3·7S and 8·1S molecules and virtually no nitrate reductase activity. The 0·4M phosphate eluate contained all species of cytochrome *c* reductase, at least some of the 8·1S being accountable for by nitrate reductase activity. The final eluate contained only the 8·1S species of cytochrome *c* reductase and associated nitrate reductase (Notton *et al.*, 1979b).

It would appear therefore that breakdown of nitrate reductase occurs during purification and that the breakdown products are separated by affinity chromatography to yield pure enzyme. No enzyme is irreversibly bound.

On the basis of the subunit composition, the metal analysis and by analogy with other purified eukaryotic nitrate reductases which all have equal molar amounts of FAD and haem (e.g. Garrett and Nason, 1969; Solomonson *et al.*, 1975), the enzyme from spinach appears to contain 2 FAD, 2 Haem and 1 molybdenum per molecule. A model of the spinach enzyme is represented (Fig. 6) but the association of the haem with a

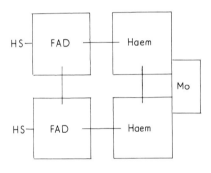

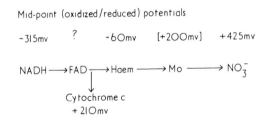

FIG. 6. Model of spinach nitrate reductase and electron flow diagram showing mid-point potentials of substrates and prosthetic groups.

particular subunit has not been shown yet and it could be present as a hinge molecule between the subunits. Also shown in Fig. 6 is an electron flow diagram of the enzyme with mid-point potentials for the various components. The mid-point potential for the molybdenum group is an estimate based on analogy with another nitrate reductase (Rosa *et al.*, 1981). The mid-point potential of the FAD is not known.

F. THE MOLYBDENUM CONTAINING COMPLEX (MCC)

Presently it is not known how Mo is bound to the protein but evidence is accumulating that all molybdoenzymes, with perhaps the exception of nitrogenase, contain a common prosthetic group termed for convenience MCC (see review Johnson, 1980). Thus, in early experiments with extracts of a mutant strain (nit 1) of *Neurospora crassa*, which contained a nitrate-inducible cytochrome *c* reductase, nitrate reductase activity was not obtained by subsequent addition of molybdate but by addition of material obtained by acid treatment of natural molybdoenzymes (nitrate reductase, sulphite oxidase, aldehyde oxidase, or xanthine dehydrogenase) (Ketchum et al., 1970; Nason et al., 1971). Similar experiments have been reported (Rucklidge et al., 1976) using acid-treated spinach nitrate reductase and extracts of molybdenum-deficient plants as an apoenzyme source (see later). An oxygen-sensitive pteridin has recently been identified as a component of the MCC (Johnson et al., 1980; Hewitt, this volume). There is also evidence that MCC may dissociate reversibly from the spinach enzyme. The use of the affinity chromatography medium 5'AMP-Sepharose to purify higher plant nitrate reductases always results in a large increase in specific activity but yields have never exceeded 30% of the applied activity. In order to show whether the enzyme was either dissociating on the column or was irreversibly bound, a semipurified preparation of the tungsten analogue, labelled with ^{185}W, was mixed with active enzyme as a marker, applied and bound to a column of AMP-Sepharose. Washing the column with 0·1M phosphate buffer released 74% of the bound radioactivity, but only 4% of cytochrome *c* reductase and no measurable nitrate reductase activity. Elution of the enzyme with 0·1 mM NADH released 21% of the bound radioactivity, 56% of the cytochrome *c* reductase and 25% of the marker nitrate reductase. When the most radioactive of the 0·1M phosphate wash fractions were bulked and submitted to ultrafiltration through molecular sieve membranes, 50% of the ^{185}W passed through membranes with nominal molecular weight cut-offs of 50 000 and 10 000 but was retained by a 2 000 membrane. When this experiment was repeated with an active nitrate reductase, the fraction passing through a 50 000 membrane was shown to be able to reconstitute nitrate reductase with apoenzyme. However, further fractionation through the 10 000 membrane resulted in complete loss of reconstituting ability even when the diffusate and retentate were remixed. These results suggested that losses which occurred during chromatography were due to release of MCC from the enzyme and that the MCC was inactivated by further ultrafiltration. It was suggested (Hewitt et al., 1977) that these results reflect an *in vivo* situation of reversible association of MCC and apoenzyme on a cell membrane.

III. Molybdenum Requirement of Plants

A. MOLYBDENUM DEFICIENCY AND NITROGEN SOURCE

Because the established role of molybdenum in higher plants is as a component of nitrate reductase, it might be expected that when plants are grown with nitrogen in a form other than nitrate, which would not require nitrate reduction before assimilation, then they would show no requirement for molybdenum. However, cauliflower plants were shown still to require molybdenum when grown with either ammonium salts, urea, glutamate, or nitrite as nitrogen sources (Agarwala, 1952; Agarwala and Hewitt, 1955; Hewitt, 1956). In the absence of molybdenum, yields were depressed and specific leaf and stem symptoms known as "whiptail" appeared, the progressive stages of which have been detailed (Hewitt and Agarwala, 1951; Hewitt and Bolle-Jones, 1952; Agarwala and Hewitt, 1954; Hewitt, 1956, 1963). The plants contained abundant free amino acids and protein indicating that nitrogen was available and was substantially absorbed in the form supplied (Hewitt et al., 1957).

Nitrate reductase was shown to be an inducible enzyme in cauliflower (Candela et al., 1957) and the formation of the enzyme depends on the presence of both nitrate and molybdenum in the living cell (Hewitt and Afridi, 1959; Afridi and Hewitt, 1964; Hewitt and Notton, 1967). It was suggested by Hewitt (1959, 1963) that the molybdenum requirement, in the presence of reduced nitrogen, may result from the derepressing effect of traces of nitrate derived from bacterial nitrification. This was confirmed by Hewitt and Gundry (1970) who showed that cauliflower plants grown under sterile conditions with ammonium sulphate and in the absence of molybdenum, showed no abnormalities. On transfer to non-sterile conditions "whiptail" symptoms appeared.

Stereoscan electron microscopy of leaf surfaces of cauliflower plants showing "whiptail" symptoms revealed isolated roughened protrusions, surrounded by radiating lines of apparent stress; small areas of underlying cells became exposed and developed into honeycomb masses with total perforation of the leaf (Fido et al., 1977).

Cells in the interveinal and marginal regions appeared in the light microscope to show two types of disorganization. One was the collapse of chloroplasts which lost their rounded outlines or appeared to fuse together and the other was the appearance of excessively expanded or collapsed cells in groups which subsequently coalesced into dense areas of shrunken cells (Hucklesby, 1961).

A study of ultrastructural features of "whiptail" symptoms (Fido et al., 1977) showed that, in samples taken progressively closer to lesion areas (Fig. 7), chloroplasts became bulbous and enlarged but with reduced grana

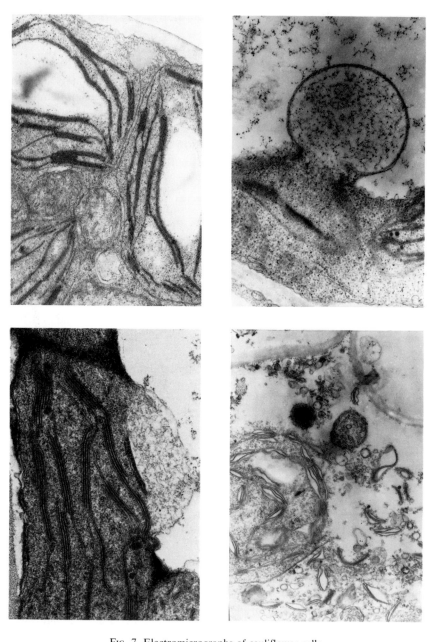

Fig. 7. Electromicrographs of cauliflower cells.
A. Section from a healthy plant grown with molybdenum and nitrate; magnification 16 K.
B. Section from a molybdenum-deficient plant grown with nitrate showing large protrusion-containing stroma and bounded by the outer chloroplast membrane; magnification 20 K.
C. Section from molybdenum-deficient plant grown with ammonium taken from non-lesion area showing loss of chloroplast membrane; magnification 60 K.
D. Section from lesion area of leaf showing complete disruption of chloroplast with intact mitochondria; magnification 22K. (Fido *et al.*, 1977).

stacking and distorted expansion of interthylakoid spaces. Decrease of electron density and appearance of cavities in the stroma were followed by spherical protrusions from the surface of the chloroplast bounded by both chloroplast and tonoplast membranes. These protrusions became filled with chloroplast debris or ruptured into the cytoplasm. Mitochondria remained relatively undamaged. The effects of molybdenum deficiency described, resembled those shown by flax plants suffering from paraquat (methyl viologen) damage (Harris and Dodge, 1972a). This was attributed by Harris and Dodge (1972b) to peroxidation of lipid components in cell membranes as a result of superoxide or peroxide produced during auto-oxidation of the reduced pyridylium radicle formed in illuminated chloro-plasts. Pederson and Aust (1972) reported that cytochrome c reductases have a role in the production of super-oxide leading to damage to lipid membranes. It is concluded therefore that "whiptail" symptoms are the result of damage to chloroplast membranes due to the activity of cytochrome c reductases induced by nitrate (Wray and Filner, 1970) in the absence of molybdenum. This is consistent with the very early conclusion of Wilson and Waring (1949) that the presence of nitrate is a contributory factor in the incidence of "whiptail" and chloroplast breakdown.

B. PRESENCE OF APO-NITRATE REDUCTASE IN MOLYBDENUM-DEFICIENT TISSUE

It seems likely that molybdenum-deficient plants grown with nitrate together with ammonium as an available nitrogen source would contain apo-nitrate reductase. This conclusion is substantiated by results of experiments using a variety of techniques to detect the presence of apoenzyme in plants grown without molybdenum.

1. Induction by Molybdenum

Plants grown without molybdenum with either ammonium nitrate or nitrate nitrogen, produced nitrate reductase activity *in vivo* when leaves were infiltrated with molybdenum. The response was biphasic and rates during the initial more rapid phase (30 min) were about four times greater for the ammonium nitrate-grown plants than for those grown with nitrate only, but rates for the second slower phase (up to 4 h) were comparable. Only this second phase was inhibited when cyclo-heximide was included in the infiltration medium to prevent protein synthesis (Afridi and Hewitt, 1965). Therefore protein synthesis was not required for the formation of the active enzyme over the first 30 min, and in the presence of ammonium more apo-protein was present.

2. Inhibition of Tungsten-analogue Formation by Puromycin

Leaves of molybdenum-deficient cauliflower plants responded to molybdate infiltration by a fourteen-fold increase in nitrate reductase activity. When stems of similar leaves were dipped for 24 h into tungstate containing ^{185}W, the isotope accumulated in fractions normally rich in enzyme activity during partial purification. When leaves were dipped into a tungstate solution containing 2 mg ml^{-1} puromycin, the formation of the analogue was reduced to 24% of the control. These plants therefore contained up to 24% of apo-nitrate reductase capable of forming the analogue. A similar experiment using the proline analogue, L-azetidine-2-carboxylic acid (1·5 mg ml^{-1}), resulted in no decrease in tungsten analogue formation (Notton et al., 1974). Both puromycin and L-azetidine-2-carboxylic acid have previously been found to inhibit the induction by molybdate of enzyme activity in molybdenum-deficient tissue (Hewitt and Notton, 1967). It therefore seems that proline is not an essential amino acid for the incorporation of molybdenum (or tungsten) into apo-nitrate reductase but that it is somehow required for the expression of enzyme activity.

3. Cross-reaction with Anti-serum to Nitrate Reductase

Antibodies to nitrate reductase, raised in rabbits, were shown to inhibit nitrate reductase activity (Notton et al., 1974). When crude extracts of plants grown without molybdenum and with either ammonium nitrate or nitrate were mixed with active enzyme, then the amount of antiserum required to inhibit the active enzyme was increased. This increase was quantitatively related to the amount of cross-reacting material in the molybdenum-deficient extracts (Graf et al., 1975). Crude extracts from both treatments were found to contain approximately 35% of the cross-reacting material found in a healthy plant. However, when these crude extracts were purified as for nitrate reductase, cross-reacting material from the nitrate-only plants was reduced to 2% of the control, whereas cross-reacting material from the ammonium-nitrate grown plants was reduced only to 20% of the control.

As nitrate reductase is composed of aggregates of subunits having cytochrome c reductase activity (Wray and Filner, 1970; Notton et al., 1976), the losses of cross-reacting material during a purification which was designed to isolate the aggregated apoenzyme, may well depend on the degree of aggregation of subunits in the preparation. Ammonium ions may be able to prevent disaggregation or to increase the rate of aggregation of subunits. In the absence of the prosthetic metal, the turnover of the apoenzyme aggregate is rapid and, even in the presence of ammonium, there is only 35% of the total cross-reacting material expected in molybdenum-sufficient tissue.

4. Reconstitution of Holo-nitrate Reductase

In spite of this accumulated evidence that molybdenum-deficient tissue, especially when grown with available nitrogen, contains up to 35% apo-nitrate reductase, addition of inorganic molybdate to crude extracts of these plants failed to produce active enzyme. Rucklidge *et al.* (1976) used *in vitro* reconstitution with MCC, obtained from acid-treated spinach nitrate reductase, to estimate the amount of apoenzyme present in molybdenum-deficient tissue grown with either nitrate or ammonium nitrate both with and without added tungsten. They compared reconstituted activity with that found in molybdenum-sufficient tissue and found that extracts of plants grown with ammonium nitrate reconstituted active enzyme up to 27% of the control but extracts of plants grown with nitrate produced no reconstitutable enzyme. When tungsten was included in the nutrient, extracts of plants grown with ammonium nitrate reconstituted up to 35% of the control while extracts of plants grown with nitrate reconstituted up to 50% of the control. It appears that tungsten stabilizes the apoenzyme by analogue formation and this reacts with MCC to produce active enzyme by exchange of metal containing components. In the absence of tungsten and particularly when ammonium nitrogen is also absent, the apoenzyme rapidly breaks down.

When a crude extract of the apoenzyme source was submitted to sucrose density centrifugation to separate the "light" 3·7S and "heavy" 8·1S cytochrome *c* reductases, both fractions were shown to be able to reconstitute nitrate reductase with MCC.

This ability of the MCC to promote aggregation of the 3·7S cytochrome *c* reductases may reflect an *in vivo* mechanism of enzyme assembly.

IV. Beneficial Effect of Tungsten

Cauliflower plants grown in the absence of molybdenum were markedly benefited by 0·25 ppm tungsten both in yield and in suppression of "whiptail" symptoms seen both visually (Gundry and Hewitt, 1974) and microscopically (Fido *et al.*, 1977). Chroroplast integrity was preserved, cavities in the grana stacking were less evident and only some areas of electron transparency persisted. Tungsten-treated cauliflower plants have been shown to produce nitrate reductase activity gradually up to 20% of normal values (Gundry and Hewitt, 1974) possibly by *in vivo* exchange between the stabilized analogue and traces of molybdenum which, however, are themselves insufficient to cause substantial aggregation of subunits.

In an investigation of the beneficial effect of small amounts of tungsten on the growth of molybdenum-deficient spinach (Notton *et al.*, 1979b;

Fido, 1980), plants were grown for 28 days in purified sand culture in the
absence of added molybdenum. Nutrient was then supplemented with
increasing concentrations of "Specpure" tungstate up to 1·0 ppm for a
further 28 days. At the end of the growth period the plants were harvested,
fresh and dry weight recorded and molybdenum content determined
(Table IV). Nitrate reductase activity and gross cytochrome c reductase

TABLE IV

Fresh weight, dry weight, and molybdenum content of spinach plants grown for 28
days with molybdenum-deficient nutrient followed by 28 days with increasing
concentrations of tungsten in the same nutrient (Fido, 1980)

Tungsten concentration (mg L^{-1})	Fresh weight (g)	Dry weight (g)	Molybdenum content (nmoles g dry wt^{-1})
0·0001	35	3·9	N/A
0·0005	138	15·2	3·52
0·001	112	13·0	3·60
0·005	268	43·3	3·00
0·01	253	30·0	3·50
0·1	235	33·3	3·14
1·0	180	22·6	2·67

N/A = not analysed.

activity were measured as well as distribution of cytochrome c reductase
activities between "light" (3·7S) and "heavy" (8·1S) fractions (Table V).
Plants receiving no tungsten did not survive the 56 days and those grown
with 0·0001 ppm tungsten provided only sufficient material for fresh and
dry weight. Maximum weight of plants occurred at 0·005 and 0·01 ppm
tungsten and 0·005 ppm tungsten plants had the maximum gross molybdenum content. However, there was no significant difference in molybdenum content g^{-1} dry wt. between 0·0005 and 0·1 ppm tungsten (3–3·6
nmoles Mo g^{-1}). The "Specpure" tungstate contained a maximum of
0·0005% molybdenum and therefore, even at the largest concentration
used, contributed negligible molybdenum to the plants. Table V shows that
nitrate reductase activity, although only approximately 2% of that found in
healthy plants, was maximal with 0·005 ppm tungsten treatment.
Cytochrome c reductase, which was predominantly in the "light" fraction
up to the 0·005 ppm tungsten treatment, had more activity in the "heavy"
fraction with the 0·01 and 0·1 ppm tungsten treatments which reflects the
increased formation of the tungsten analogue. Concentrations of tungsten
up to 0·005 ppm promoted the incorporation of molybdenum into nitrate

TABLE V

Nitrate reductase activity, gross cytochrome c reductase activity and distribution by molecular weight of cytochrome c reductase in spinach plants grown for 28 days with molybdenum deficient nutrient followed by 28 days with increasing concentrations of tungsten in the same nutrient (Fido, 1980)

Tungsten concentration (mg L^{-1})	Nitrate reductase activity[a]	Cytochrome c reductase[b]	% distribution of cytochrome c reductase (activity)[c]	
			3·7S	8·1S
0·0005	0·176	0·158	64 (0·101)	36 (0·057)
0·001	0·030	0·092	77 (0·071)	23 (0·021)
0·005	1·002	0·195	61 (0·119)	39 (0·076)
0·01	0·356	0·305	51 (0·156)	49 (0·149)
0·1	0·312	0·264	36 (0·095)	64 (0·169)
1·0	0·075	0·206	52 (0·107)	48 (0·099)

[a] Nitrate reductase activity expressed as nmoles NO_2^- produced, min^{-1}, g fresh weight^{-1}.
[b] Cytochrome c reductase activity expressed as μmoles cytochrome c reduced, min^{-1}, g fresh weight^{-1}.
[c] % distribution of cytochrome c reductases after sucrose density centrifugation.

reductase while greater concentrations were competitively antagonistic. Tungsten, by forming the analogue which has been shown to be identical to nitrate reductase in respects other than activity (Notton *et al.*, 1979a), stabilizes the aggregated form of cytochrome c reductase sufficiently for traces of molybdenum present to exchange with it to form active nitrate reductase. This enables protein synthesis to take place, producing a larger plant and the resulting increased root system would tend to accelerate the scavenging process.

V. EFFECT OF MANGANESE

A. EFFECT OF DEFICIENCY ON ACTIVITY

Soya bean nitrate reductase is influenced by manganese deficiency. Heenan and Campbell (1980) found that nitrate reductase activity measured *in vivo* was minimal under deficiency conditions and that nitrate concentrations in the tissue increased. However nitrate reductase activity in leaf extracts was not materially affected. They suggested that as photosynthesis was decreased by manganese deficiency then it was possible that the supply of reductant was insufficient to maintain the enzyme in an active state.

B. REACTIVATION *IN VITRO* OF NITRATE REDUCTASE

Nitrate reductase from *Chlorella vulgaris* (Lorimer *et al.*, 1974) and *Chlorella fusca* (Losada *et al.*, 1970) is found *in vivo* in a cyanide-bound inactive form which can be reactivated *in vitro* either completely by ferricyanide (Solomonson and Vennesland, 1972; Moreno *et al.*, 1972) or partially with manganese pyrophosphate (Funkhouser and Ackermann, 1976).

Nitrate reductase from spinach can be inactivated *in vitro* by "over-reduction" with NADH and complexing of the prosthetic molybdenum with cyanide (Relimpio *et al.*, 1971), and this has been proposed as an important regulating mechanism (see Hewitt *et al.*, 1979). Maldonado *et al.* (1980) used manganic ions, generated photochemically by illuminated whole chloroplasts, to reactivate *in vitro* inactivated enzyme as efficiently as ferricyanide (Table VI). Both pyrophosphate and excess manganous

TABLE VI

Reactivation of NADH-reduced and cyanide-inactivated nitrate reductase by photochemically oxidized manganese

System[a]	Enzyme activity[b]
Complete	107
less chloroplasts	27
less MnCl$_2$	19
less light	8

[a] Complete system as described by Maldonado *et al.* (1980).
[b] Enzyme activity expressed as nmoles NO_2^- formed, min^{-1}, cm^{-3} incubation mixture relative to ferricyanide reactivation at 100 units.

ions were found to inhibit the reactivation. It was concluded that reactivation was by direct oxidation of the molybdenum site from the "over-reduced" Mo^{iv} state to Mo^{vi} by successive external one-electron transfers from the very stable cyanide-Mo^{iv} co-ordinated state to the unstable cynanide-Mo^{vi} co-ordinated state, from which cyanide is easily and spontaneously dissociated (Notton and Hewitt, 1971b).

Evidence for "over-reduction" of spinach nitrate reductase by NADH has been obtained by electron paramagnetic studies of the pure enzyme (B. A. Notton, R. J. Fido, E. J. Hewitt, R. G. Bray, and S. Guttridge, unpublished work). The native enzyme, which displayed no signal, produced a signal ascribed to Mo^v when reduced with a molar equivalent of NADH. This signal was lost on addition of nitrate. When the native enzyme was reduced with an excess of NADH, signals ascribed to Mo^v were initially produced at two-thirds maximal intensity which slowly decreased to an assumed "silent" Mo^{iv} state. Addition of excess nitrate produced signals which then slowly decreased presumably as the molybdenum was reoxidized from Mo^{iv} through Mo^v to Mo^{vi}.

The enzyme from *Chlorella vulgaris* also produced an epr signal on reduction with NADH and was converted to an epr "silent" state by the addition of either nitrate or cyanide (Solomonson, 1979).

Manganese may therefore have an important physiological role in the regulation of nitrate reductase by reactivating the reversibly inactivated enzyme in a redox-dependent mechanism.

VI. Proposed Reaction Mechanism of Nitrate Reductase

Our studies have led to a model for the mechanism of action and reversible reactivation of eukaryotic nitrate reductase (Hewitt *et al.*, 1979). Briefly it envisages that NADH reduces one of the FAD centres to $FADH_2$ and with the second FAD generates two FADH semiquinones; these in turn reduce the two haem centres and release two protons. One haem centre then reduces $O = Mo^{vi} = O$ (the resting form of Mo in the enzyme), and with substrate and a proton produces an oxo-Mo^v-OH group liganded to nitrate. This then reacts with a second reduced haem and a proton to form oxo-Mo^{iv} with co-ordinated water and liganded nitrate. The water is immediately displaced by ligand exchange to produce a bidentate-nitrato complex in which the nitrate then undergoes rapid reduction by transfer of two electrons from the oxo-Mo^{iv} couple to yield a dioxo-Mo^{vi}-nitrito complex. Nitrite is released or displaced by nitrate to complete the cycle. In the absence of nitrate, the oxo-Mo^v-OH group is further reduced and, in the presence of cyanide, water is displaced to form an oxo-Mo^{iv}-CN complex. On reoxidation by ferricyanide or manganic ions to dioxo-Mo^{vi}, cyanide is released and the resting state of the enzyme is regained. This proposed mechanism reconciles many aspects of nitrate reductase activity and reversible inactivation. Thus it provides for a monomeric molybdenum centre participating in a two-electron transfer reaction, NADH-induced epr signals and their disappearance with nitrate as well as with cyanide, the involvement of $Mo^{vi}/Mo^v/Mo^{iv}$ oxidation states in electron transport and the formation of nitrite as the product following an oxygen transfer mechanism.

References

Afridi, M. M. R. K. and Hewitt, E. J. (1964). *J. exp. Bot.* **15**, 251–271.
Afridi, M. M. R. K. and Hewitt, E. J. (1965). *J. exp. Bot.* **16**, 628–645.
Agarwala, S. C. (1952). *Nature (Lond.)* **169**, 1099.
Agarwala, S. C. and Hewitt, E. J. (1954). *J. hort. Sci.* **29**, 278–290.

Agarwala, S. C. and Hewitt, E. J. (1955). *J. hort. Sci.* **29**, 278–290.
Candela, M. I., Fisher, E. G., and Hewitt, E. J. (1957). *Plant Physiol.* **32**, 280–288.
Fido, R. J. (1980) M.Sc. Thesis, Univ. of Bristol.
Fido, R. J., Gundry, C. S., Hewitt, E. J., and Notton, B. A. (1977). *Aust. J. Plant Physiol.* **4**, 675–689.
Fido, R. J., Hewitt, E. J., Notton, B. A., Jones, O. T. G., and Nasrulhaq-Boyce, A. (1979). *FEBS Letts.* **99**, 180–182.
Funkhauser, E. A. and Ackermann, R. (1976). *Eur. J. Biochem.* **66**, 225–228.
Garrett, R. H. and Nason, A. (1969). *J. biol. Chem.* **244**, 2870–2882.
Graft, L., Notton, B. A., and Hewitt, E. J. (1975). *Phytochemistry* **14**, 1241–1243.
Gundry, C. A. and Hewitt, E. J. (1974). *Ann. Rep. Long Ashton Res. Stn for 1973*, p. 78.
Harris, N. and Dodge, D. A. (1972a). *Planta* **104**, 201–209.
Harris, N. and Dodge, D. A. (1972b). *Planta* **104**, 210–219.
Heenan, D. P. and Campbell, L. C. (1980). *Plant and Cell Physiol.* **21**, 731–736.
Hewitt, E. J. (1956). *Soil Sci.* **81**, 159–172.
Hewitt, E. J. (1963). *In* "Plant Physiology", Vol. III, pp. 137–360, (F. C. Steward, ed.), Academic Press, London and New York.
Hewitt, E. J. (1959). *Biol. Rev.* **34**, 333–377.
Hewitt, E. J. and Afridi, M. M. R. K. (1959). *Nature (Lond.)* **183**, 57.
Hewitt, E. J. and Agarwala, S. C. (1951). *Nature, (Lond.)* **167**, 733.
Hewitt, E. J. and Bolle-Jones, E. W. (1952). *J. hort. Sci.* **27**, 257–265.
Hewitt, E. J. and Gundry, C. S. (1970). *J. hort. Sci.* **45**, 351–358.
Hewitt, E. J. and Notton, B. A. (1967). *Phytochemistry* **6**, 1329–1335.
Hewitt, E. J., Agarwala, S. C., and Williams, A. H. (1957). *J. hort. Sci* **32**, 34–48.
Hewitt, E. J., Notton, B. A., and Garner, C. S. (1979). *Proc. Biochem. Soc.* **7**, 629–633.
Hewitt, E. J., Notton, B. A., and Rucklidge, G. J. (1977). *J. Less Common Metals* **54**, 537–553.
Hucklesby, D. P. (1961). M.Sc. Thesis, University of London.
Johnson, J. L. (1980) *In* "Molybdenum and Molybdenum-Containing Enzymes", pp. 347–383, (M. Coughlan, ed.), Pergamon Press, Oxford.
Johnson, J. L., Hainline, B. E., and Rajagopalan, K. V. (1980). *J. biol. Chem.* **255**, 1783–1786.
Ketchum, P. A., Cambier, H. Y., Frazier, W. A., Nadansky, C. A., and Nason, A. (1970). *Proc. Nat. Acad. Sci. USA* **66**, 1016–1023.
Lorimer, G. H., Gewitz, H-S, Völker, W., Solomonson, L. P., and Vennesland, B. (1974). *J. biol. Chem.* **249**, 6074–6079.
Losada, M., Paneque, A., Aparicio, P. J., Vega, J. Ma., Cardenas, J., and Herrera, J. (1970). *Biochem. Biophys. Res. Commun.* **38**, 1009–1015.
Maldonado, J. M., Notton, B. A., and Hewitt, E. J. (1980). *Planta* **150**, 242–248.
Moreno, C. G., Aparicio, P. J., Palacian, E., and Losada, M. (1972). *FEBS Letts.* **26**, 11–14.
Nason, A., Lee, K. Y., Pan, S. S., Ketchum, P. A., Lamberti, A., and Devries, J. (1971). *Proc. Nat. Acad. Sci. USA* **68**, 3242–3246.
Notton, B. A. and Hewitt, E. J. (1971a). *Biochem. Biophys. Res. Commun.* **44**, 702–710.
Notton, B. A. and Hewitt, E. J. (1971b). *FEBS Letts.* **18**, 19–22.
Notton, B. A. and Hewitt, E. J. (1979). *In* "Nitrogen Assimilation in Plants", pp. 227–244, (E. J. Hewitt and C. V. Cutting, eds), Academic Press, London and New York.

Notton, B. A., Icke, D., and Hewitt, E. J. (1976). *Ann. Rep. Long Ashton Res. Stn for 1975*, p. 63.

Notton, B. A., Fido, R. J., and Hewitt, E. J. (1977). *Plant Sci. Letts.* **8**, 165–170.

Notton, B. A., Fido, R. J., and Hewitt, E. J. (1978). *Ann. Rep. Long Ashton Res. Stn for 1977*, p. 55.

Notton, B. A., Graf, L., Hewitt, E. J., and Povey, R. C. (1974) *Biochem. Biophys. Acta* **364**, 45–58.

Notton, B. A., Fido, R. J., Watson, E. F., and Hewitt, E. J. (1979a). *Plant Sci. Letts.* **14**, 85–90.

Notton, B. A., Fido, R. J., James, D. M., Watson, E. F., and Hewitt, E. J. (1979b). *Ann. Rep. Long Ashton Res. Stn. for 1978*, 73–74.

Pederson, T. C. and Aust, S. D. (1972). *Biochem. Biophys. Res. Commun.* **48**, 789–795.

Relimpio, A. Ma., Aparicio, P. J., Paneque, A., and Losada, M. (1971). *FEBS Letts.* **17**, 226–230.

Rosa, M. A. De La, Gomez-Moreno, C., and Vega, J. M. (1981). *Biochim. Biophys. Acta* **662**, 77–85.

Rucklidge, G. J., Notton, B. A., and Hewitt, E. J. (1976). *Biochem. Soc. Trans.* **4**, 77–80.

Solomonson, L. P. (1979). *In* "Nitrogen Assimilation of Plants", pp. 200–205, (E. J. Hewitt and C. V. Cutting, eds.), Academic Press, London and New York.

Solomonson, L. P. and Vennesland, B. (1972). *Biochem. Biophys. Acta* **267**, 544–557.

Solomonson, L. P., Lorimer, G. H., Hall, R. L., Borchers, R., and Bailey, J. L. (1975). *J. biol. Chem.* **250**, 4120–4127.

Wallace, W. and Johnson, C. B. (1978). *Plant Physiol.* **61**, 748–752.

Wilson, R. D. and Waring, E. J. (1949). *J. Aust. Inst. Agric. Sci.* **14**, 141–145.

Wray, J. L. and Filner, P. (1970). *Biochem. J.* **119**, 715–725.

CHAPTER 13

Selenium and Plant Metabolism

J. W. ANDERSON AND A. R. SCARF

La Trobe University, Bundoora, Victoria, Australia

I. INTRODUCTION

The study of selenium in organisms began in the 1930s with the discovery that the disorders of livestock known as alkali disease, ill thrift, and blind

---

Sulphur-containing compounds are named in this chapter in accordance with the nomenclature of Roy and Trudinger (1970). In addition, the following abreviations are used: APS, adenosine 5′-sulphatophosphate; APSe, adenosine 5′-selenophosphate; DCMU, 3-(3,4 dichlorophenyl)-1,1-dimethylurea; Fd_{red} and Fd_{ox}, reduced and oxidized forms of ferredoxin; GSH and GSSG, reduced and oxidized forms of glutathione; G-S-SO$_3^-$, S-sulphoglutathione; OAS, O-acetylserine; NEM, N-ethylmaleimide; PAPS, 3′-phosphate adenosine 5′-sulphatophosphate; PHS, phosphohomoserine.

staggers, were associated with the high selenium content of certain plant species indigenous to the seleniferous soils of western USA and Canada. These species commonly accumulate 1 to 10 mg Se g^{-1} dry weight from seleniferous soils which typically contain 2 to 14 μg Se g^{-1} (Rosenfeld and Beath, 1964; Shrift, 1969). Plants possessing these characteristics are known as selenium-accumulators (see Peterson, this volume) and have been reported in various parts of the world, e.g. *Astragalus racemosus* in USA and *Neptunia amplexicaulis* in Australia. The majority of plants, however, including many of those associated with seleniferous soils, do not accumulate high concentrations of selenium. Some species of *Astragalus* belong to this group thus permitting comparative studies of selenium metabolism in accumulator and non-accumulator species of the same genus.

Following the discovery of the selenium-accumulator plants it was soon established that selenium inhibited plant growth. It was also found that the selenium accumulators were less sensitive to selenium toxicity and that most of the selenium present in both accumulators and non-accumulators was contained in various selenoamino acids, principally as selenium analogues of intermediates or simple derivatives of the methionine biosynthetic pathway. They include Se-methylselenocysteine, selenocystathionine, and Se-methylselenomethionine. In non-accumulators, selenium also occurs in protein. Accordingly, the possibility that selenium is subject to the same (or similar) metabolic processes as sulphur has formed the working hypothesis for most studies of selenium metabolism in both accumulator and non-accumulator plants.

Following the early studies of the seleniferous vegetation and the toxicity of selenium it was found that selenium was essential for the expression of activity of several enzymes in certain anaerobic bacteria (Pinsent, 1954; Stadtman, 1979, 1980). Selenium was also found to be an essential trace element for the nutrition of animals (Schwarz and Foltz, 1957; Hartley and Grant, 1961; Stadtman, 1974, 1979). This raises the question whether plants, including accumulators and non-accumulators, also have an essential requirement for selenium. The requirement of animals for selenium also introduces important practical reasons for studying factors which affect the selenium content of non-accumulator plants since the latter are the presumed direct or indirect source of selenium for most animals.

It is evident from this introduction that selenium is metabolized in organisms in two fundamental ways. The first, which occurs in a large number of organisms, including selenium accumulator and non-accumulator plants, entails non-specific metabolism of selenium in place of sulphur, resulting in the formation of selenium isologues of sulphur-containing metabolites. This is especially evident when selenium is supplied at relatively high (millimolar) concentrations. Secondly, the requirement for trace amounts of selenium for the nutrition of animals and its incorporation into specific proteins suggests that at least some processes of some organisms are selenium specific and do not metabolize the more abundant element, sulphur. This chapter reviews the major developments in the study of selenium in plants with particular reference to the assimilation of inorganic selenium in selenium accumulator and non-accumulator plants and the specificity of the metabolic reactions of plants with respect to sulphur and selenium.

II. COMPARATIVE CHEMISTRY OF SELENIUM AND SULPHUR

Prior to the renewed interest in the chemistry of selenium, most reviews tended to stress the similarity of the two group 6 elements, selenium and sulphur. However, an increasing interest in the biological role of selenium has led to an awareness of the individual reactivity of the two elements. For example the water soluble organic thiosulphates ($R-SSO_3^-$) react with cysteine comparatively slowly and do not inhibit the enzyme glutathione reductase or the growth of various microorganisms (Scarf *et al.*, 1977, 1979, unpublished observations). Conversely, the selenosulphates ($R-SeSO_3^-$) are highly active in all three systems. These differences are important in understanding the unique and essential role of selenium in some organisms on the one hand and the toxicity of selenium on the other.

The fundamental chemical properties of selenium and sulphur are summarized in Table I. The bond strengths and electrophilicity of the two elements are similar and the larger covalent and ionic radii of selenium relative to sulphur reflect the addition of another electron shell. The pK_a

TABLE I
Some properties of sulphur and selenium[a]

Property	S	Se
Electron configuration	$[Ne]3s^2 3p^4$	$[Ar]3d^{10}4s^2 4p^4$
Covalent radius (nm)	0·103	0·117
Ionic radius (X^{2-}) (nm)	0·190	0·202
Ionic radius (X^{6+} in XO_4^{2-}) (nm)	0·034	0·040
Bond energy (X–H) (kJ mol^{-1})	368	280
Electron affinity (eV)	−3·44	−4·21
Electronegativity (Pauling)	2·44	2·48
pK_a		
XO(OH)	1·9	2·6
$XO_2(OH)_{2(aq)}$	−3	−3
$H_2X_{(aq)}$	7·0	3·8
$HX^-_{(aq)}$	12·89	11·0

[a]The symbol X refers to either S or Se as applicable.

values of the sulphur and selenium hydrides, however, are distinctly different (Table I). H_2Se is a much stronger acid than H_2S although the relative ionic character of the X—H bonds is in the reverse order. The sulphide anion has a greater charge density than the conjugate selenide anion and consequently greater proton affinity (basicity) resulting in a weaker acid. The greater acidity of H_2Se is reflected in the greater acid strength of the selenol group of selenocysteine compared with the thiol group of cysteine (Huber and Criddle, 1967a).

Sulphur and selenium compounds are known for the full range of valence states available to the Group 6 elements (II, IV, VI). With the exception of the +2 form of Se(II), all of the oxidation states are known in organisms. The redox potentials of some sulphur and selenium salts are shown in Table II. The major differences in the chemistry of selenium and

TABLE II
Reduction potentials for some forms of inorganic sulphur and selenium

Reaction	E^0 (V)
$Se_{(s)} + 2e^- \rightarrow Se^{2-}$	$-0 \cdot 78$ $(-0 \cdot 92)$
$S_{(s)} + 2e^- \rightarrow S^{2-}$	$-0 \cdot 48$
$H_2SeO_3 + 4H^+ + 4e^- \rightarrow Se_{(s)} + 3H_2O$	$0 \cdot 74$
$H_2SO_3 + 4H^+ + 4e^- \rightarrow S_{(s)} + 3H_2O$	$0 \cdot 45$
$SeO_4^{2-} + 4H^+ + 2e^- \rightarrow H_2SeO_3 + H_2O$	$1 \cdot 15$
$SO_4^{2-} + 4H^+ + 2e^- \rightarrow H_2SO_3 + H_2O$	$0 \cdot 17$

sulphur, and presumably the different roles of the two elements in organisms, originate from the easier reduction of selenium and the preferred oxidation of sulphur. Although structurally similar, the sulphate and selenate anions (+6) differ markedly in their reduction potentials with the result that selenious acid is a much stronger oxidant than the isologous sulphurous acid. Formation of the +6 oxidation state for selenium compounds requires a suitable pH environment, otherwise forcing conditions are necessary. The following equation

$$H_2SeO_3 + 2H_2SO_3 \rightarrow Se + 2H_2SO_4 + H_2O \qquad (1)$$

demonstrates the tendency of Se (+4) in selenite to undergo reduction whereas S (+4) in sulphite prefers oxidation. This characteristic of the elements predominates in their organic chemistry and is reflected in the forms of selenium present in soil. Under aerobic conditions, sulphur most commonly occurs as sulphate but selenium occurs as selenite. Thus although the particular form(s) of selenium present in soil will be a function of pH and redox potential (e.g. alkaline oxidizing conditions favour selenates, while selenides and elemental selenium are stable in reducing soils of low pH), it is assumed that selenite is the form most readily available to plants.

In plants, as in all organisms, a large proportion of the sulphur occurs as organic thiols. Much of the selenium in plants occurs as selenoamino acids and it is assumed that their synthesis involves an organic selenol (seleno-cysteine) as an intermediate. It is therefore informative to examine the reactivity of organic thiols and selenols under controlled non-biological conditions to predict likely reactions in organisms.

The pathways for the oxidation of organic thiols and their isologous selenols are shown in Fig. 1. This indicates likely metabolic differences

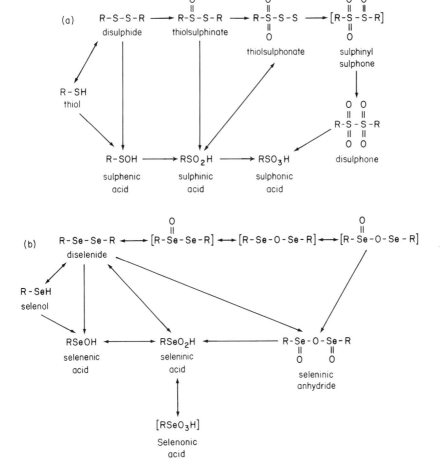

FIG. 1. Schemes for the oxidation of (a) thiols and (b) selenols. Compounds enclosed in brackets have not been isolated and are postulated intermediates. The scheme shown for (a) is adapted from Kharasch and Arora (1976).

based on the relative stabilities of the various sulphur and selenium compounds. The reversible oxidation of thiols to disulphides proceeds readily both in chemical and biological systems. On the other hand, selenols are unstable in air and are rapidly oxidized to the stable diselenides. The instability of selenols in air casts some doubt on the occurrence of free selenocysteine in aerobic biological systems. Stable selenium isologues of the thiolsulphinates, thiolsulphonates and disulphones, intermediates of disulphide oxidation, are not known (Fig. 1). Instead, diselenides are generally oxidized to the seleninic acid stage, although selective reactions can limit oxidation to either the selenenic acid or seleninic anhydride as well as the unstable selenonic acid (Klayman, 1973).

The lower portions of the schemes for the oxidation of organic thiols and selenols shown in Fig. 1 reflect the behaviour of their inorganic counterparts. Selenenic acids are more stable than selenonic acids but may be less stable than seleninic acids. The selenium-containing enzyme, glutathione peroxidase, is thought to contain selenium at the active-site in a selenenic acid group (Enz–Se–OH) which is reversibly oxidized to a seleninic acid (Enz–Se–O$_2$H) (see Section VIIB).

In addition to containing organic thiols, plants also contain various thioether amino acids. They include methionine, cystathionine, and alkylated forms of cysteine, along with their oxidation products. As mentioned previously, selenium isologues of methionine and cystathionine, and other thioether amino acids have been found in plants. The schemes shown in Fig. 2 indicate oxidation reactions for these classes of compounds which could be relevant to their behaviour in organisms.

One group of derivatives of the thio and selenoethers, which occur in

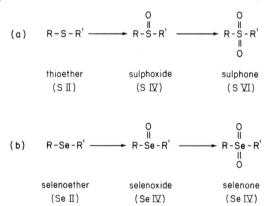

FIG. 2. Schemes for the oxidation of (a) thioethers and (b) selenoethers.

some plants, are of interest. They are the methyl sulphonium and selenonium compounds having the structure

$$CH_3 - \overset{+}{X} \overset{\displaystyle R}{\underset{\displaystyle R'}{\diagup}} \quad (X = S, Se).$$

Some compounds of this type (e.g. S-methylmethionine and Se-methylselenomethionine) serve as methyl donors in biological systems (Virupaksha *et al.*, 1966; Shrift, 1969).

In summary, this brief examination of the chemistry of sulphur and selenium demonstrates that the two elements undergo similar reactions and that the major differences in reactivity are attributed to differences in the redox characteristics of the two elements. Presumably these criteria are important in biological systems.

III. Factors Affecting the Selenium Content of Plants

The selenium content of a plant is primarily determined by the species (accumulator or non-accumulator) and whether it is grown in seleniferous or non-seleniferous soil. Seleniferous soils are defined by the seleniferous vegetation that they support rather than by their selenium content. Indeed, some seleniferous soils have lower selenium contents than many non-seleniferous soils. Readers are referred to the reviews by Rosenfeld and Beath (1964) and Shrift (1973) for a list of the primary and secondary accumulator species which comprise the seleniferous vegetation. The primary accumulators are invariably restricted to seleniferous soils; they do not accumulate selenium when established on non-seleniferous soils. Secondary accumulators are found on both types of soils but accumulate selenium only from seleniferous soils.

The selenium contents of accumulator and non-accumulator plants raised on seleniferous soils differ greatly; whereas accumulators typically contain 1 to 10 mg Se^{-1} dry wt., non-accumulators seldom contain more than 30 μg Se^{-1} g dry wt. (Watkinson, 1966). This value is of interest in relation to the minimum desirable pasture content of 30 ng Se g^{-1} dry wt. recommended by The Agricultural Research Council (1965) to prevent selenium deficiency in livestock. On the other hand, continuous ingestion of fodder containing more than 5 μg Se g^{-1} will induce selenium toxicity in stock (Underwood, 1971).

Another important factor affecting the selenium content of a given species is the sulphur/selenium nutrition. In summary the selenium content increases with the level of selenium available and decreases with the sulphur supply. This has several important practical consequences. For example, plants can tolerate much higher concentrations of selenium in the presence of high concentrations of sulphur. This is shown by the growth data in Table III for a selenium accumulator (*Neptunia amplexicaulis*) and

TABLE III
Effect of concentration of Na_2SO_4 and Na_2SeO_4 on the growth of *Trifolium repens* and *Neptunia amplexicaulis*[a]

		Dry weight (mg plant^{-1})					
	Concentration	Concentration of $Na_2SeO_4(\mu M)$					
Species	of Na_2SO_4 (μM)	0	0·32	1·6	8	40	200
T. repens	1·6	196	184	137	8	n.d.	n.d.
	8	227	239	158	8	5	n.d.
	40	449	435	520	7	7	7
	200	459	n.d.	401	19	6	4
	1000	409	n.d.	n.d.	624	405	5
N. amplexicaulis	1·6	257	289	280	281	n.d.	n.d.
	8	209	357	389	336	610	n.d.
	40	389	414	500	698	528	<10
	200	1191	n.d.	1349	1511	1876	<10
	1000	1125	n.d.	n.d.	1669	2350	2244

[a] The plants were raised in aerated water cultures containing 8·9 mM NO_3^-, 3 mM K^+, 0·1 mM $H_2PO_4^-$, 2·5 mM Ca^{2+}, 0·5 mM Mg^{2+}, and the trace elements specified by Broyer *et al.* (1972a) in addition to the concentrations of Na_2SO_4 and Na_2SeO_4 shown.

a non-accumulator (*Trifolium repens*). Another more important problem which stems from the sulphur/selenium antagonism in plants concerns sulphur-induced selenium deficiency in livestock through the application of sulphatic fertilizers to pastures and crops. This is especially common in regions with soils with inherently low levels of phosphorus and selenium where superphosphate (containing $CaSO_4$) is applied to rectify phosphorus deficiency. In a survey of some farms from an area with these characteristics we found that selenium deficiency in lambs, as judged by their glutathione peroxidase levels and increase in body weight following administration of selenium, was greatest on those farms with pastures having the highest phosphorus content and the highest rates of application of superphosphate (Paynter *et al.*, 1979, unpublished).

One group of plants of potential interest with respect to sulphur/selenium antagonisms are the gypsophilous plants. They are restricted to soils containing high concentrations of calcium sulphate and, under these conditions, contain very high concentrations of sulphur, mostly as sulphate (Parsons, 1976).

Many other factors in addition to those described above also affect the selenium content of plants. They tend to be of lesser importance and readers are referred to previous reviews for further details (Rosenfeld and Beath, 1964; Shrift, 1969, 1973).

IV. CHEMICAL FORMS OF SELENIUM IN PLANTS

Most of the selenium found in plants (accumulator and non-accumulator) occurs in selenoether amino acids of the form R–Se–R'. The identification of organoselenium compounds is complicated by the concurrence of the sulphur isologues thus requiring complex separatory procedures. Since purification techniques must distinguish between sulphur and selenium isologues, chromatographic methods, which are frequently employed for this purpose, are necessarily sensitive. This is especially relevant in view of the report that selenite binds to certain compounds non-enzymically, resulting in the formation of products which display similar chromatographic properties to the parent compound (Schwarz and Sweeney, 1964). Successful separations of selenium and sulphur isologues have been effected by several column chromatographic techniques (e.g. ion-exchange) as well as standard paper and thin layer chromatography procedures (Walter *et al.*, 1969; Benson and Patterson, 1969; McConnell and Wabnitz, 1964; Ganther, 1971). A further difficulty in characterizing the selenium-containing compounds in plants is that the selenium isologues of the sulphur amino acids are relatively labile. For example, selenocysteine and selenocystine are destroyed under conditions used for the acid hydrolysis of proteins (Huber and Criddle, 1967a); derivatization by alkylation (e.g. with iodoacetate) or some other method must be employed prior to hydrolysis to stabilize these compounds.

The selenium-containing compounds found in plants have been reviewed by Shrift (1969, 1973). Only the two main classes of compounds are summarized below.

A. DERIVATIVES OF SELENOCYSTEINE AND SELENOHOMOCYSTEINE

Free selenocysteine (I) has not been reported in plants. As noted previously, the selenol group of this amino acid would render it very susceptible to oxidation during extraction and characterization procedures.

$$
\begin{array}{ll}
\mathrm{CH_2SeH} & \mathrm{CH_2{-}SeH} \\
| & | \\
\mathrm{CH{-}NH_3} & \mathrm{CH_2} \\
\quad | \quad + & | \\
\mathrm{COO^-} & \mathrm{CH{-}NH_3} \\
& \quad | \quad + \\
& \mathrm{COO^-} \\
\text{Selenocysteine (I)} & \text{Selenohomocysteine (II)}
\end{array}
$$

The reported pK_a values of 8·25 for the thiol and 5·24 for the selenol group of cysteine and selenocysteine respectively (Huber and Criddle, 1967a) are of interest; any component containing selenocysteine in place of cysteine would thus have altered reactivity.

Although free selenocysteine has not been detected in plants, the *in vitro* synthesis of selenocysteine by pea chloroplasts has been reported (Ng and Anderson, 1978b, 1979); the product, which was formed under strictly anaerobic conditions, was trapped as the NEM adduct and shown to have the same chromatographic behaviour as selenocysteine-NEM. In addition, Brown and Shrift (1980) raised mung beans (*Vigna radiata*) in the presence of selenate and isolated the protein. After treating the protein fraction with iodoacetate and subjecting it to acid hydrolysis, they identified carboxy-methylselenocysteine. This suggests that the protein contained seleno-cysteinyl residues in peptide linkage. It remains to be determined whether the selenocysteinyl residues are formed by post-translational modification of an existing residue or by incorporation of free selenocysteine during protein synthesis.

Several other compounds structurally related to selenocysteine which have been recorded in plants are listed in Table IV. Of these, Se-methylselenocysteine is a major form of selenium in some selenium accumulators. On the other hand, the oxidized forms of selenocysteine, selenocystine, and selenocysteine seleninic acid, have been reported in trace amounts from non-accumulators (Peterson and Butler, 1962). Whilst it is possible that these could arise from oxidation of selenocysteine there is some disagreement about the occurrence of selenocystine in plants. For example, the compound from *Astragalus* tentatively identified by ion exchange chromatography as selenocystine could be an artifact resulting from a buffer change during column chromatography (Martin and Gerlach, 1969). Furthermore, it should be noted that diselenides may undergo spontaneous non-enzymic reactions with selenols and thiols over a large pH range (Walter *et al.*, 1969)

$$R-XH + R'-Se-Se-R'' \rightleftharpoons R-X-Se-R' + R''-SeH$$
$$R-X-Se-R'' + R'-SeH \qquad (3)$$

(X = S or Se)

The probability of these reactions increases with the S:Se ratio.

Free selenohomocysteine (II) has not been reported in plants. Here, the same types of problems associated with the identification of selenocysteine in plants would apply. Indeed, the reactivity of the selenol group of selenohomocysteine is exploited in the organic synthesis of seleno-cystathionine (Zdansky, 1968). Selenohomocystine has been isolated from *Astragalus crotalariae* (Virupaksha *et al.*, 1966) and is presumably subject to diselenide interchange with plant thiols or selenols as described above for selenocystine. The selenoether, selenocystathionine, is much

Some selenoamino acids found in plants, major forms of Se in accumulator plants

Derivatives of selenocysteine[a]

Se-Methylselenocysteine[a]

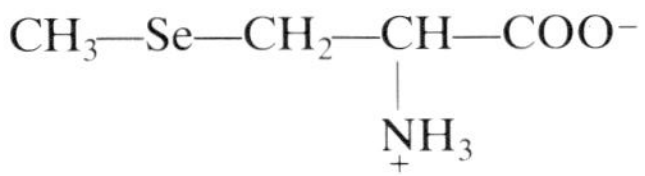

$$CH_3-Se-CH_2-CH-COO^-$$
$$|$$
$$\overset{+}{N}H_3$$

Selenocystine

$$^-OOC-CH-CH_2-Se-Se-CH_2-CH-COO^-$$
$$|\qquad\qquad\qquad\qquad\quad|$$
$$\overset{+}{N}H_3\qquad\qquad\qquad\qquad\overset{+}{N}H_3$$

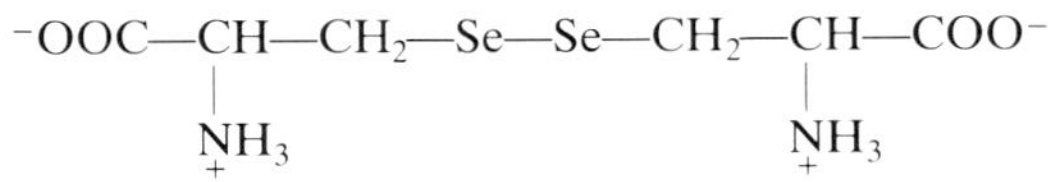

Selenocysteine seleninic acid

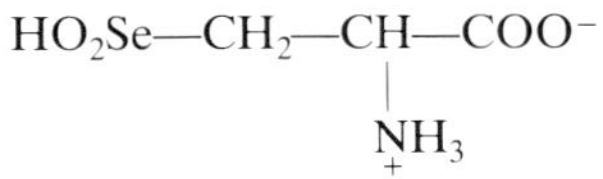

$$HO_2Se-CH_2-CH-COO^-$$
$$|$$
$$\overset{+}{N}H_3$$

Se-propenylselenocysteine selenoxide

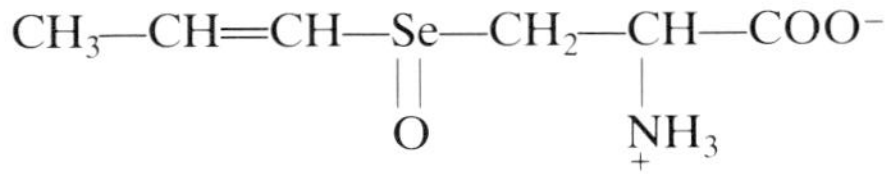

$$CH_3-CH=CH-Se-CH_2-CH-COO^-$$
$$\parallel\qquad\qquad\quad|$$
$$O\qquad\qquad\quad\overset{+}{N}H_3$$

Derivatives of selenohomocysteine

Selenohomocystine

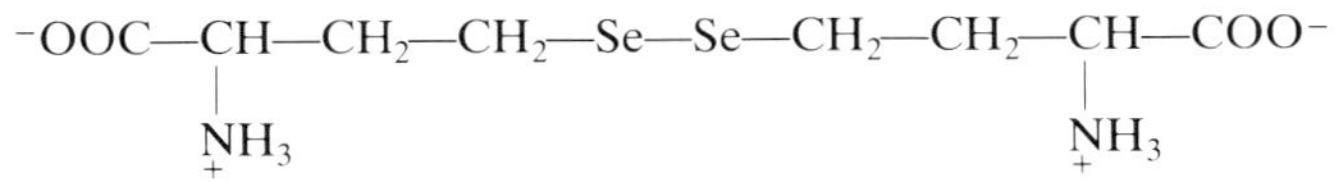

$$^-OOC-CH-CH_2-CH_2-Se-Se-CH_2-CH_2-CH-COO^-$$
$$|\qquad\qquad\qquad\qquad\qquad\qquad\qquad\qquad|$$
$$\overset{+}{N}H_3\qquad\qquad\qquad\qquad\qquad\qquad\qquad\overset{+}{N}H_3$$

Selenocystathionine[a]

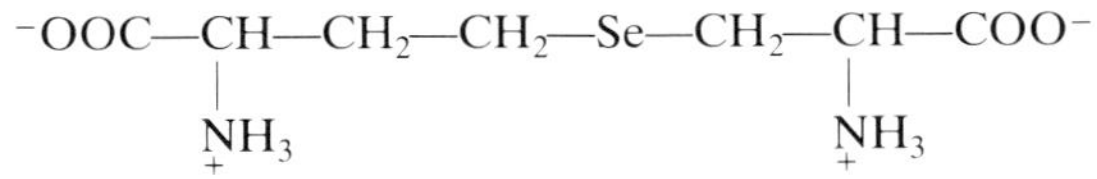

$$^-OOC-CH-CH_2-CH_2-Se-CH_2-CH-COO^-$$
$$|\qquad\qquad\qquad\qquad\qquad\quad|$$
$$\overset{+}{N}H_3\qquad\qquad\qquad\qquad\overset{+}{N}H_3$$

Derivatives of selenomethionine[b]

Selenomethionine[b]

$$CH_3-Se-CH_2-CH_2-CH-COO^-$$
$$|$$
$$\overset{+}{N}H_3$$

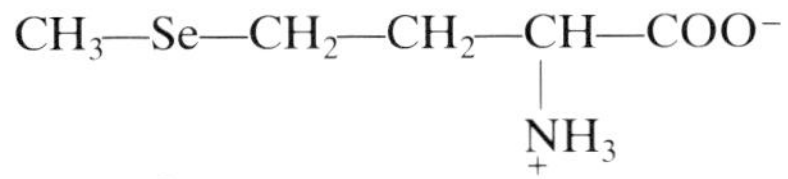

Selenomethionine selenoxide

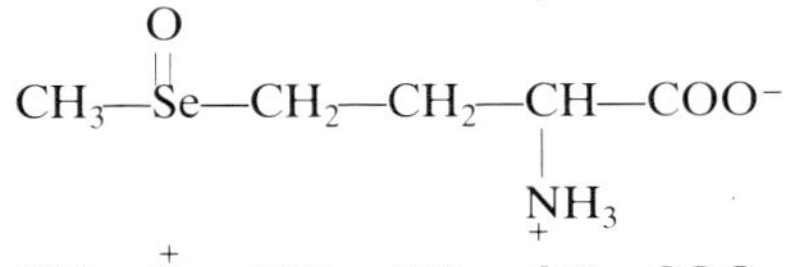

$$\overset{O}{\overset{\parallel}{CH_3-Se}}-CH_2-CH_2-CH-COO^-$$
$$|$$
$$\overset{+}{N}H_3$$

Se-Methylselenomethionine

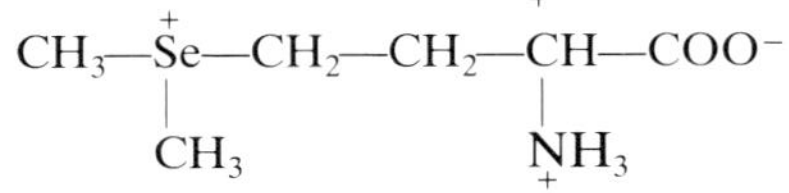

$$CH_3-\overset{+}{Se}-CH_2-CH_2-CH-COO^-$$
$$|\qquad\qquad\qquad\quad|$$
$$CH_3\qquad\qquad\quad\overset{+}{N}H_3$$

more stable than the selenol amino-acids. It has been reported in a large number of selenium accumulator plants. In some of these it constitutes the major form of selenium in the plant (Martin *et al.*, 1971; Nigam and McConnell, 1973).

B. DERIVATIVES OF SELENOMETHIONINE

Selenomethionine, like selenocystathionine is a selenoether amino-acid and is relatively stable. Presumably this is relevant to the frequent reports of selenomethionine and its derivatives (see Table IV) in non-accumulator plants supplied with exogenous inorganic selenium. However, neither selenomethionine nor its derivatives (e.g. Se-methyl-selenomethionine) constitute a very significant proportion of the selenium associated with accumulator plants.

When non-accumulators are raised on seleniferous soil or supplied with high concentrations of selenium, a considerable proportion of the selenium taken up by the plant becomes associated with the protein fraction. Following acid hydrolysis most of the selenium is found in seleno-methionine suggesting that selenomethionyl residues are found in peptide linkage (Butler and Peterson, 1967; Olson *et al.*, 1970). However, some question surrounds the stability of selenomethionine to acid hydrolysis (Shepherd and Huber, 1969) although evidence to the contrary had been advanced (Shrift, 1973).

Shepherd and Huber (1969) have reported that selenomethionine is 72% less soluble than methionine. They suggest that proteins containing selenomethionine would be less hydrophilic than their isologues. Never-theless extensive replacement of methionine by selenomethionine in β-galactosidase of *E. coli* does not significantly affect activity (Huber and Criddle, 1967b; Coch and Greene, 1971) suggesting that replacement of protein methionyl residues by selenomethionine is an unlikely explanation of selenium toxicity in non-accumulator plants.

Selenium isologues of various other sulphur-containing plant products are known. They include the isothiocyanate sinigrin in *Armoracia lapathi-folia* (Stewart *et al.*, 1974) in which presumably one or both of the sulphur atoms is replaced by selenium, and the selenium isologue of S-propenylcysteine sulphoxide, the precursor of the lachrymatory factor found in onions (Spare and Virtanen, 1964).

V. Sulphur/Selenium Antagonisms in Plants

The reversal of the toxic effects of different forms of selenium by isologous forms of sulphur has proved an informative method for elucidating many aspects of the metabolism of selenium by plants. Simple growth experiments provide elegant demonstrations of this principle. For example, selenate and selenomethionine inhibit the growth of *Chlorella*; these inhibitions are ameliorated by sulphate and methionine respectively (Shrift, 1954a, b) suggesting that selenate and selenomethionine interfere in some way with sulphate and methionine metabolism (see also Table III). Other examples include antagonisms in selenate/sulphate uptake by barley roots and cultured tobacco cells (Leggett and Epstein, 1956; Ferrari and Renosto, 1972; Smith, 1976) and antagonism in selenate/sulphate supported growth of callus cultures of accumulator and non-accumulator species of *Astragalus* (Ziebur and Shrift, 1971). Collectively, these experiments demonstrate that the toxic effects of selenium are caused, at least in part, by interference with sulphur metabolism. It has been suggested that the components of the cell most sensitive to selenium toxicity are those in which sulphur atoms are required for some specific and essential reactivity, e.g. a catalytic -SH group of an enzyme; activity is lost if sulphur is replaced by selenium.

Further evidence for the existence of sulphur/selenium antagonisms within plants has been obtained by examining the effect of sulphur/selenium nutritional regimes on the synthesis and accumulation of various sulphur-containing metabolites and their selenium isologues. For example the selenium accumulator species *Astragalus bisulcatus* accumulates both S-methylcysteine and Se-methylselenocysteine. Chow *et al.* (1971) found that the relative proportions of these two amino acids, synthesized by the plant, depended on the sulphate/selenate nutritional regime; sulphate enhanced the amount of S-methylcysteine and decreased the amount of Se-methylselenocysteine. Conversely, selenate had the opposite effect. These results imply that at least some parts of the pathway involved in the incorporation of sulphur and selenium from sulphate and selenate into methylcysteine and its selenium isologue are common for both elements. Although inorganic selenium is reportedly incorporated into the selenium isologues of a number of other sulphur-containing compounds the effect of inorganic sulphur/selenium on the relative amounts of the sulphur and selenium products has not been investigated. However, it is frequently assumed that such incorporations are competitive, with respect to sulphur and selenium, even though, as reviewed by Shrift (1969), there is some evidence to the contrary especially regarding selenium-specific reactions in selenium accumulator plants. Here, however, it should be noted that many of the sulphur isologues of selenoamino acids, previously thought to be absent from certain accumulator species, have now been detected by improved techniques (Chen *et al.*, 1970; Martin *et al.*, 1971; Chow *et al.*, 1972; Peterson and Robinson, 1972). Furthermore, as noted previously, the relative amounts of the sulphur and selenoamino acids in plants are highly dependent on the sulphur/selenium nutrition.

VI. METABOLISM OF SELENIUM IN PLANTS

A considerable amount of evidence discussed above suggests that macro concentrations of selenium are metabolized by at least some parts of the pathways of sulphur metabolism. This generalization is probably not correct for all aspects of selenium metabolism, expecially the assimilation of inorganic selenite. Moreover, free selenocysteine, a presumed key intermediate in the assimilation of inorganic selenium via the sulphate assimilatory pathway, has not been demonstrated in plants. Nevertheless, the sulphur assimilation pathway provides a framework for discussing selenium metabolism in accumulator and non-accumulator plants.

A. ASSIMILATION OF INORGANIC SELENIUM BY PLANTS AND PLANT TISSUES

Studies of the incorporation of [^{75}Se]selenate and [^{75}Se]selenite into the various selenium-containing metabolites of whole plants have confirmed that macro concentrations of these forms of inorganic selenium are subject to similar metabolic fates as inorganic sulphur. Our current knowledge of the pathway for the assimilation of inorganic sulphate into methionine is summarized in Fig. 3. It shows that cysteine is a key intermediate in the biosynthesis of methionine. Cysteine is also a precursor of GSH, S-methylcysteine, and many other compounds (Giovanelli et al., 1980). Peterson and Butler (1962) examined whether inorganic selenite was incorporated into the selenium isologues (and/or their derivatives) of the intermediates of the pathway shown in Fig. 3. They raised Neptunia amplexicaulis, Triticum aestivum, and the pasture plants Trifolium repens, T. pratense, and Lolium perenne for 10 days in nutrient solution containing [^{75}Se]selenite. Most of the label associated with the wheat and pasture plants (60–80%) was associated with protein in the material insoluble in ethanol. Approximately 20–30% of the ^{75}Se-label in the pasture plants was associated with selenocystine, selenocysteic acid, selenomethionine, and selenomethionine selenoxide in the ethanol-soluble fraction, although in wheat the label in this fraction was associated with selenite. On the other hand, very little label was associated with the protein fraction of N. amplexicaulis; most of the ^{75}Se-label in this species was associated with selenocystathionine in the ethanol-soluble fraction (Peterson and Butler, 1962, 1967). Butler and Peterson (1967) also established that Spirodela oligorrhiza (non-accumulator) incorporated [^{75}Se]selenate and selenite

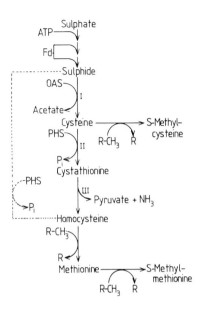

Fig. 3. Pathway of incorporation of sulphate into methionine and other amino acids. Details of the reduction of sulphate to sulphide are as shown in Fig. 4. Reactions I, II, and III are catalysed by cysteine synthase, cystathionine γ-synthase and β-cystathionase respectively. The pathway of methionine biosynthesis via reactions I to III is referred to as transulphurylation and is probably the major route of methionine biosynthesis (Giovanelli et al., 1980). The alternative route shown by the dotted line is known as direct sulphurylation.

into the compounds described above for the non-accumulator pasture species, and that the selenium associated with the protein fraction occurred as selenomethionine. Nigam and McConnell (1973) also reported that leaves (but not pods) of *Phaseolus lunatus* (non-accumulator) incorporate selenate into selenomethionine. On the other hand, maturing seeds of this species incorporate selenate into Se-methylselenocysteine. Most of the general conclusions from these early experiments were subsequently confirmed by Nigam and McConnell (1976) in a study of [75Se]selenate metabolism in wheat, *Phaseolus lunatus* (non-accumulators), and *A. bisulcatus* (accumulator). However, they found relatively similar proportions of 75Se-label in the soluble protein fractions of all three species and concluded that differences in the sensitivity of the three species to selenium could not be attributed to differences in the incorporation of selenium into soluble protein.

Shrift and Virupaksha (1963, 1965), Virupaksha and Shrift (1963, 1965), and Martin et al. (1971) investigated the labelling patterns of a large number of accumulator and non-accumulator plants indigenous to seleniferous soils when supplied with 75Se-labelled selenate or selenite. Of the selenium isologues of the intermediates shown in Fig. 3 the only compound labelled was selenocystathionine; it was found in some accumulators and non-accumulators. However, ^{75}Se-label was incorporated into Se-methyl-selenocysteine in all of the accumulators and Se-methylselenomethionine in all of the non-accumulators.

In summary, these experiments suggest that in the non-accumulator species *T. repens*, *T. pratense*, *L. perenne*, and *S. oligorrhiza*, selenite is assimilated via a pathway analogous to that shown for sulphur (Fig. 3) and that selenocysteine is a likely key intermediate. However, these experiments provide no information on how selenite or selenate are reduced. The incorporation of selenite into selenocystathionine and Se-methylseleno-cysteine in accumulators is also consistent with the pathway of sulphur assimilation, but in the absence of ^{75}Se-labelling of any other intermediates of the pathway (especially selenocysteine or its oxidation products), this possibility remains uncertain. Similar remarks apply to the incorporation of selenite and selenate into Se-methylselenomethionine and seleno-cystathionine in non-accumulator species indigenous to seleniferous soils. There is therefore a requirement for further labelling experiments with accumulators and non-accumulators using [^{75}Se]selenite and selenate. In particular, there is a need for experiments designed to determine the order of labelling of metabolites in the ethanol-soluble fraction.

B. ASSIMILATION OF INORGANIC SELENIUM INTO SELENOCYSTEINE BY CELL-FREE SYSTEMS

The principal form of sulphur incorporated into intermediary sulphur metabolism is inorganic sulphate. It is taken up via a specific uptake mechanism (Smith, 1976; Anderson, 1980) and activated by ATP to form APS. Although plants contain an active APS kinase (Burnell and Anderson, 1973) it is now generally agreed that APS rather than PAPS serves as the precursor for reduction to oxidation state -2. This is achieved via the reaction sequence shown in Fig. 4. Since the sulphur from sulphate is attached to a carrier molecule, possibly GSH (Tsang and Schiff, 1978), through the action of an APS specific sulphotransferase (Schmidt, 1973, 1975) and remains attached throughout the pathway, it is known as the bound pathway. Plants also reduce and assimilate inorganic sulphite readily. However, free sulphite is reduced to free sulphide by an independent mechanism involving sulphite reductase. It is known as the free pathway and is presumed not to be involved in the assimilation of sulphate since *Chorella* mutant Sat_2^-, which contains sulphite reductase but lacks thiosulphonate reductase (the principal reductive enzyme of the bound pathway) does not grow on sulphate. Accordingly, it has been proposed

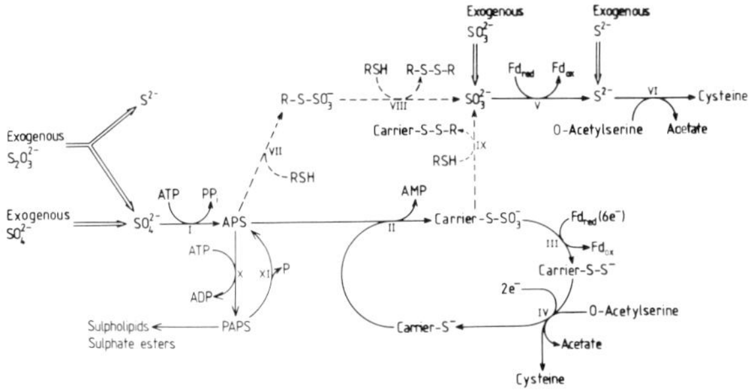

FIG. 4. Summary of the free and bound pathways for the assimilation of inorganic sulphur in chloroplasts. The reaction sequence catalysed by (I) ATP sulphurylase, (II) APS sulphotransferase, (III) thiosulphonate reductase, and (IV) cysteine synthase constitutes the bound sulphate assimilation pathway. The reaction sequence catalysed by (V) sulphite reductase and (VI) cysteine synthase constitutes the free pathway. It is not clear whether reactions (IV) and (VI) are catalysed by the same enzyme. Reactions shown by dotted lines represent various side reactions of the bound pathway which could give rise to free sulphite; reactions (VIII) and (IX) are non-enzymic but reaction (VII) is catalysed by APS sulphotransferase. APS and PAPS are interrelated via (X) APS kinase and (XI) bisphosphonucleoside phosphohydrolase (adapted from Anderson, 1980, 1981).

that the free pathway serves to reduce, and so detoxify, free sulphite taken up from the environment or produced by the plant itself by side reactions of the bound pathway (Schiff and Hodson, 1973; Schmidt et al., 1974). It should be noted that the sulphur flux capacity of the free pathway is far in excess of the bound pathway and that both pathways are associated with chloroplasts and use light-generated Fd_{red} and ATP as their sources of reducing equivalents and energy (Anderson, 1980, 1981).

1. Metabolism of Selenate

Several reactions of the bound and free pathways have been examined with respect to their specificity towards the appropriate selenium isologue. The reaction catalysed by the enzyme ATP sulphurylase, which catalyses the first reaction of sulphate metabolism:

$$ATP + SO_4^{2-} \rightleftharpoons APS + PP_i \qquad (4)$$

has a highly positive free energy change ($+40$ kJ mol^{-1}) thus permitting a study of the reaction by sulphate-dependent [^{32}P]PP$_i$–ATP exchange (Shaw and Anderson, 1972). The enzyme has been purified from various plant species, three of which are accumulators. All of the enzymes studied to date catalyse both sulphate and selenate-dependent PP$_i$–ATP exchange (Shaw and Anderson, 1972, 1974b; Burnell, 1981). This suggests that selenate enters metabolism by forming the selenium analogue of APS in both accumulator and non-accumulator plants. Studies of the ATP sul-phurylases from yeast and *Neptunia amplexicaulis* are consistent with this view. Wilson and Bandurski (1958) reported that the yeast enzyme, in the presence of pyrophosphatase, catalysed the incorporation of [^{75}Se]selenate into a compound which had many of the characteristics of APS. More recently Dilworth and Bandurski (1977) reported that the yeast enzyme in the presence of pyrophosphatase and a thiol, catalysed the reduction of selenate to elemental selenium (Se^{o}) with the concomitant production of AMP and P_i. They attributed this to the enzyme-dependent formation of a selenate anhydride (probably APSe) which then successively reacted with the thiol to form a thiolselenic acid, selenite and finally Se^{o}. However, Shaw and Anderson (1972, 1974b) found that whereas the plant enzymes in the presence of pyrophosphatase catalysed the incorporation of sulphate into APS, the incorporation of selenate into a compound having the chromatographic properties of APSe could not be demonstrated. ATP is the first substrate to react with ATP sulphurylase, and APS the last product released (Shaw and Anderson, 1974a; Farley *et al.*, 1976):

$$E \xrightarrow{ATP} E\text{-}ATP \xrightarrow{SO_4^{2-}} E \underset{SO_4^{2-}}{\overset{ATP}{\rightleftharpoons}} E \underset{APS}{\overset{PP_i}{\rightleftharpoons}} E\text{-}APS \xrightarrow{PP_i} E \underset{}{\overset{APS}{\longrightarrow}} E \qquad (5)$$

(i)   (ii)   (iii)   (iv)   (v)

Accordingly, Shaw and Anderson (1974b) concluded that selenate partici-pated in this reaction in lieu of sulphate but that the dissociation constant of the fifth partial reaction was so small that it effectively blocked the formation of free APSe. This problem has been investigated using the ATP sulphurylase from *Neptunia amplexicaulis*. Burnell (1981) was unable to identify free APSe as a reaction product but found that ^{75}Se-label was bound to charcoal when [^{75}Se]selenate was supplied as substrate. Further, the *Neptunia* enzyme catzlysed the reduction of selenate to Se^{o} in the presence of pyrophosphatase and GSH, with the production of 2 mol of P_i per mol of Se^{o} formed. This suggests that the yeast and plant enzymes are similar with respect to the production of free APSe, and, in the presence of thiols, the subsequent non-enzymic production of selenite. Since ATP sulphurylase in plants is associated with chloroplasts (Burnell and Ander-son, 1974) and chloroplasts contain a light-coupled glutathione reductase (Jablonski and Anderson, 1978), the reduction of selenate to selenite in plants may be a light-dependent process.

    In a different approach to this problem, Reuveny (1977) found that selenate derepressed the ATP sulphurylase of cultured tobacco cells grown on sulphate but not cells grown with cysteine as the sulphur source. Since selenate was more toxic towards cells grown under conditions which derepressed ATP sulphurylase (i.e. with sulphate) than under conditions which repressed the enzyme (with cysteine), Reuveny (1977) proposed that selenate entered metabolism as an alterantive substrate of sulphate via the reaction catalysed by ATP sulphurylase and that the selenium isologue of the sulphur-containing compound responsible for end product-represession of ATP sulphurylase is ineffective (i.e. depression ensues).

2. Reduction of Selenite

As noted in Section II of this chapter, selenite rather than selenate is the presumed source of selenium taken up by plants from aerobic soils. By analogy with the sulphate assimilation pathway, selenite could be reduced via the free pathway (sulphite reductase) or undergo exchange with the carrier–S–SO_3^- intermediate of the bound pathway (Fig. 4). Another possibility involves the mechanism proposed by Hsieh and Ganther (1975) for the reduction of selenite in yeast. It involves reduction of selenite by GSH and NADPH in the presence of glutathione reductase. Since chloroplasts contain a highly active glutathione reductase (Foyer and Halliwell, 1976; Schaedle and Bassham, 1977) which utilizes light-generated reducing equivalents (Jablonski and Anderson, 1978), then selenite reduction in chloroplasts could proceed by the mechanism shown in Fig. 5.

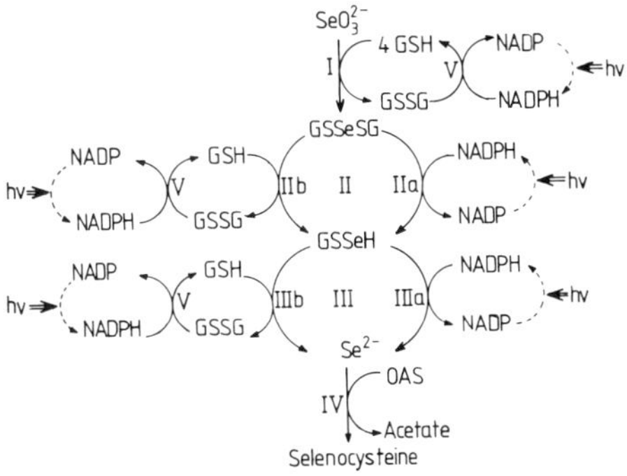

FIG. 5. Proposed pathway for the incorporation of selenite into selenocysteine by illuminated chloroplasts. Reactions I–III are as proposed by Hsieh and Ganther (1975) and reaction IV involves cysteine synthase (Ng and Anderson, 1978b). Reaction I is non-enzymic. Reactions II and III can proceed enzymically in the presence of glutathione reductase using NADPH as reductant (IIa and IIIa) or non-enzymically using GSH as reductant (IIb and IIIb). Light-coupled glutathione reductase (reaction V) catalyses reduction of GSSG; 0·5 mol of O_2 are evolved per mol of GSSG reduced by this process (Jablonski and Anderson, 1978). Light is also required for the production of NADPH for use in the enzyme-catalysed reactions in II and III (from Ng and Anderson, 1979).

Isolated pea chloroplasts contain the enzyme cysteine synthase (Fank-hauser et al., 1976; Ng and Anderson, 1978a). In addition to catalysing the synthesis of cysteine from sulphide in the presence of OAS, it also supports the reaction

$$Se^{2-} + OAS \rightarrow selenocysteine + acetate \qquad (6)$$

Thus the reduction of selenite by chloroplasts can be monitored by the formation of selenocysteine in the presence of OAS. Ng and Anderson (1979) found that intact chloroplasts supported the formation of cysteine and selenocysteine from sulphite and selenite respectively. Both reactions required light and OAS and were inhibited by DCMU, indicating that light was required for the formation of the electron donor. However, whereas the incorporation of sulphite was sensitive to KCN, an inhibitor of sulphite reductase, and insensitive to $ZnCl_2$, an inhibitor of glutathione reductase, the incorporation of selenite exhibited the opposite characteristics (Table V). Sonicated chloroplasts, supplemented with GSH and NADPH

TABLE V

Incorporation of [^{35}S]sulphite and [^{75}Se]selenite into cysteine and selenocysteine by isolated pea chloroplasts in the light[a]

| | | Rate of synthesis (%)[b] | |
Chloroplasts	Treatment	Cysteine	Selenocysteine
Intact	Complete[c]	100	100
	Without light	0	2
	With 2 μM DCMU	0	7
	Without OAS	0	0
	With 1 mM $ZnCl_2$	97	76
	With 1 mM KCN	0	96
Sonicated	Complete[d]	100	100
	With 1 mM $ZnCl_2$	103	28
	With 1 mM KCN	0	93

[a] Recalculated from Ng and Anderson (1979).
[b] All rates expressed as % of appropriate complete system.
[c] With 10 mM OAS and Na_2SO_3/Na_2SeO_3.
[d] With 1 mM NADPH, 4 mM GSH, 10 mM OAS and Na_2SO_3/Na_2SeO_3.

also incorporated sulphite and selenite in the presence of OAS. With this system, however, it was noted that whereas GSH and NADPH enhanced the incorporation of sulphite by 3 to 4-fold, these reagents enhanced selenite incorporation by 17 to 34-fold suggesting that GSH and NADPH fulfil different roles for the two substrates. Incorporation of selenite and sulphite by sonicated chloroplasts exhibited qualitatively similar sensitivity to $ZnCl_2$ and KCN as intact chloroplasts, although quantitatively the incorporation of selenite by sonicated chloroplasts was more sensitive to $ZnCl_2$ (Table V). Ng and Anderson (1979) concluded from these observations that the incorporation of sulphite into cysteine was consistent with the free pathway of sulphur assimilation, whereas the incorporation of selenite was consistent with the scheme shown in Fig. 5. The latter conclusion was verified by examining the synthesis of selenocysteine from *OAS* and selenite in reaction mixtures containing purified cysteine synthase from clover and glutathione reductase from peas. The reaction was dependent on the addition of the two enzymes, GSH and NADPH.

Jablonski and Anderson (1978) demonstrated that ruptured chloroplasts support GSSG-dependent O_2 evolution in the presence of catalytic concentrations of NADPH. This reaction was attributed to light-coupled glutathione reductase activity.

$$GSSG + NADPH + H^+ \longrightarrow 2GSH + NADP^+ \qquad (7)$$

$$NADP^+ + H_2O \xrightarrow{hv} NADPH + H^+ + \tfrac{1}{2}O_2 \qquad (8)$$

$$GSSG + H_2O \xrightarrow{hv} 2GSH + \tfrac{1}{2}O_2 \qquad (9)$$

It follows that if GSH is oxidized by selenite as shown in Fig. 5, then ruptured chloroplasts should support selenite-dependent O_2 evolution in the presence of GSH (or GSSG) and catalytic concentrations of NADPH. The data in Fig. 6 show that under these conditions, repeated additions of 0·2 mM selenite elicit the evolution of 0·76 to 0·46 mol of O_2 per mol of selenite supplied. Although the ratio is less than the theoretical O_2:selenite stoichiometry predicted from Fig. 5, the experiment nevertheless indicates that selenite can serve as a final electron acceptor for reducing equivalents emanating from water. Further details of these experiments will be reported elsewhere.

3. Assimilation of Selenide

As noted previously, the cysteine synthase of pea chloroplasts supports the synthesis of selenocysteine from selenide and *OAS*. Ng and Anderson (1978b) also found that the cysteine synthases from six additional species, three of them selenium accumulators, also support this reaction. This would afford an explanation for the entry of inorganic selenium into the selenoamino acids in accumulator and non-accumulator plants. However, they failed to detect any consistent differences in the properties of the cysteine synthases from accumulator and non-accumulator species. Thus, if inorganic selenium enters intermediary metabolism via the reaction catalysed by cysteine synthase *in vivo*, then the studies of Ng and Anderson

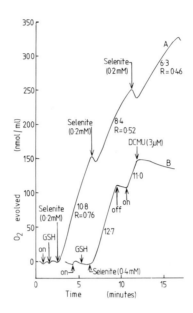

FIG. 6. Some properties of selenite-dependent oxygen evolution by ruptured pea chloroplasts. Chloroplast preparations A and B (94% and 85% intact respectively) were sonicated and incubated at a chlorophyll concentration of 200 μg ml^{-1} in the presence of 50 μM NADPH. The reaction mixtures were illuminated and GSH (0·4 mM), DCMU (3 μM) and Na$_2$SeO$_3$ (0·2 mM or 0·4 mM) were added as shown. The values beside the curves denote the rate of O$_2$ evolution in μmol h^{-1} mg^{-1} Chl^{-1}. The ratio of O$_2$ evolved:selenite suppled (R) for each addition of selenite is also shown (data per courtesy of P. P. Jablonski).

(1978b) suggest that the differences in the metabolism of selenium in accumulator and non-accumulator plants cannot be traced to the assimilation reaction.

The data in Table VI show that the selenium flux capacities of the various reactions of chloroplasts are lower than those for the sulphur flux of the analogous reaction of the free pathway of sulphur assimilation. Nevertheless, the selenium fluxes, which are almost certainly underestimated (Ng and Anderson, 1978b, 1979), are not less than the theoretical sulphur flux (i.e. flux from SO$_4^{2-}$ and S^{2-} required to account for reduced sulphur incorporated by the plant) of the bound pathway. Given that sulphate and selenite are the likely forms of sulphur and selenium available to plants under aerobic conditions and that the mechanism of selenite assimilation described in Fig. 5 occurs *in vivo*, the data in Table VI imply that sulphate and selenite could, in theory, be assimilated at approximately equal rates.

TABLE VI

Sulphur and selenium fluxes for some reactions of sulphur and selenium assimilation in intact chloroplasts

Element	Species	Reaction	Flux (μg atoms h^{-1} mg^{-1} Chl)	Reference
S	Theoretical value	$SO_4^{2-} \rightarrow$ cysteine[a]	0·4	Anderson (1980)
	Spinach	$S^{2-} + OAS \rightarrow$ cysteine[b]	123	Fankhauser et al. (1976)
	Pea	$S^{2-} + OAS \rightarrow$ cysteine[b]	88	Ng and Anderson (1978a)
	Pea	$SO_3^{2-} \rightarrow S^{2-}$[b]	5·8	Ng and Anderson (1979)
	Pea	$SO_3^{2-} + OAS \rightarrow$ cysteine[b]	6	Ng and Anderson (1979)
Se	Pea	$Se^{2-} + OAS \rightarrow$ selenocysteine	22–26	Ng and Anderson (1978b)
	Pea	$SeO_3^{2-} + OAS \rightarrow$ selenocysteine	0·36	Ng and Anderson (1979)

[a] Denotes a reaction of the bound pathway of sulphur assimilation.
[b] Denotes a reaction of the free pathway of sulphur assimilation.

Although some subcellular organelles and individual enzymes of sulphur assimilation and glutathione metabolism have been examined *in vitro* with respect to the assimilation of inorganic selenium there is currently very little evidence to suggest that these processes operate *in vivo*. In particular, an analysis of the assimilation of selenate and selenite in mutants of a photosynthetic organism with deficiencies in such enzymes as ATP sulphurylase, cysteine synthase, glutathione reductase, sulphite reductase etc., such as conducted on *Chlorella* mutants with respect to sulphur assimilation (Schiff and Hodson, 1973; Schmidt *et al.*, 1974) has not been reported. Such information is all the more desirable in view of the absence of any reports of free selenocysteine in plants and its presumed central role in selenium assimilation.

C. THE PRESUMED METABOLISM OF SELENOCYSTEINE

In general, most of the inorganic selenium incorporated by selenium accumulators, occurs as selenocystathionine and/or Se-methylselenocysteine. In non-accumulators most of the selenium is incorporated into protein where it is associated with selenomethionyl residues or is found as free selenomethionine or its Se-methylated derivative. Assuming that these compounds are derived from selenocysteine by mechanisms similar to those shown for sulphur (Fig. 3), this suggests that accumulators and non-accumulators differ in their pattern of metabolism of selenocysteine.

Several lines of evidence suggest that those accumulators which contain Se-methylselenocysteine, synthesize this compound from selenocysteine in a manner analogous to the synthesis of S-methylcysteine from cysteine. Most species which contain Se-methylselenocysteine also contain S-methylcysteine which is synthesized by methylation of cysteine (Thompson, 1967). The reciprocal relationship between the synthesis of the two compounds under varying conditions of sulphate/selenate nutrition (Chow *et al.*, 1971) suggests that they are synthesized by a common pathway. The studies of Chow *et al.* (1972) and Chen *et al.* (1970) support this conclusion. When [U-^{14}C]cysteine was supplied to *Astragalus bisulcatus*, relatively little label was incorporated into Se-methylselenocysteine; the ratio of the specific radioactivity of the Se:S isologues was in the range 0·05–0·1. However, when [U-^{14}C]serine, a precursor of OAS was supplied, relatively more ^{14}C-label was incorporated into Se-methylselenocysteine than S-methylcysteine; the ratio of the specific radioactivities for the Se:S isologues was between 5 and 10. Notwithstanding the limitations of using specific radioactivities of the compounds rather than the rate of synthesis of the two products, these data are consistent with the following sequence:

$$\text{Serine} \longrightarrow OAS \underset{\text{Selenide}}{\overset{\text{Sulphide}}{\Large\langle}} \overset{\displaystyle R\text{–}CH_3}{\underset{\displaystyle R\text{–}CH_3}{\begin{array}{l} \text{Cysteine} \xrightarrow{} \text{S-methylcysteine} \\ \text{Selenocysteine} \xrightarrow{} \text{Se-methylselenocysteine} \end{array}}} \qquad (10)$$

The synthesis of S-methylselenocysteine has been examined by other approaches. Chow *et al.* (1972) supplied [Me-^{14}C]methionine and [Me-^{14}C]S-adenosylmethionine to *A. bisulcatus* to trace the origin of the methyl group. Both substrates caused labelling of S-methylcysteine and Se-methylselenocysteine suggesting that the S/Se-methyl groups of the products arose by transfer of a C_1 unit to simple unmethylated precursors such as cysteine and selenocysteine.

One other point of interest is that some species, which are regarded as non-accumulators since they are not indigenous to seleniferous soils, readily synthesize S-methylcysteine. Nigam and McConnell (1973) reported that *Phaseolus lunatus*, a plant of this type, readily incorporates selenate into Se-methylselenocysteine. This raises the question whether these plants, like the selenium accumulators, are relatively insensitive to selenium toxicity. Whatever the answer to this question, it appears that plants capable of synthesizing S-methylcysteine, be they accumulators or non-accumulators, are capable of synthesizing the selenium isologue and that the formation of the two products are competing processes. In summary, it seems likely that the products arise by methylation and that the relative amounts formed depend on the relative amounts of sulphur and selenium available for metabolism.

The high concentrations of selenocystathionine in several accumulator species are presumed to arise through the action of cystathionine γ-synthase using selenocysteine as substrate in lieu of cysteine (Fig. 3). The relative absence of selenomethionine and its derivatives is surprising, for these could in theory be formed from selenocystathionine by the transulphurylation pathway (Fig. 3). One possibility which we are currently pursuing concerns the specificity of the reaction catalysed by β-cystathionase

$$\text{Cystathionine} \rightarrow \text{homocysteine} + \text{pyruvate} + NH_3 \qquad (11)$$

towards the selenium analogue, selenocystathionine. The enzyme from accumulator species could be specific for cystathionine thus excluding selenium from methionine metabolism, causing selenocystathione to accumulate. Conversely, the β-cystathionase of non-accumulators could be non-specific towards the selenium isologue, thus permitting the synthesis of selenomethionine and its eventual incorporation into protein. To date, we have achieved the chemical synthesis of L-selenocystathionine by the method of Zdansky (1968) and established that it is cleaved by purified β-cystathionase from spinach. However, we have yet to establish whether it serves as a substrate for the enzymes from accumulator species. Nevertheless, other mechanisms for excluding selenocystathionine from methionine metabolism in accumulators are possible. One possibility is that selenocystathionine is accumulated in a non-metabolic compartment(s) of the cell. Another is that accumulators lack β-cystathionase activity thus blocking the transulphurylation pathway; homocysteine and methionine could be synthesized by an alternative (sulphur specific?) pathway (e.g. direct sulphurylation?).

Very little is known about the presumed methylation reactions leading to the synthesis of selenomethionine and Se-methylselenomethionine from selenohomocysteine. Selenite and selenate are incorporated into these compounds in non-accumulator species of *Astragalus* and in the roots of pasture plants (Virupaksha and Shrift, 1965; Peterson and Butler, 1962). Non-accumulator species of *Astragalus* also incorporate [^{35}S]sulphate into S-methylmethionine (Virupaksha and Shrift, 1965) suggesting that sulphur and selenium are metabolized via similar mechanisms. It is probable that Se-methylselenomethionine is formed by methylation of selenomethionine since label from [Me-^{14}C]selenomethionine and [^{75}Se]selenomethionine is incorporated into Se-methylselenomethionine (Virupakash *et al.*, 1966).

D. INCORPORATION OF SELENIUM INTO PROTEIN

As noted previously, inorganic selenium is incorporated into the protein of non-accumulators where it occurs as selenomethionine, and, at least in *Vigna radiata*, selenocysteine. On the other hand only trace amounts of selenium are incorporated into the protein of the accumulator *Neptunia amplexicaulis* (Peterson and Butler, 1962). One explanation could be that accumulators possess mechanisms which select the sulphur amino acids for incorporation into protein but not their selenium isologues. Pursuing this line of enquiry, Shrift *et al.* (1976) and Burnell and Shrift (1977, 1978) investigated the substrate specificity of cysteinyl t-RNA synthetase, the enzyme catalysing the first step in the incorporation of cysteine into protein, using the enzymes from various accumulator and non-accumulator plants. Activity was measured by amino acid-dependent [^{32}P]PP$_i$-ATP exchange:

$$\text{Cysteine} + \text{ATP}^* \rightleftharpoons \text{Cysteinyl-AMP} + \text{PP}_i^* \qquad (12)$$

The enzymes from *Phaseolus aureus* (now *Vigna radiata*) and *A. lentigino-*

sus catalysed both cysteine- and selenocysteine-dependent PP_i-ATP exchange; the V_{max} and K_m of the enzymes for selenocysteine were only slightly less than for cysteine (Table VII). Thus, provided that the later stages of amino acid incorporation do not distinguish between the two substrates, these results demonstrate that selenocysteine competes with cysteine for incorporation into protein. The enzyme from *A. bisulcatus* (accumulator), on the other hand, did not catalyse selenocysteine-dependent PP_i-ATP exchange thus demonstrating that selenocysteine cannot enter protein in this species. However, the enzymes from *A. crotalariae* and *A. racemosus* catalysed both selenocysteine- and cysteine-dependent PP_i-ATP exchange and the kinetics of these reactions were similar to those for the enzymes from the non-accumulators (Table VII).

TABLE VII

Some properties of the cysteinyl t-RNA synthetases from accumulator and non-accumulator plants as measured by amino acid-dependent PP_i-ATP exchange[a]

Species	K_m (μM)		V_{max} (selenocysteine) / V_{max} (cysteine)
	Cysteine	Selenocysteine	
Non-accumulators			
Phaseolus aureus	62	50	0·75
Astragalus lentiginosus	60	50	0·90
Accumulators			
A. bisulcatus	170	b	b
A. racemosus	60	50	0·80
A. crotalariae	43	63	0·90

[a] From Burnell and Shrift (1977, 1979).
[b] Synthetase from *A. bisulcatus* did not catalyse selenocysteine-dependent PP_i-ATP exchange.

Thus some other mechanism must exist in *A. crotalariae* and *A. racemosus* for preventing the incorporation of selenocysteine into protein in these species. By analogy with other studies (Peterson, 1967) it is possible that cysteinyl t-RNA might fail to accept the selenocysteinyl adenylate synthesized by the cysteinyl t-RNA synthetase.

Analogous studies with the methionyl t-RNA synthetases from accumulator and non-accumulator plants have not been reported but studies with the enzyme from various non-accumulator organisms have shown that selenomethionine will serve as a substrate (Hahn and Brown, 1967; Hoffman *et al.*, 1970).

The metabolism of [^{75}Se]selenomethionine (mixed with [Me-^{14}C] selenomethionine in some cases) has been examined in some accumulator and non-accumulator species of *Astragalus* (Virupaksha *et al.*, 1966; Chow *et al.*, 1972). This substrate would seem an unlikely source for the synthesis of ^{75}Se-labelled isologues of intermediates in earlier parts of the pathway shown in Fig. 3 since the pathway is considered to be irreversible (Thompson, 1967; Giovanelli *et al.*, 1980). Nevertheless, in *A. crotalariae* and *A. bisulcatus* (accumulators), ^{75}Se-label was associated with one or both of Se-methylselenocystine and selenohomocystine, although in *A. lentiginosus* (non-accumulator) ^{75}Se-label was associated with Se-methylselenomethionine and a selenopeptide (Virupaksha *et al.*, 1966; Chow *et al.*, 1972). This raises the question whether accumulators (but not non-accumulators) have a mechanism(s) for the catabolism of seleno-methionine analogous to that for methionine catabolism in animals.

VII. Is Selenium Essential for Plant Growth?

A. SELENIUM REQUIREMENTS OF ORGANISMS OTHER THAN PLANTS

The essential nutritional requirements of the rat, birds, and various livestock for selenium has been reviewed by Stadtman (1974, 1979). In summary, selenium is required in only trace amounts and can be supplied in a variety of inorganic or organic forms. In animals, four gram atoms of selenium are associated with each mol of the enzyme glutathione peroxi-dase which is comprised of four subunits and catalyses the reactions:

$$2GSH + H_2O_2 \rightarrow GSSG + 2H_2O \tag{13}$$

$$2GSH + R-OOH \rightarrow GSSG + H_2O + ROH \tag{14}$$

Since conditions such as necrosis of the liver and sensitivity of erythrocyte membranes to oxidative damage (haemolysis) are associated with selenium deficiency in animals, it is assumed that the reduction of lipid peroxides by glutathione peroxidase (Eq. 14) forms the basis of essentiality. Active glutathione peroxidase is undetectable in the erythrocytes of selenium-deficient organisms but increases with the administration of selenium and as selenium becomes associated with the enzyme. It is therefore apparent that selenium, which is present as selenocysteine (Forstrom *et al.*, 1978), fulfils an essential role in the catalytic function of glutathione peroxidase and that the requirement for selenium is specific, i.e. it cannot be replaced by the more abundant element, sulphur. This is supported by the observed valence state changes of the selenium atom during enzyme functioning (Ladenstein and Wendel, 1976; Ladenstein et al., 1979).

Selenium is also required for the expression of several enzymes in some anaerobically grown bacteria. They include glycine reductase from several species of *Clostridium* which catalyses the reaction

$$\text{Glycine} + \text{ADP} + \text{P}_i + \text{R} \overset{-\text{SH}}{\underset{\text{SH}}{\diagup}} \rightarrow \text{acetate} + \text{ATP} + \text{NH}_3 + \text{R} \overset{-\text{S}}{\underset{\text{S}}{\diagup}} \quad (15)$$

and formate dehydrogenase from *Escherichia coli, Methanococcus vannielii* and several species of *Clostridium*. Three other selenium-containing enzymes are also known in Clostridia (Stadtman, 1980).

B. THE FUNCTION OF SELENIUM IN SELENIUM-SPECIFIC PROTEINS

Glutathione peroxidase has been purified to homogeneity and its three-dimensional structure determined by X-ray analysis. The enzymes from the bacterial sources have been studied less extensively presumably because of their extreme sensitivity to oxygen. Nevertheless, it seems that the selenium-containing enzymes studied to date undergo redox changes of the selenium atom during enzyme catalysis.

In rat liver glutathione peroxidase, the selenium has been derivatized, isolated, and identified as a selenocysteine moiety (Forstrom et al., 1978; Forstrom and Tappel, 1979). On the other hand X-ray studies of the bovine erythrocyte enzyme led to the conclusion that selenium undergoes reversible oxidation from selenol via a selenenic acid to a seleninic acid. The inability of the oxidized or reduced enzyme to bind mercury was evidence against a selenol moiety but was compatible with the selenenic/seleninic acid transition (Ladenstein et al., 1979):

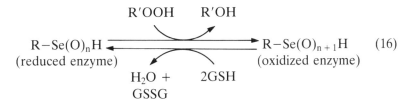

where $n = 0$, selenol; $n = 1$, selenenic acid; $n = 2$, seleninic acid.
Further evidence that selenocysteine may not be the active form of selenium in glutathione peroxidase is the observation that free selenocysteine was a poor source of selenium for incorporation into the protein (Sunde and Hoekstra, 1980). However, the identification of selenocysteine residues in the enzyme and their ability to undergo a valence change to a

derivative other than one containing oxygen means that with a peroxide substrate, more alternatives are possible than selenenic and seleninic acids (Stadtman, 1980).

The mode of action of clostridial glycine reductase is thought to parallel that of *E. coli* ribonucleoside diphosphate reductase which involves the interaction of redox-active disulphides with iron and a tyrosinyl free radical. For glycine reductase on the other hand, Stadtman (1979) has proposed that the two redox thiol groups and the single selenol group (Cone *et al.*, 1976, 1977) react with the iron which is known to be associated with the enzyme. The inhibition of the enzyme by hydroxylamine, an inhibitor of free radical enzymes, is consistent with this proposal. It has also been suggested that one of the redox-active thiol or selenol groups ($R(XH)_2$) may participate directly in the phosphate esterification reaction catalysed by glycine reductase. The report of a phosphate group in ester linkage with a cysteinyl sulphur in thioredoxin from *E. coli* (Pigiet and Conley, 1978) is consistent with this proposal.

C. STUDIES OF SELENIUM ESSENTIALITY IN PLANTS

The most direct and definitive method of determining whether selenium is required for plant growth is to determine whether omitting selenium from the nutrient solution impairs growth. Broyer *et al.* (1966) applied this technique to the non-accumulators *Medicago sativa* and *Trifolium subterraneum*. The growth of plants supplied with 0·025 to 2·5 μM selenite was not distinguishable from control plants grown without added selenium; a concentration of 25 μM inhibited growth. However, some selenium was present as a contaminant in the materials used for the control cultures (e.g. water, nutrient salts, seeds etc.). This amounted to 50 ng of Se per culture but at the conclusion of the experiment, Broyer *et al.* (1966) found that the control plants contained 1·5 μg of Se per culture. They suggested that the control plants acquired selenium from the air. Their experiments therefore do not rule out the possibility that selenium is essential for plant growth. If selenium is essential, their results imply that under suitable conditions, gaseous forms of selenium could be suitable sources.

The effect of inorganic selenium on the growth of *Astragalus* species is more controversial. Trelease and Trelease (1939) reported that selenium enhanced the growth of *Astragalus racemosus* (accumulator) and inhibited the growth of *A. succulentus* (non-accumulator). These experiments were conducted in the presence of 1 to 2 mM phosphate. Broyer *et al.* (1972a, b) repeated the experiments of Trelease and Trelease (1939) using accumulator and non-accumulator species of *Astragalus* grown in 2 to 10 μM phosphate. With one exception, Broyer *et al.* (1972a) found no evidence for selenium-enhanced growth of accumulators. They concluded that the contrary observations described by Trelease and Trelease (1939) were due to selenium alleviation of phosphate toxicity. Here, the results shown in Table III on the effect of selenate on the growth of *Trifolium repens* and *Neptunia amplexicaulis* at 0·1 mM phosphate are of interest; if the level of phosphate employed was toxic to *N. amplexicaulis*, then it is evident that adding selenate equivalent to only 8% of the concentration of phosphate has a marked effect on growth at most sulphate concentrations. Examples of phosphorus/selenium interactions in other organisms are discussed by Shrift (1969).

An alternative technique for investigating the essentiality of selenium is to determine whether any of the proteins reported to contain selenium in other organisms are present in plants. It would be necessary to establish also that the selenium cannot be replaced by sulphur and that the protein(s) do not function without selenium. In animals, glutathione peroxidase has a similar distribution to superoxide dismutase (Stadtman, 1979). Since illuminated chloroplasts photoreduce O_2 to superoxide (O_2^-) which is dismutated to H_2O_2 through the action of chloroplast superoxide dismutase (Asada *et al.*, 1973, 1974) or reduced by Fd_{red} (Allen, 1977), then the association of glutathione peroxidase with chloroplasts forms an attractive hypothesis, especially since chloroplasts exhibit light-coupled glutathione reductase activity (Jablonski and Anderson, 1978) for the regeneration of GSH:

$$2O_2^- + 2H^+ \rightarrow H_2O_2 + O_2 \tag{17}$$

$$H_2O_2 + 2GSH \rightarrow 2H_2O + GSSG \tag{18}$$

$$GSSG + H_2O \xrightarrow{h\nu} 2GSH + \tfrac{1}{2}O_2 \tag{19}$$

Glutathione peroxidase activity has also been postulated as a component process in the dark deactivation of light-modulated enzymes in chloroplasts (Wolosiuk and Buchanan, 1977).

Flohe and Menzel (1971) and Neubert *et al.* (1962a, b) have reported some tentative and indirect evidence for the occurrence of glutathione peroxidase activity in plants. However, Smith and Shrift (1979) found no evidence of this enzyme in a wide range of plants. In our laboratory P. P. Jablonski (unpublished results) has found that crude extracts of pea shoots catalyse H_2O_2-dependent oxidation of GSH but that a factor which is removed by gel filtration on Sephadex G-25 is essential for the expression of this activity. This factor was shown to be a flavonoid. The reaction cannot therefore be attributed to glutathione peroxidase activity. As regards the H_2O_2 produced in chloroplasts (Eq. 17), an alternative mechanism for its reduction (detoxification) is that proposed by Halliwell and Foyer (1978) involving ascorbate as reductant. This scheme has received support from reports that the enzymes ascorbate peroxidase, glutathione dehydrogenase, and glutathione reductase are associated with chloroplasts and form a coupled system for light-dependent reduction of H_2O_2 (Groden and Beck, 1979; Jablonski and Anderson, 1978, 1981, unpublished results).

It is clear that a search for specific selenium-containing proteins in plants on the basis of known selenoproteins in other organisms has been unrewarding to date. In continuing the search it is evident that if selenium is required by plants it is needed in minute amounts since plant material which causes selenium deficiency in stock (e.g. *Trifolium repens*) contains as little as 6–10 ng of Se g^{-1} dry wt. (Watkinson, 1966). It would therefore seem appropriate to supply selenium to culture solutions at very low concentrations. This would also diminish the likelihood of non-specific selenium incorporation through replacement of sulphur. It will also be necessary to examine a large number of the individual proteins present in crude extracts for selenium by high resolution techniques such as electrophoresis, isoelectric focusing, etc. Another possibility is that selenium could comprise a specific constituent of a component nucleotide of a species of t-RNA (Hoffman and McConnell, 1974; Young and Kaiser, 1979). Should a selenium-containing protein or t-RNA be found in plants in which the selenium cannot be replaced by sulphur, then the exciting task of finding the function of the protein and the function of the selenium atom(s) can begin. Here, the report of selenocysteinyl residues in the protein fraction of *Vigna radiata* by Brown and Shrift (1980) gives some cause for encouragement, for although it is likely that the selenocysteinyl residues represent non-specific replacement of cysteine, the selenium present in the selenium-specific proteins examined to date occurs in this form (Cone *et al.*, 1976; Forstrom *et al.*, 1978; Jones *et al.*, 1979).

ACKNOWLEDGEMENTS

We wish to express our gratitude to P. P. Jablonski for his comments and for allowing us to refer to his data.

REFERENCES

Agricultural Research Council (1965). *The Nutrient Requirements of Farm Livestock. No. 2, Ruminants* ARC, London.
Allen, J. F. (1977). *Cur. Adv. Plant Sci.* **9,** 459–469.
Anderson, J. W. (1980). *In* "The Biochemistry of Plants", Vol. 5, pp. 203–223, (B. J. Miflin, ed.), Academic Press, New York and London.

Anderson, J. W. (1981). *In* "The Biochemistry of Plants", Vol. 8, pp. 473–500, (M. D. Hatch and N. K. Boardman, eds.), Academic Press, New York and London.

Asada, K., Kiso, K., and Yoshikawa, K. (1974). *J. Biol. Chem.* **249**, 2175–2181.

Asada, K., Urano, M., and Takahashi, M. (1973). *Eur. J. Biochem.* **36**, 257–266.

Benson, J. V. and Patterson, J. A. (1969). *Anal. Biochem.* **29**, 130–135.

Brown, T. A. and Shrift, A. (1980). *Plant Physiol.* **66**, 758–761.

Broyer, T. C., Johnson, C. M., and Huston, R. P. (1972a). *Plant Soil* **36**, 635–649.

Broyer, T. C., Johnson, C. M., and Huston, R. P. (1972b). *Plant Soil* **36**, 651–669.

Broyer, T. C., Lee, D. C., and Asher, C. J. (1966). *Plant Physiol.* **41**, 1425–1428.

Burnell, J. N. (1981). *Plant Physiol.* **67**, 316–324.

Burnell, J. N. and Anderson, J. W. (1973). *Biochem. J.* **134**, 565–579.

Burnell, J. N. and Shrift, A. (1977). *Plant Physiol.* **60**, 670–674.

Burnell, J. N. and Shrift, A. (1979). *Plant Physiol.* **63**, 1095–1097.

Butler, G. W. and Peterson, P. J. (1967). *Aust. J. Biol. Sci.* **20**, 77–86.

Chen, D. N., Nigam, S. N., and McConnell, W. B. (1970). *Can. J. Biochem.* **48**, 1278–1283.

Chow, C. M., Nigam, S. N., and McConnell, W. B. (1971). *Phytochemistry* **10**, 2693–2698.

Chow, C. M., Nigam, S. N., and McConnell, W. B. (1972). *Biochim. Biophys. Acta* **273**, 91–96.

Coch, E. H. and Greene, R. C. (1971). *Biochim. Biophys. Acta* **230**, 223–236.

Cone, J. E., Martin del Rio, R., Davis, J. N. and Stadtman, T. C. (1976). *Proc. Natl. Acad. Sci. U.S.A.* **73**, 2659–2663.

Cone, J. E., Martin del Rio, R., and Stadtman, T. C. (1977). *J. Biol. Chem.* **252**, 5337–5344.

Dilworth, G. L. and Bandurski, R. S. (1977). *Biochem. J.* **163**, 521–529.

Fankhauser, H., Brunold, C., and Erismann, K. H. (1976). *Experientia* **32**, 1494–1497.

Farley, J. R., Cryns, D. F., Yang, Y. H. J., and Segel, I. H. (1976). *J. Biol. Chem.* **251**, 4389–4397.

Ferrari, G. and Renosto, F. (1972). *Plant Physiol.* **49**, 114–116.

Flohe, L. and Menzel, H. (1971). *Plant Cell Physiol.* **12**, 325–333.

Forstrom, J. W. and Tappel, A. L. (1979). *J. Biol. Chem.* **254**, 2888–2891.

Forstrom, J. W., Zakowski, J. J. and Tappel, A. L. (1978). *Biochemistry* **17**, 2639–2644.

Foyer, C. H. and Halliwell, B. (1976). *Planta* **133**, 21–25.

Ganther, H. E. (1971). *Biochemistry*, **10**, 4089–4098.

Giovanelli, J., Mudd, S. H. and Datko, A. H. (1980). *In* "The Biochemistry of Plants", Vol. 5, pp. 453–505, (B. J. Miflin, ed.), Academic Press, New York and London.

Groden, D. and Beck, E. (1979). *Biochim. Biophys. Acta* **546**, 426–435.

Hahn, G. A. and Brown, J. W. (1967). *Biochim. Biophys. Acta* **146**, 264–271.

Halliwell, B. and Foyer, C. H. (1978). *Planta* **139**, 9–17.

Hartley, W. J. and Grant, A. B. (1961). *Fed. Proc.* **20**, 679–688.

Hoffman, J. L. and McConnell, K. P. (1974). *Biochim. Biophys. Acta* **336**, 109–113.

Hoffman, J. L., McConnell, K. P., and Carpenter, D. R. (1970). *Biochim. Biophys. Acta* **199**, 531–534.

Hsieh, H. S. and Ganther, H. E. (1975). *Biochemistry* **14**, 1632–1636.

Huber, R. E. and Criddle, R. S. (1967a). *Arch. Biochem. Biophys.* **122**, 164–173.

Huber, R. E. and Criddle, R. S. (1967b). *Biochim. Biophys. Acta* **141**, 587–599.

Jablonski, P. P. and Anderson, J. W. (1978). *Plant Physiol.* **61**, 221–225.

Jablonski, P. P. and Anderson, J. W. (1981). *Plant Physiol.* **67**, 1239–1244.

Jones, J. B., Dilworth, G. L. and Stadtman, T. C. (1979) *Arch. Biochem. Biophys.* **195**, 255–260.

Kharasch, N. and Arora, A. S. (1976). *Phosphorus and Sulfur* **2**, 1–50.

Klayman, D. L. (1965). *J. Org. Chem.* **30**, 2454–2456.

Klayman, D. L. (1973). *In* "Organic Selenium Compounds: Their Chemistry and Biology", pp. 67–171, (D. L. Klayman and W. H. H. Gunther, eds.), John Wiley, New York.

Ladenstein, R. and Wendel, A. (1976). *J. Mol. Biol.* **104**, 877–882.

Ladenstein, R., Epp, O., Bartels, K., Jones, A., Huber, R., and Wendel, A. (1979). *J. Mol. Biol.* **134**, 199–239.

Leggett, J. E. and Epstein, E. (1956). *Plant Physiol.* **31**, 222–226.

Martin, J. L. and Gerlach, M. L. (1969). *Anal. Biochem.* **29**, 257–264.

Martin, J. L., Shrift, A. and Gerlach, M. L. (1971). *Phytochemistry* **10**, 945–952.

McConnell, K. P. and Wabnitz, C. H. (1964). *Biochim. Biophys. Acta* **86**, 182–185.

Neubert, D., Rose, T. H., and Lehninger, A. L. (1962a). *J. Biol. Chem.* **237**, 2025–2031.

Neubert, D., Wojtczak, A. B., and Lehninger, A. L. (1962b). *Proc. Natl. Acad. Sci. U.S.A.* **48**, 1651–1658.

Ng, B. H. and Anderson, J. W. (1978a). *Phytochemistry* **17**, 879–885.

Ng, B. H. and Anderson, J. W. (1978b). *Phytochemistry* **17**, 2069–2074.

Ng, B. H. and Anderson, J. W. (1979). *Phytochemistry* **18**, 573–580.

Nigam, S. N. and McConnell, W. B. (1973). *Phytochemistry* **12**, 359–362.

Nigam, S. N. and McConnell, W. B. (1976). *J. Exp. Bot.* **27**, 565–571.

Olson, O. E., Novacek, E. J., Whitehead, E. I., and Palmer, I. S. (1970). *Phytochemistry* **9**, 1181–1188.

Parsons, R. F. (1976). *Am. Mid. Nat.* **96**, 1–20.

Paynter, D. I., Anderson, J. W., and McDonald, J. W. (1979). *Aust. J. Agric. Res.* **30**, 703–709.

Peterson, P. J. (1967). *Biol. Rev.* **42**, 552–613.

Peterson, P. J. and Butler, G. W. (1962). *Aust. J. Biol. Sci.* **15**, 126–146.

Peterson, P. J. and Butler, G. W. (1967). *Nature* **213**, 599–600.

Peterson, P. J. and Robinson, P. J. (1972). *Phytochemistry* **11**, 1837–1839.

Pigiet, V. and Conley, R. R. (1978). *J. Biol. Chem.* **253**, 1910–1920.

Pinsent, J. (1954). *Biochem. J.* **57**, 10–16.

Reuveny, Z. (1977). *Proc. Natl. Acad. Sci. U.S.A.* **74**, 619–624.

Rosenfeld, I. and Beath, O. A. (1964). "Selenium: Geobotany, Biochemistry, Toxicity and Nutrition", Academic Press, New York and London.

Roy, A. B. and Trudinger, P. A. (1970). "The Biochemistry of Inorganic Compounds of Sulphur", Cambridge University Press, London and New York.

Scarf, A. R., Cole, E. R., and Southwell-Keely, P. T. (1977). *Phosphorus and Sulfur* **3**, 285–291.

Scarf, A. R., Cole, E. R., and Southwell-Keely, P. T. (1979). *Aust. J. Pharm. Sci.* **8**, 125–127.

Schaedle, M. and Bassham, J. A. (1977). *Plant Physiol.* **59**, 1011–1012.

Schiff, J. A. and Hodson, R. C. (1973). *Annu. Rev. Plant Physiol.* **24**, 381–414.

Schmidt, A. (1973). *Arch. Mikrobiol.* **93**, 29–52.

Schmidt, A. (1975). *Planta* **124**, 267–275.

Schmidt, A., Abrams, W. R., and Schiff, J. A. (1974). *Eur. J. Biochem.* **47**, 423–434.

Schwarz, K. and Foltz, C. M. (1957). *J. Am. Chem. Soc.* **79**, 3292–3293.
Schwarz, K. and Sweeney, E. (1964). *Fed. Proc.* **23**, 421.
Shaw, W. H. and Anderson, J. W. (1972). *Biochem. J.* **127**, 237–247.
Shaw, W. H. and Anderson, J. W. (1974a). *Biochem. J.* **139**, 27–35.
Shaw, W. H. and Anderson, J. W. (1974b). *Biochem. J.* **139**, 37–42.
Shepherd, L. and Huber, R. E. (1969). *Can. J. Biochem.* **47**, 877–881.
Shrift, A. (1954a). *Am. J. Bot.* **41**, 223–230.
Shrift, A. (1954b). *Am. J. Bot.* **41**, 345–352.
Shrift, A. (1969). *Annu. Rev. Plant Physiol.* **20**, 475–495.
Shrift, A. (1973). *In* "Organic Selenium Compounds: Their Chemistry and Biology", pp. 763–814, (D. L. Klayman and W. H. H. Gunther, eds.), John Wiley, New York.
Shrift, A. and Virupaksha, T. K. (1963). *Biochim. Biophys. Acta* **71**, 483–485.
Shrift, A. and Virupaksha, T. K. (1965). *Biochim. Biophys. Acta* **100**, 65–75.
Shrift, A., Bechard, D., Harcup, C., and Fowden, L. (1976). *Plant Physiol.* **58**, 248–252.
Smith, I. K. (1976). *Plant Physiol.* **58**, 358–362.
Smith, J. and Shrift, A. (1979). *Comp. Biochem. Physiol.* **63B**, 39–44.
Spare, C. G. and Virtanen, A. I. (1964). *Acta Chem. Scand.* **18**, 280–282.
Stadtman, T. C. (1974). *Science* **183**, 915–922.
Stadtman, T. C. (1979). *Adv. Enzymol.* **48**, 1–28.
Stadtman, T. C. (1980). *Annu. Rev. Biochem.* **49**, 93–110.
Stewart, J. M., Nigam, S. N., and McConnell, W. B. (1974). *Can. J. Biochem.* **52**, 144–145.
Sunde, R. A. and Hoekstra, W. G. (1980). *Biochem. Biophys. Res. Commun.* **93**, 1181–1188.
Thompson, J. F. (1967). *Annu. Rev. Plant Physiol.* **18**, 59–84.
Trelease, S. F. and Trelease, H. M. (1939). *Amer. J. Bot.* **26**, 530–535.
Tsang, M. L-S, and Schiff, J. A. (1978). *Plant Sci. Lett.* **11**, 177–183.
Underwood, E. J. (1971). "Trace Elements in Human and Animal Nutrition", Academic Press, New York and London.
Virupaksha, T. K. and Shrift, A. (1963). *Biochim. Biophys. Acta* **74**, 791–793.
Virupaksha, T. K. and Shrift, A. (1965). *Biochim. Biophys. Acta* **107**, 69–80.
Virupaksha, T. K., Shrift, A., and Tarver, H. (1966). *Biochim. Biophys. Acta* **130**, 45–55.
Walter, R., Schlesinger, D. H., and Schwartz, I. L. (1969). *Anal. Biochem.* **27**, 231–243.
Watkinson, J. H. (1966). *Anal. Chem.* **38**, 92–97.
Wilson, L. G. and Bandurski, R. S. (1958). *J. Biol. Chem.* **233**, 975–981.
Wolosiuk, R. A. and Buchanan, B. B. (1977). *Nature* **266**, 565–567.
Young, P. A. and Kaiser, I. I. (1979). *Plant Physiol.* **63**, 511–517.
Zdansky, G. (1968). *Ark. Kemi* **29**, 449–453.
Ziebur, N. K. and Shrift, A. (1971). *Plant Physiol.* **47**, 545–550.

CHAPTER 14

A Perspective of Mineral Nutrition: Essential and Functional Metals in Plants

E. J. HEWITT

University of Bristol, Long Ashton Research Station, Bristol, UK

I. INTRODUCTION

About 40 years ago, when I first learned the distinction between trace elements and major elements as then understood, the list of known essential elements appeared "complete", well-documented, and subject to little argument (see Hewitt, 1952). However, some doubts were already appearing as to the importance of additional elements notably chlorine and silicon, and about the rigid application and relevance of the well-known

Criteria of Essentiality {Arnon and Stout, 1939) which had clarified much of the thought at the time. The term micronutrient element (Loomis and Shull, 1937; Arnon, 1950) was becoming widely accepted in place of trace element, minor element, etc. By 1966 evidence regarding the limited importance of sodium, vanadium, and silicon became stronger though still confusing. The role of cobalt and the circumstances of its requirement were defined, and nickel was suspected as possibly important (Roach and Barclay, 1946). New ideas had emerged suggesting that changing metabolic pathways, or species differences, could affect qualitative as well as quantitative mineral requirements (see Hewitt, 1958, 1966).

My intention now is to amplify some of these ideas about metabolic dependence and species requirements. I shall examine how far we can account for the utilization of some mineral micronutrients and try to quantify the present limits of search for evidence of essential or functional elements. I shall also discuss ways in which elements normally considered as antagonistic to corresponding essential nutrients may show beneficial effects on account of this relationship.

II. Accountable Proportions of Some Micronutrients

The principal micronutrients are generally present in, or required by plants in a fairly wide range of concentrations (see Chapman, 1966), but for many plants the respective limits for normal (sufficient) or deficient growth states, can be specified for a particular element in terms of typical values which are quite well established (see Hewitt, 1966, 1979; Chapman, 1966). The typical values for sufficient concentrations of different micronutrients span a wide range from not less than $200 \mu M$ (fresh cell volume) for iron, to possibly less than $3 \mu M$ for molybdenum, and still less ($0 \cdot 1 \mu M$) for cobalt, whilst manganese, zinc, and copper show intermediate values; representative values have been assigned in the tables below to help quantify functional proportions.

Little information seems to be available showing the proportions of the principal micronutrients which are identifiable in terms of known enzymes or which represent their abundance as atoms per cell. I have therefore attempted some tentative calculations of these quantities which are certainly subject to errors of magnitude due to large differences between sources, and may also be subject to some mistakes and guesses when several parameters are necessary to deduce the required values. Nevertheless, as will appear below, there are large and surprising differences between the extents to which different elements can be accounted for in specific known functional states, and these differences are clearly not all due to random choice of data.

In order to make some calculations I have had to use representative, experimentally determined or sometimes arbitrary parameters for cell volumes, protein contents, properties of organelles, enzyme activities or amounts of metallo-proteins. Some of the principal parameters are summarized below, and have been collated from diverse sources (Kirk and Tilney-Bassett, 1967; Opik, 1968; Bonner, 1976; Wilson, 1963; Agarwala, 1952; Hewitt, 1951).

Leaf water content: 90%
Leaf protein (cauliflower, tomato or spinach): 3–4% fr. wt. max.
Cell volume (tomato or brassica spp.): 4–$17 \times 10^4 \, \mu m^3$
Chloroplast protein: 70% of cell protein, 65–70% protein in dry wt.
Chlorophyll concentration (spinach, beet): 1·8 mM.
Mitochondria: 3–10 (characteristically 5)% of cell volume, 15% of cell protein, 700 per cell, volume 5–50 (characteristically 10) μm^3, 40% protein in dry wt.

In most cases the above values will tend to overestimate the calculated percentage recoveries of metals in specific proteins, and so emphasize in general the remarkable gaps between total leaf content and identified forms found for some of the elements considered below.

A. IRON

Three main forms of iron in proteins are considered in Tables I and II and here. Table I shows calculations for the recovery of iron in known

TABLE I

Proportions of leaf iron in some haem proteins

Protein	μM in cell	Fe atoms mol^{-1}	Fe concn in Cell μM	Fe as total of cell Fe (%)
Cytochrome f	2	1 or 2	2 or 4	1·0 or 2·0
Cytochrome b_6	4	1	4	2
Cytochrome b_{559}	5	1	5	2·5
Cytochrome $a + a_3$	1·0	2	2	1·0
Cytochrome c	0·5	1	0·5	0·25
Cytochrome b	0·5	1	0·5	0·25
Nitrate reductase	0·1	2	0·2	0·1
Nitrite reductase	0·1(?)	1	0·1(?)	0·05
Sulphite reductase	0·1(?)	1	0·1(?)	0·05
Catalase	0·2(?)	1	0·2(?)	0·1(?)
Peroxidase	0·5(?)	1	0·5(?)	0·25(?)
			Total about	8·5

The total iron content of leaf is taken as $200 \, \mu M$.
Data for mitochondrial complexes taken from Green and Silman (1967) and for chloroplast haem-proteins from Cramer and Whitmarsh (1977).

TABLE II

Proportions of leaf iron in some non-haem iron proteins

Protein	Protein concn in cell (μM)	Fe atoms mol^{-1}	Fe concn in cell (μM)	Fe as % of total in cell
Ferredoxin 2FeS	2·5	2	5	2·5
Thylakoid complexes:				
PSI 4FeS	2·5(?)	4	10(?)	5(?)
Rieske Centre 2Fe?	1(?)	2(?)	2(?)	1
Mitochondrial complexes:	1·0	18	18	9·0
2FeS and 4FeS	each	total		
Aconitase 3FeS	0·1(?)	3	0·3(?)	0·15(?)
Nitrite reductase 4FeS	0·1(?)	4	0·8(?)	0·4(?)
Sulphite reductase 4FeS	0·1(?)	4(?)	0·8(?)	0·4(?)
			Possible total	18·5%

Total iron concentration of fresh leaf is 200 μM.

haem-proteins and Table II shows values for non-haem iron proteins containing acid labile sulphur.

The total iron recovered in all these classes amounts to about 28% of that present in many plants, though total iron concentrations are quite commonly 50–100% greater than the 200 μM shown. It will be seen that non-haem iron proteins probably account for more than the haem proteins. Much of the remainder can be accounted for as ferritin in chloroplasts which are estimated to be about 35% of the total (see Seckbach, 1969 and H. F. Bienfait, this volume). Thus about 65% of total leaf iron can be related to known iron proteins in normal leaves. There is little corresponding information for deficiency conditions where chlorosis is visible, but total iron may be half of that found in normal leaves (Chapman, 1966).

B. ZINC

Two important zinc enzymes have been fairly accurately quantified, carbonic anhydrase in parsley (Tobin, 1969, 1970) and superoxide dismutase in spinach (Asada et al., 1973). These are probably by far the most abundant, making together about 17% of cell zinc (Table III) and accounting for 1·6% of total or >7% of extractable protein (Tobin, 1970; Kandel et al., 1978). The data for an RNA polymerase in wheat germ, based on results of Petranyi et al. (1977) and Jendrisak and Burgess (1975), are more tentative and may not reflect the amounts in leaf cells. Aspartate

TABLE III
Proportions of leaf zinc in some plant metallo-proteins

Protein	Enzyme concn in cell (μM)	Zn atoms mol^{-1}	Zn concn in cell μM	Enzyme-Zn as % of total
Superoxide dismutase (Spinach)	0·7 (0·6% of cell protein)	2	1·4	1·9
Carbonic anhydrase (Parsley)	1·8 (1% of cell protein)	6	11	14·5
RNA polymerase	0·2(?) (0·025% of "germ" protein)	8	1·6(?)	2(?)
Aspartate trans carbamylase	0·2(?)	6	1·2(?)	1·5(?)
			Total	19·9

Total leaf Zn is assumed to be 75 μM.

transcarbamylase is quite abundant in pea seedlings, and alcohol de-hydrogenase is abundant in maize. Probably 20 more additional zinc-dependent enzymes occur in plant tissues. If these others are present in $0 \cdot 1 \mu M$ concentrations and contain, on average, 4 zinc atoms per mole as might be expected for zinc proteins, then their zinc concentration might amount to 8 μM. In all, the known and predicted zinc proteins may thus account for about 30% of total cell zinc when this is 75 μM. In plants having as little as 40 μM total zinc, a large part of it (60%) might be accounted for by these expected forms in similar concentrations, and so the discrepancy is reasonable as zinc is bound also by other macromolecules and cell surfaces.

Our investigations on the effects of zinc supply at sufficient or deficient rates to tomato and radish plants under various environmental conditions (Wilson, 1963) provide some accurate estimates of zinc atoms per cell (Table IV), and are compared with the estimated values of zinc contained in superoxide dismutase and carbonic anhydrase.

The agreement between two alternative methods of calculations for two enzymes of known concentration is encouraging considering the uncertainties involved in different species and growth conditions. The moderate discrepancy between total and identifiable zinc is therefore probably realistic if not yet understood.

C. COPPER

The data for copper in Table V present quite a different picture. Here only two copper proteins, plastocyanin and superoxide dismutase, account for 60 to >90% of the total copper content of leaf cells. The values for superoxide dismutase are based on data of Asada et al. (1973). For plastocyanin our own estimations for vegetable marrow (C. pepo), spinach and barley (Scawen et al., 1975) generally agree with the values of Plesničar and Bendall (1970) for barley and other species. Thus there appears to be little surplus copper in leaf cells of many normal plants. Although chlorophyll content is not appreciably decreased, and often appears even to be temporally increased in concentration during onset of acute copper deficiency, our measurements of plastocyanin in copper-deficient spinach and barley (Fig. 1) and C. pepo (Table V), showed that the concentration on a chlorophyll basis was decreased about ten-fold whereas the concentration of total copper was decreased to about one-third that found in normal plants. There is now, therefore, quite a large discrepancy between total and identifiable functional copper, and it seems that other cell constituents compete for limited supplies of copper at the expense of the principal functional proteins. This observation explains the sensitivity of photosynthesis to impaired copper nutrition first noted in

TABLE IV
Zinc concentrations and distribution in leaf cells

		Cell vol. $\mu m^3 \times 10^{-3}$	Zn atoms cell^{-1} $\times 10^{-9}$	Zn in carbonic anhydrase and superoxide disinutase*	
				Atoms cell^{-1} (+Zn) $\times 10^{-9}$	% of Total
RADISH Cotyledon (25 days)	Short day +Zn	170	3·8		
	−Zn	50	1·8		
	Long day +Zn	150	4·0	1·1	28
	−Zn	120	0·3		
First leaf (30 days)	Short day +Zn	42	1·1	0·31	28
	−Zn	8	0·25		
	Long day +Zn	90	3·2	0·67	21
	−Zn	12	0·16		
TOMATO First leaf (29 days)	Long day +Zn	69	4·2	0·52	13
	−Zn	20	0·47		

* Notional atoms per cell and percentage of total Zn as for sum of maximal values of carbonic anhydrase and superoxide dismutase in Table III.

TABLE V

Proportions of copper in some plant metallo-proteins

Copper status of plant	Cu concentration in cell μM		Protein concn in cell (μM)	Cu atoms mol^{-1}	Cu concn in cell (μM)	Protein Cu as % of total Cu
NORMAL	10	Superoxide dismutase	0·7	2	1·4	14
		Plastocyanin	5–8	1	5–8	50–80
		Cytochrome $a + a_3$	1 (Table 1)	2	2	20
DEFICIENT (C. pepo)	3	Plastocyanin	<0·5–1	1	0·5–1	15–30

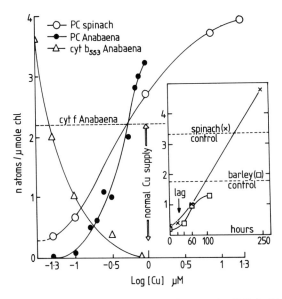

FIG. 1. Effects of copper supply on production of plastocyanin (PC) in higher plants and blue-green algae. *Main* figure shows effects of copper concentrations in nutrient media on plastocyanin content in spinach and *Anabaena* and the reciprocal changes in cytochrome b_{553}. *Inset* figure shows formation of plastocyanin with time after restoring copper supply to spinach or barley plants given smallest copper supply in main figure.

tung trees (Loustalot *et al.*, 1945). I suggest that measurements of plastocyanin activity or copper content of appropriately isolated chloroplasts might provide some of the most sensitive, simple and relevant means of identifying and quantifying copper status in crops.

Figure 1 compares the effects of copper supply on the formation of plastocyanin in two dissimilar plant types, a blue-green alga *Anabaena variabilis* studied by Sandmann and Böger (1980), and two higher plants, spinach and barley grown at Long Ashton (M. D. Scawen and E. J. Hewitt, unpublished data). The relationships on a chlorophyll basis are generally similar but *Anabaena* seems possibly more sensitive to omission of copper and requires less copper for maximal plastocyanin formation than the plants. The relationship is sigmoidal over several decades of copper concentration. Figure 1 shows also a curious effect of copper in repressing the formation of a haem protein, cytochrome b_{553}, in *Anabaena* chloroplasts, an effect found also in *Plectonema boryanum* (Sandmann and Böger, 1980). There was almost a reciprocal relationship between plastocyanin and cytochrome b_{533} formation but the most severe repression coincided with the first, scarcely detectable, increases in plastocyanin formation. The time course of plastocyanin formation in response to added copper is shown in Fig. 1 for spinach and barley. There is a lag period in the formation of recognizable amounts of the protein, lasting about 20–40 h after giving copper, followed by its synthesis to different extents by the two species. In barley, production attained that found in normal plants after about 200 h, and then continued at a rate decreasing to half the maximal value. In spinach, production exceeded the amount found in normal spinach plants after the same period and continued at a linear rate faster than in barley.

With plastocyanin from *Cucurbita pepo* (Scawen *et al.*, 1975) we were able quite easily to remove and replace copper from the protein *in vitro*. However, applying the reconstitution procedure to plastocyanin-type preparations from copper-deficient but non-necrotic leaves of marrow, failed to produce any copper protein. We therefore concluded that absence of copper prevents synthesis of apo-protein. This result contrasts with conclusions regarding the effects of molybdenum on nitrate-induced formation of apo-protein of nitrate reductase in spinach and in other plants. The specific location of plastocyanin as the single copper protein in chloroplasts, its sensitivity to copper supply and the sensitive procedures available to measure it in the metal-containing state, suggest that it may provide an ideal model for studying aspects of protein synthesis in chloroplasts. As several milligrams of homogeneous plastocyanin can be obtained, the production of monospecific antibodies, the isolation and localization of its RNA, and the identification of the sites of its synthesis should be feasible. These experiments can be related to environmental and also to pathological factors such as the effects of amphotericin B, a polyene antibiotic which causes dissociation of plastocyanin from the lipo-protein environment of the thylakoid membranes (Nolan and Bishop, 1974).

D. MOLYBDENUM

Three proteins containing molybdenum are known or reasonably inferred in non-leguminous plants. One, sulphite oxidase, is so far reported only from wheat germ embryo (Kessler and Rajagopalan, 1972), where it may be an important constituent present in amounts sufficient to allow its isolation in milligram quantities. Another, assumedly, is xanthine dehydrogenase. This can only be detected with radio-active assay methods, and so probably contributes negligibly to the identifiable total cell molybdenum (Nguyen and Feierabend, 1978). The third, nitrate reductase, is the principal molybdenum protein, and in leaves and most other vegetative tissues (Rijven, 1958), is certainly the most abundant. Our experience of isolating nitrate reductase from plants and measurement of its molecular weight, specific activity, initial enzyme activity, molybdenum content, and

TABLE VI
Proportions of molybdenum in some plant metallo proteins

Plants and Mo-status	Concn of Mo in cell (μM)	Protein	Concn of protein in cell (μM)	Mo atoms mol^{-1}	Concn of protein Mo in cell (μM)	Protein Mo as % of total Mo
NORMAL						
Spinach leaf	3·0	Nitrate reductase	0·1	1(?)	0·1	3·3
Cauliflower leaf	3·0	Nitrate reductase	0·1	1	0·1	3·3
Wheat embryo (50% d. wt.)	6·0	Nitrate reductase	0·02	1	0·02	0·33
DEFICIENT						
Spinach	0·3	Nitrate reductase	0·0003	1	0·0003	0·1
Cauliflower	0·03	Nitrate reductase	0·0003	1	0·0003	1·0
NORMAL						
Wheat embryo	6·0	Sulphite oxidase	0·3(?)	1(?)	0·3(?)	5·0(?)
Tobacco leaf	1·0	Xanthine dehydrogenase	0·002(?)	1(?)	0·002(?)	0·2(?)

yield from a known fresh weight, shows that in spinach the normal enzyme concentration is about $0.1\,\mu M$ assuming one atom Mo per mole. Spinach is among the more active producers of nitrate reductase but is exceeded by wheat, maize, and soybean to the extent of about 50%. Table VI shows that for a normal spinach leaf with a concentration of $3\,\mu M$ total molybdenum, nitrate reductase represents only about 3% of its total molybdenum. In severely molybdenum deficient plants the estimated enzyme concentration is only $0.0003\,\mu M$ while the total cell molybdenum may be between $0.3\,\mu M$ and $0.03\,\mu M$. In these plants nitrate reductase represents between 0.1% and 1% of total molybdenum, i.e. less than that in the normal plant. Thus recovery of molybdenum in identifiable proteins shows the largest discrepancy of all the micronutrients reviewed above and seldom exceeds 6%. The sulphite oxidase of rat liver again accounts for only about 5% of the total tissue molybdenum (Johnson *et al.*, 1977).

The necessary concentrations of molybdenum in cells of various microorganisms and plants and the probable values for deficiency conditions are shown in Table VII, based on data of Ichioka and Arnon (1955), Arnon *et al.* (1955), Arnon (1958), Agarwala and Hewitt (1954), C. S. Gundry and E. J. Hewitt (unpublished work), and on my calculations and notional values applied to these.

Cauliflower cells are larger than those of *Scenedesmus* or *Anabaena* and so more atoms per cell are necessary to produce similar concentrations, but requirements are clearly greater for plants on concentration or on atoms per cell bases, regardless of molybdenum supply.

The estimates of nitrate reductase molecules per cell show the same proportion of total Mo bound as reductase-Mo as in Table VI, but are not independent estimates (unlike those for zinc enzymes). They do, nevertheless, indicate the limits in cell requirements for amounts of nitrate reductase. For *Azotobacter*(s) with much smaller cells ($8\,\mu m^3$) where the optimum concentration for nitrogen fixation is equal to 10^4 atoms molybdenum per cell (Burk, 1934) the aqueous concentration would be possibly 700 times greater than for *Scenedesmus* growing with nitrate according to Arnon *et al.* (1955).

E. MANGANESE

Unlike the elements already considered, where probably all the specifically metal-dependent enzymes are metallo-proteins with small dissociation constants ($\ll 10^{-6}\,M$), manganese may activate several enzymes whose metal-dissociation constants are much larger, corresponding to K_m values between 10^{-4} and $10^{-5}\,M$; and in many cases the requirement is not specific, and can be replaced usually by magnesium. Therefore, an

TABLE VII

Estimated cellular concentrations of molybdenum in cells

Species	Source of N	cell vol μm^3	Growth response	Mo concn in medium (μM)	Estimate of Mo atoms cell^{-1}	Nitrate reductase molecules cell^{-1}
Scenedesmus obliquus	NO_3^-	$1800\ \mu m^3$	Severe deficiency	10^{-5}	$1 \cdot 6 - 2 \cdot 7 \times 10^3$	
			50% growth	10^{-4}	$1 \cdot 8 - 2 \cdot 2 \times 10^3$	
			Full growth	10^{-3}	$4 \cdot 2 \times 10^3 - 1 \cdot 4 \times 10^4$	
			Full growth		$<10^2(?)$	
Cauliflower	NH_4^+ *or* Urea NO_3^-	$20\ 000\ \mu m^3$	Severe deficiency	5×10^{-5}	$2 \cdot 5 \times 10^5$	$3 \cdot 6 \times 10^3$
		$50\ 000\ \mu m^3$	50% growth	5×10^{-4}	6×10^6	
		$125\ 000\ \mu m^3$	Full growth	10^{-2}	$2 \cdot 5 \times 10^8$	$7 \cdot 5 \times 10^6$
Azotobacter sp.	Nitrogen	$8\ \mu m^3$	Full growth	—	10^4	
Anabaena cylindrica	NO_3^- or Nitrogen	$100\ \mu m^3$	Severe deficiency	10^{-3}	2×10^5	
	Nitrogen		50% growth	10^{-2}	—	
			Full growth	10^{-1}	5×10^5	

important proportion of total cell manganese, where it is essential or needed to achieve balanced rates in sequentially linked systems, is involved in dissociable metal-activated systems requiring fairly large (0·1 to 1 mM) concentrations of ionic metal.

Only perhaps in chloroplasts is manganese present in a tightly bound state which amounts to about 6 atoms per 400 chlorophyll molecules (Cheniae, 1970). On the basis of 1·8 mM chlorophyll, the bound manganese is equivalent to 28 μM and so about 28% of average total cell concentration (100 μM). About two-thirds of the chloroplast-bound manganese is held in a less tightly combined state and seems to be most closely involved in oxygen evolution, whereas the smaller fraction may be more directly involved in thylakoid structure or stability (Weiland et al., 1975; Takahashi and Asada, 1977). Manganese-containing proteins which restore oxygen evolution to depleted chloroplasts or thylakoid preparations have been isolated (Lagoutte and Duranton, 1975; Spector and Winget, 1980) and contain two manganese atoms per 65 000 mol wt. (see Hipkins, this volume). A recently described superoxide dismutase may contribute about 0·8 μM manganese in a bound form (Sevilla et al., 1980). The proportion of unidentified manganese is much less than that for molybdenum or zinc, more than that for copper, and comparable to that for iron.

F. SUMMARY

The general conclusion seems to be that for all metal micronutients, in some circumstances there are substantial discrepancies between total and identifiable amounts. For molybdenum especially and also for zinc, the discrepancy is very large. Presumably either competing metal-binding factors, possibly proteins, cell walls, nucleic acids, or smaller chelating compounds are involved; or else quite large proportions of these metals are present in ionic states. This second possiblility seems unlikely, because acute deficiency effects clearly occur when total cell concentrations are reduced to amounts which very often are only about one-half or one-third of the normal amounts, and so still much in excess of the identified enzymic requirements in normal plants; moreover, in these conditions the amounts of metallo-enzymes are much reduced.

III. MOLYBDENUM–TUNGSTEN INTERACTIONS: PROPOSAL FOR COMPETITION BETWEEN STORAGE AND ENZYME METAL COMPLEXES

It was deduced that in bacteria a metal-binding component is present which accounts for at least 90% of total cell molybdenum, possibly at the expense of enzymic requirements (Elliott and Mortenson, 1975, 1976). Tungsten (W) is probably similarly sequestered and the kinetics of uptake are similar in *Clostridium pasteurianum* (Elliott and Mortenson, 1975, 1976). Storage proteins, which are able to hold more than 80% of cell tungsten and having molecular weights of about 88 000, have been found in *Clostridium formicoaceticum* (Leonhardt and Andreesen, 1977). Other proteins which may bind W and Mo are thought to occur in *Escherichia coli* and *C. thermoaceticum* (see Ljungdahl, 1980).

It has been found that giving 0·005 ppm "Specpure" W to nutrient media providing sub-optimal concentrations of Mo (0·00005–0·0001 ppm) is remarkably beneficial to growth and nitrate reductase activity of spinach. Somewhat larger concentrations (0·1–0·25 ppm W providing < 0·000003 ppm Mo) are similarly beneficial to cauliflower (see Notton, this volume and Hewitt, 1979). The first interpretation (Hewitt, 1979) was that W was able to induce or stabilize a cell receptor. Because Mo and W are competitive antagonists it was thought that the receptor would act as a scavenging agent for Mo, and facilitate its subsequent transfer to the apo-protein or apo-cofactor of nitrate reductase. This view has now been modified based on the notional proportions and properties of a postulated intracellular non-enzymic Mo-binding component (A) present in excess of apo-protein or apo-cofactor, and on some established parameters for metal and enzyme concentrations. Table VIII shows the notional values for the dissociation constant (K) in the expression [Mo] [A]/[MoA] where [Mo] is the concentration of Mo in the cell able to form cofactor in active nitrate reductase (0·1 μM), [A] is the concentration of the Mo-binding component and [MoA] is a monomolecular complex. The thermodynamic equilibria for such chelating systems where ligand competition is involved have been described in general terms by Da Silva (1978).

As nitrate reductase does not increase appreciably with large increases in Mo supply beyond 0·05 ppm, when total Mo is 33 times enzyme Mo, it has been assumed that total A is not less than total Mo (i.e. 3μM) or possibly more (e.g. 5μM) in normal plants. The computed values of K for these notional values of A are respectively $3\cdot45 \times 10^{-3}$ and $72\cdot4 \times 10^{-3}$. In Mo-deficient plants enzyme bound Mo was determined to be 0·0003 μM whereas total Mo was 0·3 μM (Table VI). Therefore, either the concentration of A may have changed or K has altered or both. Either change is conceivable and may have a precedent. Thus, formation of the binding component was repressed by metal uptake and de-repressed by metal deficiency in *C. pasteurianum* (Elliot and Mortenson, 1976).

Alternatively, if A is related to the cofactor molecule, as postulated below, it may exist in two states. The Mo cofactor which is regarded as common to several molybdoenzymes (see Johnson, 1980) is suspected to be a pterin derivative (Johnson *et al.*, 1980) which is "active" when reduced but "inactive" when autoxidized to the tautomeric quinone forms depicted in Fig. 2. These tautomers are readily interconverted and are reversibly reduced non-enzymatically by NADH or ascorbate (Benkovic, 1980). I suggest for the purpose of my hypothesis that chelation of Mo is

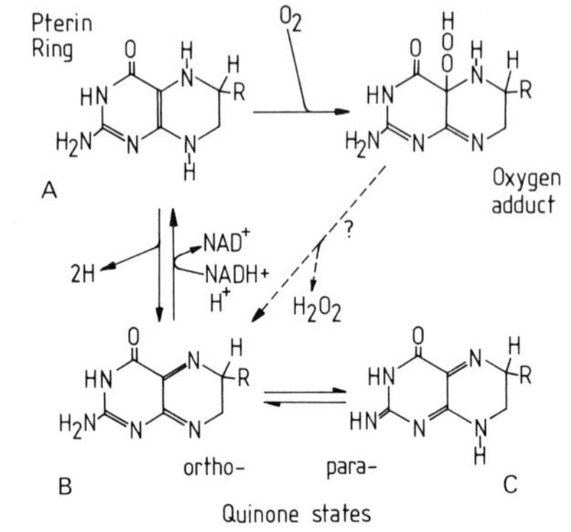

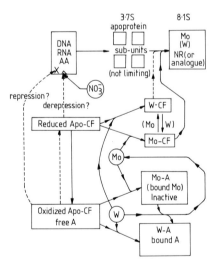

FIG. 2. Structures of pterin ring compounds. A, basic tetrahydro pterin structure. B, *ortho* "quinone" configuration of oxidized product; C, *para* "quinone" configuration in equilibrium with B; D, molecular oxygen adduct and possible decomposition into B and hydrogen peroxide; redrawn from Benkovic (1980).

FIG. 3. Schematic representation of the role of tungsten in availability of molybdenum for nitrate reductase (NR). Molybdenum and tungsten are supposed to react equally and competitively with the apo-cofactor (Apo–CF) represented in hypothetical oxidized and reduced states. The reduced state of the complex is supposed to react with enzyme subunits to unite them into the holoenzyme. Tungsten may equilibrate with the molybdenum complexes. Oxidized Apo–CF is suggested to be equivalent to free A and may combine with either Mo or W to form bound A.

stronger with the quinone state, that oxidation of the reduced "active" site occurs more readily in the metal free state so that the active cofactor is stabilized by ionic Mo as reported by Lee *et al.* (1974) and Johnson *et al.* (1980). Therefore, in the absence of metal the amount of the oxidized cofactor will increase and it may also undergo irreversible oxidation. This will intensify the sequestration of molybdenum in deficient plants as appeared in Table VI and is illustrated in Fig. 3. Thus the alternatives mentioned above, namely that the Mo-binding component might increase in amount or in binding ability when the metal is deficient may both operate.

The notional values in Table VIII for K or [A] are shown for convenience on the basis of one or other as exclusive alternatives. If both changes occur, as seems quite possible, then intermediate values of both of those parameters will occur. If A is assumed to be $3\,\mu M$ in both normal and deficient cells, the corresponding values of K are similar ($3 \cdot 45 \times 10^{-3}$ and $2 \cdot 7 \times 10^{-3}$ respectively). If K is now assumed to be independent of the nutritional status and A is $3\,\mu M$ in the normal cell, then A is found to increase slightly to $3 \cdot 75\,\mu M$ in deficient cells. However if A is fixed at 5 μM, Table VIII shows that the respective values of K and concentration A predicted with each alternative are markedly dependent on nutritional status.

The consequences of this hypothesis for the uptake of a small amount of W by Mo-deficient plants are shown in Table IX for the notional concentrations of Mo in the reactive, and ultimately enzyme-bound, state and in the sequestered state (MoA), which has been calculated assuming for simplicity that Mo and W have similar chelation stabilities and therefore attain mixed equilibria with A determined by the same value of K. For these calculations it is assumed that K remains constant and that A changes in response to the total metal status. For the specimen parameters shown in Table IX the results are identical no matter whether A in normal plants is assumed to be $3\,\mu M$ or $5\,\mu M$. As total A and unbound A supposedly change with [Mo] or [Mo + W] the appropriate, graphically interpolated values of A have been used in the calculations for the ratio Total Metal/Total A in Table IX. The values show that a tenfold excess of W increases the amount of enzyme-bound Mo by a factor of 33.

Tungsten competition with Mo uptake was not noticeable in our experiments; but even if total Mo uptake were notionally decreased 66% by W to $0 \cdot 1\,\mu M$ in the presence of 29 times as much W in the cell, the Mo free for enzyme formation would still exceed that available in the absence of W by ten times. The extent to which W may antagonize Mo uptake in plants is not well known. Quin and Hoglund (1976) found that both 5 ppm and 50 ppm W added to a nutrient containing 0·024 ppm ($0 \cdot 25\,\mu M$) Mo increased Mo uptake by white clover grown with atmospheric nitrogen,

TABLE VIII

Notional parameters for partition of molybdenum between Mo-binding component(A) and nitrate reductase

Status and Assumption	Total Mo (μM)	Enzyme-bound Mo (μM)	Apo-protein (μM)	Concentration of Mo-binding component (μM)		K ($\times 10^3$) for	
				(i)	(ii)	(i)	(ii)
I Mo Sufficient Concn of A. assumed	3·0	0·1	>0·1	3·0	5·0	3·45	72·4
II Mo deficient Concn of A as in I	0·3	0·0003	0·03	3·0	5·0	2·7	4·7
III Mo deficient K as in I	0·3	0·0003	0·03	3·75	72·7	3·45	72·4

TABLE IX

Effect of tungsten on available molybdenum

Mo μM in cell	W	Total metal/Total A	Enzyme-bound Mo (μM)	Enzyme-bound W (μM)	% Increase in enzyme bound Mo
0·3	0	0·08	0·0003	0	
0·3	1·0	0·38	0·00053	0·0017	180
0·3	2·0	0·72	0·0012	0·0078	400
0·3	2·5	0·91	0·0032	0·027	1070
0·3	2·7	1·0	0·010	0·090	3330
0·1	2·9	1·0	0·0033	0·097	1100

nitrate, or ammonium fertilizer. Growth was increased by 40–100% by 5 ppm W respectively in the presence of nitrate or ammonium, but not with atmospheric nitrogen. Tungsten uptake (given at 5 ppm) was only one-quarter or one-eighth that of Mo uptake in the presence of nitrate or atmospheric nitrogen indicating little antagonism between the elements. In earlier work on leguminous pastures in New Zealand, Davies and Stockdill (1956) observed that yields of dry matter were increased by W in the absence but not with additions of molybdate, which was also more effective alone.

There are two other essential factors involved in the manifestation of predicted and observed beneficial W effects, namely the abundance of apo-protein, and its affinity for the W cofactor. As W forms a metal-analogue of nitrate reductase (Notton and Hewitt, 1971; see also Hewitt and Notton, 1980) there must be some competition between the metals for apo-protein as shown by the now widely used practice of blocking nitrate reductase activity with W. Apo-protein formation apparently not only increases in the cell to match increasing supplies of either metal, until optimal but not excessive concentrations resembling those in the normal Mo-grown plants are reached, but surplus free apo-protein is also present in cells of Mo-deficient plants given W (Rucklidge *et al.*, 1976; Hewitt *et al.*, 1977; Notton, this volume). A similar situation occurs with rat liver sulphite oxidase following excess W administration which causes the production of both W analogue and free apo-protein in proportions of about 1:2 (Johnson *et al.*, 1974; Jones *et al.*, 1977).

Serological measurements (Notton *et al.*, 1974) and *in vivo* responses to Mo by excised leaf tissues or cell cultures in the presence of inhibitors of protein synthesis (Hewitt *et al.*, 1977; Jones *et al.*, 1978) show that apo-protein is limited by Mo deficiency to between 20% and 40% of that present as nitrate reductase in normal cells. There is therefore a quite large increase in apo-protein formation, or less turnover of metallo-protein, in the presence of adequate amounts of Mo or W. For this reason the additional presence of W does not depresss nitrate reductase formation so long as adequate apo-protein is also available as in our experiments, although as shown in Table VI normal Mo concentrations are at least ten times those of corresponding protein concentrations.

The possible relationships between Mo and W when the latter can be beneficial are emphasized in Fig. 3. Here I have depicted A, the Mo-binding component, as the oxidized metal-free cofactor (oxidized apo-CF) which combines with the metal more tightly (perhaps ten times) than the reduced cofactor so that it competes effectively for Mo. This model corresponds more closely with the alternative in which the value of K changes in relation to molybdenum supply. However, if total A comprises both oxidized (strong binding) and reduced (weak binding) states such that total A is nearly equivalent to normal molybdenum concentrations and the value of K for the oxidized state is, say, two decades greater than for the reduced state, the scheme also predicts that decrease in metal concentration will promote reciprocal increase in oxidized A. One additional assumption must nevertheless be made, namely that once the active metal cofactor combines with the apoenzyme the value of K for the reduced (active state) cofactor now becomes much smaller so that active cofactor is stabilized in the enzyme-bound state as observed. For the reasons argued here, that formation of a more tightly binding component is promoted by absence of metal, the kinetics can be described essentially in terms of the change in total A, as adopted here, and also a change in K of two interchangeable states of A.

The outcome of tungsten–molybdenum interactions cannot, however, be predicted without some knowledge of competitive parameters. Thus, although the ratio 10^2–10^3 : 1 for effective W antagonism of Mo utilization is observed in some fungi, *Azotobacter* spp. and plants (see Hewitt, 1979; Quin and Hoglund, 1976) the distinction between antagonism at the uptake stage or competition for an apo-protein or other constituent within the cell, was not generally recognized in the earlier work. In some plants W uptake is small compared to Mo (Quin and Hoglund, 1976) but in *C. pasteurianum* the metals compete on a nearly equal basis, slightly favouring W (Elliott and Mortenson, 1975). However, final incorporation into apo-proteins of nitrate reductase or nitrogenase appears more efficient for Mo or conversely the molybdo-form is more stable (see Hewitt, 1979). In *C. formicoaceticum* by contrast, W is by far the favoured metal, and the functional ratio is about 10^3 in faviour of W for formation of the formate-carbon dioxide reduction system in this organism (Leonhardt and Andreesen, 1977) and several clostridial species appear to produce W-containing enzymes (see Ljungdahl, 1980). There is no evidence at present for the type of cofactor present in these systems. Perhaps the frequent association between carbon dioxide reducing systems and folate metabolism indicates some additional or specific link with the probable pterin nature of the cofactor in some molybdo-proteins.

IV. Efffect of Alternative Metabolic Pathways on Metal Requirements

A. MOLYBDENUM REQUIREMENTS

The requirements for molybdenum in nitrate reductase seem so large compared to those for xanthine dehydrogenase (Nguyen and Feierabend, 1978) and probably for sulphite oxidase (Kessler and Rajagopalan, 1972), that it is possible to explore the need for molybdenum when nitrate metabolism is by-passed by use of other nitrogen sources.

When cauliflower plants are grown in molybdenum-deficient nutrient with ammonium sulphate or other non-nitrate sources of nitrogen and exposed to the air, they initially remain dark-green instead of becoming chlorotic, and growth is scarcely reduced compared to that with molybdenum. However, after about six to eight weeks the young leaves begin to collapse in a very characteristic manner known as "whiptail" (Agarwala and Hewitt, 1955; Hewitt, 1956). When plants are grown in sterilized media to minimize bacterial nitrification of ammonium, "whiptail" symptoms are less frequent or absent (Hewitt and Gundry, 1970 and unpublished work). Contrary to earlier conclusions (Agarwala & Hewitt, 1955), we now believe that molybdenum is of very little importance for some plants if nitrate reduction is not necessary for nitrogen assimilation, although this only holds if nitrate is also rigidly excluded from the nutrient medium. Thus there are two conditions necessary for inducing negligible dependence on Mo. The first is elimination of metabolic dependence. The second one, the exclusion of nitrate, involves the role of nitrate in the induction of nitrate reductase apo-protein in absence of the metal. For various reasons discussed by B. A. Notton (this volume), presence of the apo-enzyme uncombined with Mo or W, leads to damage of chloroplasts (Fido *et al.*, 1977). Thus there is another beneficial effect of W, which appears to be distinct from the ligand competition described above.

The effect of Mo on the synthesis and stability of nitrate reductase apo-proteins contrasts markedly with that of copper on plastocyanin synthesis.

B. VANADIUM REQUIREMENTS

1. Biological Aspects

Another example of different metal requirements in relation to different metabolic pathways is provided by vanadium. Meisch and his associates re-opened an old topic of vanadium requirements of some green algae first reported by Arnon and Wessel (1953) for *Scenedesmus obliquus*, where growth responses to adding $0.2 \mu M$ vanadium in the medium ($10 \mu g$ L^{-1}) had been found, although maximal growth required up to $2 \mu M$. An effect on photosynthesis in saturating light intensity was also observed both for this organism (Arnon, 1958), and *Chlorella* sp. (Warburg *et al.*, 1955). These early results were confirmed and extended for green algae including *S. obliquus*, *Chlorella pyrenoidosa*, *C. fusca*, *C. vulgaris* (yellow mutant 211–11h/20) by Payer and Trultsch (1972), Meisch and co-workers (1975, 1977, 1978a, b, 1980a, b, c), Wilhelm and Wild (1980), and possibly also for *Bumulariopsis filiformis* (Xanthophyceae) in terms of photosynthesis of isolated chloroplasts (Böger, 1969) or growth (Hesse, 1974). In most of these experiments, vanadium contamination was minimized by purification of nutrients (Meisch and Bielig, 1975; Meisch *et al.*, 1977) and with analytical checks. In corresponding experiments with various higher plant species, where comparable and perhaps even more rigorous procedures were used, no evidence whatsoever could be obtained for a need for vanadium (see Abbott and Hewitt, 1966; Welch and Huffman, 1973).

I have estimated (Table X) the probable concentrations of vanadium as atoms per cell in lettuce or tomato and in some green algae grown without

TABLE X
Tentative threshold concentrations for vanadium in plants grown with nitrate

Species	Growth response	V concn in medium (μM)	Estimate of V atoms cell^{-1}
Scenedesmus	50% growth	10^{-4}(?)	10^4-10^5(?)
obliquus	Full growth	4×10^{-3}	10^6-10^7(?)
Chlorella	20–50% growth	2×10^{-5}	5×10^3
pyrenoidosa	70% growth	2×10^{-4}	—
	Full growth	$0\cdot1-1\cdot0$	5×10^6
Chlorella	50% chlorophyll	10^{-2}(?)	5×10^5(?)
fusca	(50% growth ?)		
	Max. chlorophyll and cytochrome *f*	$0\cdot4$	4×10^6
Tomato, lettuce	full growth	8×10^{-4}	$2\cdot5 \times 10^5$

or with the element. In plants the minimal requirements, if any, are much less than $2\cdot5 \times 10^5$ atoms cell^{-1}. Any possible threshold for deficiency is estimated to be between one-quarter and one per cent of the above, say, one-tenth or $2\cdot5 \times 10^4$ atoms cell^{-1}. This value is about the same as for the green algae where obvious deficiency effects were observed. However, plant cell volumes are about 500 times those of the vanadium-deficient algae, and for full growth the algae contained about ten times more than the minimal concentrations found in the plants. The minimal cell concentration for 50% growth of *Chlorella pyrenoidosa* is therefore at least 250 times greater than that present in lettuce and tomato leaves.

Typical cell sizes of tomato plants (e.g. $10^5 \mu m^3$; Wilson, 1963) and *Chlorella* (< 500 to $> 2000 \mu m^3$; Meisch and Benzschawel, 1978) differ by about two decades. Cell sizes of the vanadium deficient algae (about 200 to $300 \mu m^3$) are more closely distributed and are about one-fifth of the average value of cell sizes in the presence of 10μM vanadium (about $1500 \mu m^3$). Cells numbering about 4×10^{10} L^{-1} suspended in 2μm vanadium absorbed all of it in 24 h so that an estimate of atoms per cell is possible in these conditions. For a cell volume of about 10^{-12} L the absorbed vanadium is present at 3×10^7 atoms per cell, and cell concentrations producing near maximal sizes and growth rates, may then be about $50 \mu M$. The smaller V-deficient cells might contain about 3×10^4 atoms per cell and have a cell concentration of about $1 \mu M$. Vanadium concentrations in the larger plant cells are about $4 \times 10^{-3} \mu M$.

There are therefore quite good reasons to suppose that higher plants may not require vanadium at all when grown with normal complete nutrient media, whereas several green algae certainly do. Moreover, the green algae show increased requirements either when iron is omitted or is supplied in a poorly utilized form, e.g. $FeCl_3$ (Meisch et al., 1980c; Wilhelm & Wild, 1980); V requirements in these conditions have not been tested with higher plants.

The reasons for the differences between these two groups of chlorophyllous plants may conceivably be related to the predominant metabolic pathways involved in the formation of δ amino laevulinic acid (ALA) which is needed in the formation of protoporphyrin IX, the common precursor of both chlorophyll and haem pigments.

The first step in this pathway for ALA synthesis established in animal tissues and bacterial cells involves ALA synthetase which catalyses the conjugation between glycine and succinyl CoA with elimination of CO_2, and this reaction is important over a wide phylogenetic range (see Shemin, 1956):

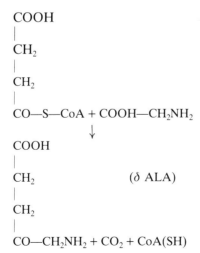

It is often assumed that this pathway also operates in higher plants, although some of the evidence is controversial (see Harel, 1978). However, another pathway is known which may possibly be restricted to certain phylogenetic groups or species and which might involve vanadium. This involves a five-carbon precursor and leads finally to the transamination of 4,5-dioxovaleric acid (DOVA) at the terminal carbon:

$$
\begin{array}{ccc}
\text{COOH} & & \text{COOH} \\
| & & | \\
\text{CH}_2 & & \text{CH}_2 \\
| & & | \\
\text{CH}_2 & \xrightleftharpoons{\text{transamination}} & \text{CH}_2 \\
| & & | \\
\text{CO} & & \text{CO} \\
| & & | \\
\text{CHO (DOVA)} & & \text{CH}_2\text{—NH}_2 \ (\delta\,\text{ALA})
\end{array}
$$

Although a specific transaminase which catalyses this reaction has been isolated from mammalian and bacterial tissue, DOVA is also transaminated non-enzymically in the presence of pyridoxal phosphate and V; V can be replaced, although less effectively, by Fe or Al (Meisch et al., 1978c).

There is good evidence to suggest that in several higher green plants, some green algae, Euglena and in a photosynthetic bacterium Rh. spheroides, the five-carbon DOVA transaminase route is either the physiologically more important one, or has some significance (Gassman et al., 1968; Wider de Xifra et al., 1971; Ramaswamy and Madhusudanan-Nair, 1973; Beale and Castelfranco, 1974; Beale, 1978; Porra and Grimme, 1974, 1978; Hampp et al., 1975; Gough and Kannangara, 1976, 1977; Salvador, 1978; Weinstein and Castelfranco, 1978; Klein and Senger, 1978; Lohr and Friedmann, 1976; Ford and Friedmann, 1979; Kipe-Nolt and Stevens, 1980). If this route is more important in some species, and if the DOVA is transaminated in a specific V-dependent mechanism, a restricted requirement for vanadium would occur. However, a metal requirement in a non-enzymic metabolic reaction would be unique, and there are other problems. Thus in Scenedesmus obliquus and Chlorella fusca where there seems to be a clear need for vanadium, the formation of ALA is achieved comparably by the synthetase and transamination routes (Doernermann and Senger, 1980; Porra and Grimme, 1974; Klein and Senger, 1978a, b); but if both routes are functional, no need for vanadium would be expected. However, an involvement of vanadium in chlorophyll synthesis might explain the increased requirement for vanadium by algae growing with little Fe and at higher light intensities (Meisch and Bielig, 1975; Meisch et al., 1980c; Wilhelm and Wild, 1980). As chlorophyll synthesis greatly exceeds haem synthesis, vanadium requirements might not be detected under dark heterotrophic conditions. It is difficult to interpret the apparently reciprocal substitution effects of iron and vanadium as observed in C. fusca (Meisch et al., 1980c) because there is still no evidence regarding the role of iron for chlorophyll formation, and as iron is so abundant in chloroplasts it may be involved in protecting the organelles from photo-chemical injury. Under these conditions with strong light the need to repair the damage occurring with restricted iron would be emphasized and so vanadium might become a necessary specific substitute here. Corresponding experiments on iron and vanadium interactions are needed with higher plants, but absolute vanadium requirements, if any, are very much smaller than for green algae (Welch and Huffman, 1973). Small beneficial effects of vanadium have been reported for *Candida sloofii* (Roitman *et al.*, 1969) and *Aspergillus niger* (Bertrand, 1966).

A recent and alternative interpretation of V activity is that vanadium accelerates electron transport between photosystems II and I (Meisch and Becker, 1981). The addition of $20 \mu M$ V in either the four or five valency states increased activity four-fold in spinach choroplasts, while $100 \mu M$ V doubled it in *Chlorella* chloroplasts. The response to V here appeared to depend on the provision also of iron during growth, and was increased by high light intensities. Thus V may have a role in chloroplast functioning besides, or instead of, a role in chlorophyll synthesis. Nevertheless chloroplasts are the probable organs of ALA formation (see Harel, 1978).

2. Chemical Reactions

The chemical aspects of vanadium catalysis of transamination are discussed by Meisch *et al.*, 1978c). The vanadyl ([IV] oxidation state; Fig. 4) and the vanadato ([V] oxidation state) both appear possible candidates for chelation by pyridoxal and the amino or oxo acids in a planar system. Vanadyl chelates often tend to be easily oxidized to the vanadato state so the vanadyl pyridoxal derivative was used under nitrogen. By comparing the transamination rates from pyridoxamine and DOVA, or from alanine with pyridoxal phosphate and DOVA, Meisch *et al.* concluded that vanadium is involved mainly or only in the first part of the reaction forming pyridox-amine and pyruvate from alanine and pyridoxal phosphate, and there was little effect on the conversion of DOVA to ALA by pyridoxamine. Both V^{IV} and V^{V} states appeared able to catalyse the reaction. It is difficult to see how vanadium should be important *in vivo* for the DOVA transamina-tion, because it is the step involving pyridoxamine formation common to familiar transaminase activity that is stimulated chemically. However, in a possibly faster and specific enzymic system the effects may be different; but a hitherto undetected general role of vanadium in transaminases seems very unlikely.

There is an interesting similarity between the postulated vanadyl pyri-doxal chelate of Fig. 4 and amavadin (Fig. 5), a natural vanadium compound found in large amounts (60–330 mg kg^{-1} dry wt.) as the laevo

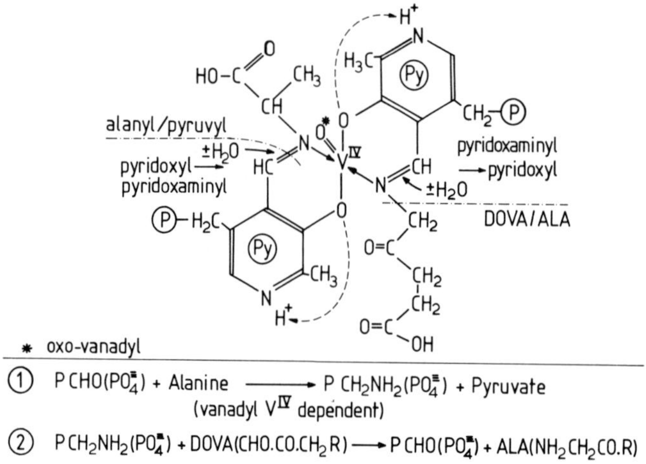

FIG. 4. Possible structure of pyridoxyl-amino acid chelate of vanadium and its role in non-enzymic transamination. The alanyl α amino group reacts with pyridoxal phosphate to produce an intermediate complex able to release pyruvate and pyridoxamine (reaction 1) whilst dioxovalerate (DOVA) reacts with pyridoxamine phosphate to produce a complex able to release δ amino laevulinate (ALA) and pyridoxal (reaction 2). Pyridoxal residues therefore cycle between amino and aldehyde states on opposite molecules in a reciprocal manner. Modified structure redrawn from Meisch et al. (1978c).

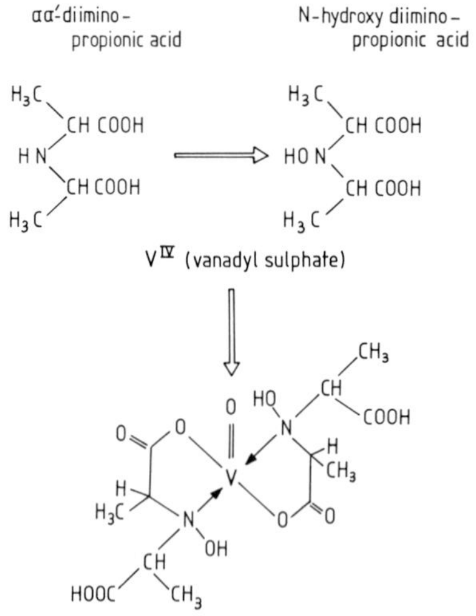

FIG. 5. Synthesis and structure of amavadin. Redrawn from Kneifel and Bayer (1973).

isomer only in *Amanita muscaria* (Meisch *et al.*, 1978d). This compound has been synthesized by Kneifel and Bayer (1973) in the racemic form. Amavadin is pale blue with a stable vanadyl (V^{IV}) state which is not autoxidizable unlike many other V^{IV} chelates. The *bis* N-hydroxy-α δ'-diiminopropionic acid ligand quite closely resembles alanine. Amavadin differs from the pyridoxyl-amino acid complex of Fig. 4 in having hydroxyl oxo groups on the nitrogen atom(s) instead of the pyridoxyl side chain(s). It resembles the 2N-substituted hydroxamates used for V analyses (Stary, 1964).

C. NICKEL REQUIREMENTS

The importance of nickel has been shown beyond doubt only recently for a few micro-organisms and in one specific context for certain plants. Several years previously, however, Roach and Barclay (1946) reported improved growth of barley and potatoes in field conditions in response to a nickel containing spray.

1. Nickel in Microorganisms

In three microorganisms shown in Table XI, nickel is needed for growth under chemolithotrophic conditions depending on carbon dioxide and hydrogen; in two others the activity of a hydrogenase system or of a carbon monoxide dehydrogenase system are dependent on the presence of nickel. In these examples the necessary concentrations are mainly between 0·1 and 1 μM in the medium and so rather more than the usual requirements for molybdenum or cobalt. In the case of *Oscillatoria* sp. (Van Baalan and O'Donnell, 1978) where nickel requirements are rather less than the rest (0·05 μM), I have very tentatively estimated that concentrations in the cells might be 2×10^5 atoms per cell for 40% growth and 5×10^5 atoms per cell for full growth (see Table XIII), values similar to those for molybdenum in blue-green algae (Table VII) and about three decades less than for zinc requirements in tomato (Table IV).

Three other reports are interesting. The possible importance of nickel in nitrogen fixation (Bertrand and De Wolf, 1973; Bertrand, 1974) suggests by analogy with examples of the chemolithotrophic organisms that hydrogen metabolism may be involved in some way, even though very indirectly. I do not believe that this concerns the proton reducing activity of the nitrogenase complex, which independently evolves hydrogen. Nevertheless nickel may be beneficial in those organisms which have an independent hydrogenase system, which in some views is thought to be able to recycle any gaseous hydrogen formed by proton reduction and so re-utilize its

TABLE XI

Evidence for functional nickel in microorganisms

Species	Activity	Optimal concn in medium (μM)	Reference
Rhizobia and soil bacteria	N_2 fixation	—	Bertrand and De Wolf (1973) Bertrand (1974)
Hydrogenomonas sp. (2 strains) H. eutropha	Chemolithotrophic $H_2 + O_2 + CO_2$	0·3	Bartha and Ordel (1965) Repaske and Ambrose (1970)
Alcaligenes eutrophus (2 strains)	Chemolithotrophic only No response with fructose	0·1	Tabillon and Kaltwasser (1977)
Methanobacterium thermoautrophicum	Chemolithotrophic $CO_2 + H_2$	1·0	Schonheit et al. (1979)
Clostridium pasteurianum	Carbon monoxide dehydrogenase Prevented by Ni chelators. No response to Co	0·1	Diekert et al. (1979)
Nocardia opaca	Hydrogenase ? (specific Mg also needed)	500	Aggag and Schliegel (1974)
Oscillatoria sp.	Growth on NO_3^- or NH_4^+ or Amino acids	0·05	Van Baalen and O'Donnell (1978)
Aspergillus nidulans (mutant)	Urease (specific)	100	Mackay and Pateman (1980)

energy. The possible importance of nickel specifically for hydrogenase of *Nocardia opaca* is some reason for considering this idea, although this organism does not fix nitrogen (Hill and Postgate, 1969). Just as ferrous ions activate aconitase (having also three non-haem iron with three acid-labile sulphur atoms), so nickel may stimulate some hydrogenases which have four non-haem iron and acid-labile sulphur atom-clusters. Probably only anaerobic (including facultative) types of nitrogen fixing organisms have this separate hydrogenase and hence any effects of nickel would depend on conditions. It is less easy to see why fixation by rhizobial systems should be similarly stimulated because free hydrogenase is of very doubtful and controversial occurrence in rhizobial systems (see Postgate, 1971). An alternative possibility is that other soil organisms remove hydrogen through nickel-dependent systems and so reduce the competitive inhibitory effects of molecular hydrogen on the rhizobial nitrogenase. *Desulfovibrio gigas* hydrogenase has one atom of nickel (Peck and Leball, 1982). Another interesting example is that a species of *Oscillatoria* appears to need small amounts of nickel for autotrophic growth with either nitrate or ammonium used as nitrogen source (Van Baalen and O'Donnell, 1978).

2. Urease Systems

The third report links microorganisms of Table XI to plants shown in Table XII where the clearly identified role of nickel in urease is now established beyond doubt following its original discovery (two atoms per mole) in urease of jack bean (Dixon *et al.*, 1975). Now it is clear that other plants have nickel-dependent urease activity and so growth on urea but not ammonium is limited by nickel deficiency (Polacco, 1977a, b). The discovery that large amounts (0·1 mM) of nickel are needed for growth on urea by a mutant of *Aspergillus nidulans* (Table XI; Mackay and Pateman, 1980) suggests that this fungus has a similar type of urease and it is thought that the mutant shows the need very clearly because its enzyme may be less able to bind nickel.

Urease does not occur ubiquitously in plants. It is abundant in some Cucurbitaceae (*C. pepo* water melon, and *C. moschata*), some legumes (Table XII), and in *Citrullus* spp.; but many plants have no detectable urease, and metabolize ureides by the ornithine cycle (Bollard, 1959). Plants grown with nitrate or ammonium are of course unlikely to have nickel requirements for urease, but it is possible that they need it for phenylalanine ammonia lyase (PAL). The presence of a broad absorbance band at 310–340 nm in purified PAL from potatoes and maize is suggested as a possible indication of a nickel component (Dixon *et al.*, 1976) by analogy with the light absorbance of urease.

TABLE XII

Evidence for functional nickel in higher plants

Species	Activity	Observation	Reference
Barley	Field Growth	Beneficial response	Roach and Barclay (1946)
Jack bean	Urease	2 Ni atoms	Dixon et al. (1975)
(Canavallia ensiformis)		present per monomer	Fishbein et al. (1976)
Cell suspension of	Inhibition of both	Addition of nickel salt	Polacco (1977a,b)
soybean, tobacco, rice	growth on urea and	($10 \mu M$) overcomes	
	synthesis of urease	inhibition	
	by $10 \, mM$ citrate		

3. Nickel (and other Metal) Accumulator Plants

A separate aspect of nickel nutrition, which is related to the general topic of metal accumulation in plants may involve specialized use of carboxylic acids (see also Peterson, this volume).

The nickel accumulator species *Sebertia acuminata* forms a complex with nickel containing two molecules of citrate (Lee *et al.*, 1977). This may explain why the addition of citrate to nutrient media is effective in revealing the need for the element in non-accumulator species (Polacco, 1977a, b). At least two tridentate citrato-nickel complexes can be imagined in *cis* or *trans* configurations (Fig. 6). It is possible that in *S. acuminata* and

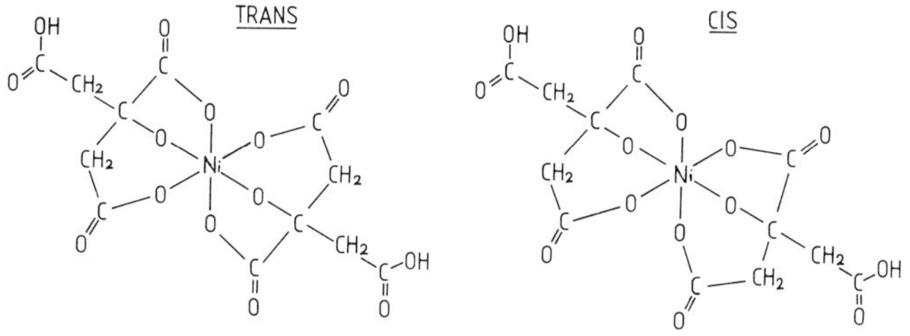

FIG. 6. Co-ordination of nickel by citrate. Hypothetical configurations of hexa co-ordinate nickel-citrate chelates showing possible *cis* and *trans* forms based on evidence for structures of tridentate citrate-metal chelates deduced by Glusker (1968).

similar plants an '"enzyme" induces a special configuration of the citrate, which confers maximal stability on the complex. In *Alyssum bertholoni*, another nickel accumulator, malate and malonate, and perhaps other organic acids, together with magnesium and calcium are associated with nickel chelation and detoxication (Pelosi *et al.*, 1976). Again the formation of *cis* or *trans* type complexes seems possible and may influence stability or involve a conformational action by a protein to induce a stable complex.

Chromium accumulation by *Leptospermium scoparium* is associated with the formation of a trioxalato-chromium complex and also some diaquo dioxalato complexes (Lyon *et al.*, 1969). While the formation of the dioxalato diaquo complexes of chromium in *cis* (violet) and *trans* (pink) forms have been described (Palmer, 1954) and involve different procedures, the production of a trioxalato complex *in vitro* is not similarly recorded to my knowledge. Trivalent chromium has hexacoordinate capability which would be exploited in a trioxalato complex but the dioxalato complexes would be also hexacoordinate with the two aquo groups. Again biological activity might select especially stable forms of the possible chelates, and the trioxalato chelate may be very stable. In zinc-tolerant ecotypes, e.g. *Thlaspi alpestre*, malate and sometimes oxalate is specifically increased by zinc (Mathys, 1977).

V. Comparative Aspects of Nutritional Functions

The chemical properties of the inorganic elements are immutable but their exploitation may change and extend kinetically when combined with different proteins, though only so long as the overall reactions catalysed are thermodynamically feasible. The remarkable range ($+360$ to -550 mv) of mid-point redox potentials spanned by proteins containing identical four non-haem iron and acid-labile sulphur complexes shows how flexibly these systems may behave (Sweeney and Rabinowitz, 1980). Newly evolved proteins may thus catalyse metabolic reactions by new routes so that thermodynamically unfavourable steps are avoided or alternative energy sources exploited.

I shall not attempt to enlarge on the strictly chemical aspects of this subject because a superficial treatment would be of little value. Extensive treatments have been presented by Williams and Da Silva (1978) and Pope *et al.* (1980) regarding the electronic and chemical properties of metal elements and how they match the metabolic activities mediated by their associated proteins. Within these chemical limitations, different proteins can mediate very diverse reactions with new substrates, either to form new products or to achieve familiar endpoints by new routes.

The evolution of a protein can therefore influence the chemical activity of potential substrates. Less often a particular reaction can be mediated by different metals in different proteins or in the same protein as for example in the catalysis of superoxide anion disproportionation by superoxide dismutases (SOD). Here, either zinc and copper, together or iron alone, or manganese alone are functional in different SOD systems (Fridovich, 1974). Furthermore, copper, alone or with cobalt, mercury, or cadmium can replace zinc in the zinc plus copper enzyme *in vitro* with retention of substantial (60–90%) activity (Beam *et al.*, 1974; Moss and Fee, 1975). Possibly, though arguably, vanadium may replace molybdenum in the reactions of nitrogenase without change in the protein. Alternatively, reactions occur which are quite specific to a phylogenetic class and involve the participation of an element not apparently important in other classes, as for example sodium in blue-green algae and C_4-type plants. Examples of these limited or exceptional functions are given below.

A. VANADIUM

The possible role of vanadium in the synthesis of chlorophylls by some organisms has already been discussed. In a separate context, vanadium stimulates nitrogen fixation in *Azotobacter* spp. (Horner *et al.*, 1942; Becking, 1962) and the original conclusions of Bortels and Jensen (see

Hewitt, 1966) on this subject have been confirmed. Vanadium has large effects on the apparent activity and some properties of nitrogenase preparations obtained from *Azotobacter chroococcum* and *A. vinelandii* grown without molybdenum (McKenna *et al.*, 1970; Burns *et al.*, 1971; Benemann *et al.*, 1972). Vanadium was incorporated into protein isolated as, or with, nitrogenase. One interpretation was that vanadium produced an altered and less efficient type of nitrogenase. An alternative view was that vanadium "induced" and stabilized a protein like nitrogenase into which free molybdenum, present only in very restricted concentrations, could be subsequently inserted, so increasing the amount of true nitrogenase by a sparing or scavenging effect (Benemann *et al.*, 1972); a process closely resembling the interaction which we have proposed for tungsten and nitrate reductase. It differs perhaps in conception in that for tungsten a competition with a storage system or inert cofactor is now visualized rather than a direct exchange into the respective active proteins (Hewittt, 1979). Nevertheless, this alternative mechanism may also be involved or be even more representative. There is evidence that *in vivo* substitution of one metal by the other can occur in existing proteins as distinct from alternative incorporation during protein synthesis. The evidence is not unequivocal because the participation of pre-existing apo-protein in induced cells has not been excluded (Heimer and Filner, 1971). The view that vanadium does not produce an alternative nitrogenase (Benemann *et al.*, 1972) must also be qualified by the evidence that the vanadium analogue probably retains substantial proton reducing (hydrogen evolution) ability, more than that expected from small amounts of the molybdenum protein produced as a scavenging product. As the protonation reaction and H_2/N_2 exchange may be an essential feature of all nitrogenase activity whereby molybdenum trihydride formation precedes reduction of dinitrogen by the same system (Chatt, 1980), the vanadium analogue may be very useful in studying this process.

A curious feature of vanadium "replacement" for molybdenum in nitrogenase is that strains and species of *Azotobacter* and *Beijerinkia* differ greatly in the extent to which vanadium can substitute for molybdenum (Becking, 1962). Only one variant strain of *A.* (Syn. *Beijerinkia*) *agile* out of ten tested and none of *B. indicum* could utilize vanadium while out of 30 strains of *A. chroococcum* and *A. vinelandii*, 26 were able to do so. Where does this difference in vanadium utilization have its origin? The *in vivo* nitrogenase systems of different species in each genus differ by two decades in their requirements for molybdenum, so that either the synthesis of the Fe–Mo–S cofactor or its affinity for apo-protein differs between species. Generic differences suggest that vanadium can be used for the synthesis of a Fe–V–S cofactor in some cases (e.g. *Azotobacter*), or that alternatively this cofactor analogue may be generally formed but that it is not always compatible with apo-protein. Additionally the Fe–V–S cofactor may de-repress the synthesis of apo-protein or stabilize it, and so enhance Mo-utilization.

B. TUNGSTEN

The occurrence of mutants showing altered requirements for tungsten and molybdenum has been reported. Nitrogen fixation or nitrate reduction, measured as growth, appeared to utilize tungsten instead of molybdenum in a mutant of *Nostoc muscorum* (Singh *et al.*, 1978). Both metals were apparently equally effective and slightly synergistic for both activities in a mutant of *Anabaena doliolum* (Tyagi, 1974). Two quite different structures of metal cofactor are, however, involved in the two types of nitrogen metabolism (see discussion by Pienkos *et al.*, 1977). Therefore rather unusual mutations must have occurred independently in two distinct organisms in order that tungsten could replace molybdenum in two different proteins mediating two quite different chemical reactions. Perhaps in these organisms or generally in blue-green algae, a common polypeptide sequence is involved in binding the cofactor in nitrate reductase and the cofactor in nitrogenase. A mutation in this part of the gene complex could then have effects common to both proteins. It is necessary, however, to assume that the mutation was of the sort which allowed both very different types of activity to occur, and in one case exclusively with tungsten instead of molybdenum.

In several clostridial species, the differences between specificity for molybdenum or tungsten in the CO_2-formate oxido-reductase reaction, are subtle: exclusive specificity, substitution or obligatory incorporation of both metals are all known (Ljungdahl, 1980). In addition, the mutually antagonistic effects of the metals on incorporation or formation of inert analogue proteins vary between different species and the enzymes they produce. The importance of tungsten for some reactions involving carbon dioxide reduction possibly depends on the generally more negative redox potentials of tungsten compared to molybdenum (Pope *et al.*, 1980) and this property may be important in the mutants.

C. COBALT

The familiar role of cobalt is in the activity of the corrinoid cofactors present in vitamin B_{12} and and coenzyme B_{12} and the reactions they mediate (Lester Smith, 1965). Cobalt is needed by Cyanophyceae, propionic and lactic acid bacteria, *Rhizobium* spp., and the non-legume symbionts in plants able to fix nitrogen; it is also needed by *Euglena*, *Ochromonas malhamensis*, and related genera (Holme-Hansen *et al.*, 1954; Reisenauer, 1960; Delwiche *et al.*, 1961; Lowe *et al.*, 1961; Lowe and Evans, 1962; Bond and Hewitt, 1962; Hewitt and Bond, 1966; Ahmed and Evans, 1960, 1961; Kliewer and Evans, 1962; Nicholas *et al.*, 1962; Hutner *et al.*, 1950; Hutner, 1972). In propionic acid bacteria and Rhizobia, the B_{12} coenzyme is essential for the methylmalonyl CoA-succinyl CoA isomerase system which controls the formation of succinyl CoA and so indirectly the utilization of propionate (Hertogh *et al.*, 1964a, b). Therefore, cobalt may control the formation of haemoglobin in the root nodules of both legumes (Ahmed and Evans, 1961) and non-legumes indirectly via the succinyl-CoA and glycine pathway.

There is doubt about the possible cobalt requirements of higher plants, green algae, and fungi, but some exceptional reports must be considered. These include requirements by filamentous red algae (A. Provasoli, cited by Hutner, 1972), a diploid hybrid *Saccharomyces* sp. (Nicholas *et al.*, 1962), cocoa (Maskell *et al.*, 1953), rubber, tomato (Bolle-Jones and Mallikarjuneswara, 1957a, b), subterranean clover (Wilson and Hallsworth, 1962; Hallsworth *et al.*, 1965) red clover and wheat (Wilson and Nicholas, 1967; see Table XIII). The last two of these plants were grown in enclosed environments, and omission of cobalt caused chlorosis in the young leaves as well as growth reduction. The earlier reports by Maskell *et al.* (1953) were most remarkable. As seen in retrospect, after experience of producing deficiencies of micronutrients, they appear to me to provide important evidence for some role of cobalt in cocoa. Symptoms which appeared in young leaves were quite distinct from those produced by copper, zinc, or molybdenum deficiencies. The most exacting techniques which are usually necessary for these elements were employed and results were in all respects characteristic of acute deficiencies and competent technology. Plants deprived of cobalt recovered when injected with cobalt chloride. The symptoms were reproduced on separate occasions and cobalt injections had beneficial effects under field conditions. Chlorine deficiency seems impossible in that situation.

Since these reports an enzyme, transcarboxylase, has been found in propionic acid bacteria to contain both bound cobalt (two atoms) and zinc (four atoms) per mole of 670 000 m. wt. (Northrop and Wood, 1969). The reaction it catalyses is:

Methylmalonyl CoA + Pyruvate $\rightleftharpoons$ Propionyl CoA + oxaloacetate.

It is of special interest as being the first enzyme protein known to bind cobalt naturally in a form different from the B_{12} compounds, and because the reaction involves methylmalonyl CoA the substrate for the B_{12} coenzyme-dependent isomerase system. It is also noteworthy that alcohol

TABLE XIII

Threshold concentrations in medium for cobalt, nickel or other elements

Element	Species and growth conditions	Growth response	Concn in medium (μM)	Atoms cell^{-1}
Cobalt	Nostoc muscorum	50% growth	3×10^{-5}	10^3(?)
	Nitrogen	Full growth	3×10^{-3}	10^5(?)
	Lucerne (Medicago sativa) Nitrogen nodulated	Full growth	$<0 \cdot 1$	
	Wheat, clover: Nitrate	Full growth	$<0 \cdot 1$	
	Soybean (Glycine max) Nitrogen nodulated	Full growth	$<1 \cdot 6 \times 10^{-2}$	$<10^5$(?)
	Nitrate	Full growth	$<10^{-4}$(?)	
	Rhizobium meliloti*	Full growth	$1 \cdot 6 \times 10^{-3}$	
	R. leguminosarum*	Full growth	$1 \cdot 6 \times 10^{-3}$	
	R. trifolii*	50% growth	$\sim 10^{-4}$	10^2(?)
		Full growth	$\sim 10^{-2}$	
	R. japonicum*	50% growth	$1 \cdot 6 \times 10^{-4}$	10^2(?)
		Full growth	$1 \cdot 6 \times 10^{-2}$	10^4(?)
Nickel	Oscillatoria sp.	50% growth	2×10^{-2}	2×10^5(?)
	Nitrate or ammonium	Full growth	5×10^{-2}	5×10^5(?)
	Alcaligenes eutrophus	40% growth	10^{-2}	
	Chemolithotrophic	Full growth	$0 \cdot 1$	
Chromium (Huffman and Allaway, 1973)	Lettuce (Lactuca sativa) Tomato (Lycopersicon esculentum) Wheat (Triticum aestivium) Beans (Phaseolus sp.)	Full growth	$<3 \cdot 8 \times 10^{-4}$	$<10^5$(?)
Any element (Dinman, 1972)	Lemna sp.	Full growth	$<3 \cdot 8 \times 10^{-5}$	$<10^4$(?)
	e.g. Liver hepatosomes	Minimal requirement	$0 \cdot 1$ ng g^{-1} fr. wt.	$\lesssim 10^{-4} > 10^4$(?)

* All on nitrate. 50% or less growth, where not shown, probably on 10^{-4} μM metal.

dehydrogenase and D-lactate dehydrogenase which are naturally zinc proteins, form functional cobalt analogues when zinc deficient organisms are given cobalt (see Curdel, 1966; Curdel and Iwatsubo, 1968). In the transcarboxylase system it seems that the two metals are incorporated but the chelation mechanisms *in vivo* may be specific and distinct for each. Perhaps similar CoA metabolism occurs in higher plants.

D. SODIUM

Higher plants which have the C_4 types of photosynthesis mostly have an absolute requirement for sodium (see Brownell and Crossland, 1972; Brownell, 1979). Some species, however, including maize and sugar cane, show rather doubtful responses to sodium. In these the anatomical features with rather loose bundle sheaths may allow diffusion of carbon dioxide into the Calvin cycle sufficiently fast to make the C_4 trapping system and transport mechanisms much less important (P. F. Brownell, personal communication). When the partial pressure of carbon dioxide is increased from 300 to 1500μl litre^{-1}, responses to sodium shown by such C_4 plants are then greatly increased.

There is, however, another evolutionary link found in some C_4 plants where the appearance of true Krantz type bundle sheath anatomy is correlated with one type of aspartate pathway in C_4 metabolism. In these plants, CO_2 fixed into oxaloacetate is converted into aspartate instead of malate in mesophyll cells before transport into the bundle sheath cells where oxaloacetate is regenerated by transamination. In plants which then form pyruvate and CO_2 via the malate enzyme system (instead of by pyruvate carboxykinase), this malate enzyme seems specifically to require manganese for activation and uses only NAD$^+$ (see Table XIV based on work of Hatch and Kagawa, 1974; Hatch *et al.*, 1975; Gutierrez *et al.*, 1974; P. F. Brownell, personal communication; and S. Boag, unpublished work).

In the "malate-C_4" species the malate enzyme uses NADP$^+$, and either manganese or magnesium are equally good activators; this malate system is characteristic also of C_3 plants. In one species, *Portulaca grandiflora*, which is a C_4 malate type, both the NAD$^+$–Mn specific and NADP$^+$– Mg/Mn type systems are present in similar activities. Thus apart from anatomical changes the distinctions between C_3 and C_4 types of photosynthesis are large involving several differences and two metals.

The biochemical requirement of sodium in C_4 photosynthesis may be related to the discovery that pyruvate and alanine accumulate whilst phosphoenolypyruvate is depleted in sodium-deficient plants, possibly indicating a role in the activation of pyruvate-phosphate dikinase (P. F. Brownell, personal communication). Shomer-Ilan and Waisel (1973) sug-

TABLE XIV

Relationships between Mn and Na requirements in C_4 and C_3 photosynthetic types

Species	PS type	Malate enzyme type	Sodium growth effect
Panicum miliaceum and several *P.* spp.	C_4 aspartate	NAD$^+$ Mn	Not tested
Atriplex spongiosa	C_4 aspartate	NAD$^+$ Mn specific	Large
Atriplex hortensis	C_3 type	NADP$^+$ Mg = Mn	Small ('nil')
Amaranthus edulis	C_4 aspartate	NAD$^+$ Mn specific	Large
Amaranthus tricolor	C_4 aspartate	NAD$^+$ Mn specific	Large
Amaranthus palmeri	C_4 aspartate	NAD$^+$ Mn specific	Not tested
Cynodon dactylon	C_4 aspartate	NAD$^+$ Mn specific	Large
Portulaca oleracea	C_4 aspartate	NAD$^+$ Mn specific	Not tested
Portulaca grandiflora	C_4 malate	NAD$^+$ Mn specific and NADP$^+$ Mg = Mn	Large
Kochia childsii	C_4 malate	NADP$^+$ Mg = Mn	Large
Kochia pyramidata	C_3 type	NADP$^+$ Mg = Mn	Small ('nil')
Zea mays	C_4 malate	NADP$^+$ Mg = Mn	Small[a]
Sorghum vulgare	C_4 malate	NADP$^+$ Mg = Mn	Small[a]
Sorghum sudanense	C_4 malate	NADP$^+$ Mg = Mn	Small[a]
Saccharum officinale	C_4 malate	NADP$^+$ Mg = Mn	Small[a]
Pennisetum typhoides	C_4 malate	NADP$^+$ Mg = Mn	Small[a]
Chloris barbata	C_4 aspartate[b]	NADP$^+$ Mg = Mn	Large
Chloris gayana	C_4 aspartate[b]	Not tested	Large
Eleucine indica	C_4 aspartate[b]	NADP$^+$ Mg = Mn	Large
Mesophytic spp.	C_3 types	NADP$^+$ Mg = Mn	Small ('nil)

[a] Bundle sheath leaky to CO_2?
[b] Pyruvate carboxykinase sub-type.

gested that sodium activates phosphoenolpyruvate carboxylase in *Aeleuropus litoralis* though scarcely in *Chloris gayana* or *Zea mays*; but their results were based on effects of 100 mM Na, which does not reflect the quantitative sodium requirements for C_4 plants in general. Sodium influences carboxylic acid metabolism in diverse species of bacteria, Cyanophyceae, and green plants, either being necessary for particular oxaloacetate decarboxylases or in diverting metabolic behaviour in the direction of C_4 type amino acid or carbon metabolism. The sodium requirements for C_4 type plants may therefore be similarly derived suggesting a basic relationship between sodium and some aspects of carboxylic acid metabolism. The evidence for specific changes in the malate enzyme in some C_4 species suggests that specific changes in sodium requirements for the other proteins are similarly likely. The Cyanophyceae also show general requirements for small amounts of sodium (Allen, 1952; Allen and Arnon, 1955; Kratz and Meyers, 1955; Ward and Wetzel, 1975) and carbon metabolism is possibly more directly affected than nitrate metabolism or nitrogen fixation studied by Brownell and Nicholas (1967).

The significant connection between these diverse groups may be that the Cyanophyceae are essentially C_4 type photosynthesis organisms and possibly "archetypal" for this activity according to Raven and Glidewell (1978); these authors also include the alga *Hydrodictyon* sp. in this metabolic class and so its sodium requirements would be of special interest. Sodium stimulation of carbon fixation into C_4 carboxylic or amino acids may be quite general, occurring in spinach and in red and brown algae (Joshi *et al.*, 1962).

E. ZINC

The presence of zinc in the aldolase of yeast and its replacement by cobalt or manganese, and the absence of metals in the different (Class Two) aldolases of plants and animals are well known (see Harris *et al.*, 1969). These two classes of aldolases show that the same reaction can be achieved by different proteins and without obligatory participation of a metal.

However, aldolases of wheat and clover were considerably reduced by zinc deficiency (Quinlan-Watson, 1953) and zinc may be present in the enzyme from onions (T. Walsh, unpublished work). More recently Agarwala *et al.* (1976) have shown that total aldolase activity of radish is decreased by deficiencies of iron, copper, zinc, or manganese in decreasing order of importance. In the control plants a single aldolase band appeared during electrophoresis, but two bands appeared with all deficiency treatments. Agarwala *et al.* reviewed evidence suggesting the possible presence of two aldolases in normal plants of other species, and considered that the micronutrients probably limited protein synthesis for which one aldolase type was a sensitive indicator. It is therefore possible that both classes of aldolases occur in some plants, one normally to a greater extent.

The ubiquitous presence of zinc in carbonic anhydrases may also need re-consideration (H. A. O. Hill, personal communication). The carbonic anhydrases of pea and parsley seem undoubtedly to be zinc proteins containing six atoms zinc per mole of six sub units (Tobin, 1970; Kiliel and Graf, 1972). As shown in Table III, carbonic anhydrase may account for about one per cent of cell protein (Tobin, 1969, 1970) and removal of zinc causes loss of activity. Very active and quite pure carbonic anhydrase preparations from spinach obtained by Pocker and Ng (1973) and Kandel *et al.* (1978) respectively contained six or eight atoms of zinc and as many subunits. Zinc deficiency in many plants causes loss of carbonic anhydrase activity (Day and Franklin, 1946; Bradfield, 1947; Wood and Sibly, 1952). Zinc excess induces more carbonic anhydrase in zinc-tolerant, but not in sensitive ecotypes (Mathys, 1977). Animal carbonic anhydrases and the parsley enzyme are inhibited by sulphonamides, azide, cyanide, and sulphide. However, a crude preparation of carbonic anhydrase from parsley obtained by Fellner (1963) did not appear to contain zinc. A supposedly homogeneous carbonic anhydrase of spinach with specific activity comparable to purified enzyme from animal sources, was devoid of zinc, and did not react with the above inhibitors (Kondo *et al.*, 1952a, b), and later independent work confirmed this finding (Rossi *et al.*, 1969). Corresponding results have been reported for carboxy peptidase normally regarded as a zinc enzyme. Zuber (1968) could not detect Zn in the enzyme from orange peel.

Thus there seem to be well-substantiated reports that some proteins characteristically contain one atom of zinc per protein subunit, while functionally similar enzymes lack zinc. Copurification or serology methods to test homology more strictly have not been reported and so the significance of the differences cannot yet be interpreted. Perhaps, as for aldolases, there are actually two classes of carbonic anhydrases, one rather rare and containing no metal.

VI. CONCLUSIONS

I have shown that for some metallic micronutrients, only very small proportions of their total concentrations in plants are identified in well-known metabolic forms; this holds especially for molybdenum in nitrate reductase. On the other hand, almost all the copper content can be accounted for in two proteins. Other elements show intermediate types of distribution. Much of the unidentified fraction(s) seems to be located in unknown combinations. These combinations may be storage forms under regulatory control, or just immobilized waste, and various macro-molecules (pectins, proteins, nucleic acids) may bind quite large propor-tions of metals inertly. Additionally, sequestration in vacuoles as chelates of small molecules, e.g. carboxylic acids (Woolhouse, 1966; Mathys, 1977; and above) may serve for detoxication or storage. Perhaps a quarter of the total manganese is involved in the oxygen evolving reactions and in the thylakoid membranes. Much of the rest may be ionic and involved in the specific activation of some enzymes in dissociable systems. In such systems the K_m values are about 10–50 μM. Thus 100 μM manganese in cytoplasm would be necessary for adequate enzyme activity. Nucleic acids also probably bind manganese in ribosomal structures and manganese dioxide also occurs.

Although possibly all organisms when properly tested seem to require iron, manganese, copper, zinc, and molybdenum, there are occasionally some unusually large quantitative differences. In some cases the operation of alternative metabolic pathways, e.g. nitrate or ammonium nutrition, provide a probable explanation, but in others mutations may have altered metal requirements qualitatively so that partial substitutions occurred. It seems likely that for some metals, e.g. nickel, vanadium, sodium, tungsten, and cobalt, protein evolution and the existence of alternative metabolic pathways explain both the apparently limited essentiality of these elements, and also the difficulties in proving such special requirements when alternatively functional elements are present.

It is not possible to say that an element is not required by a plant. The limits of present techniques may allow estimates of requirements as being possibly less than a certain concentration. The limits at present are indicated by minimal numbers of atoms per cell. For a higher plant grown with nitrate, 10^5 atoms molybdenum per cell, i.e. about $1.6\,nM$, produces severe deficiency. In a green alga a concentration of 10^3 atoms per cell produces a similar effect at a similar $(0.8\,nM)$ cytoplasmic concentration. For ammonium or urea nutrition, any deficiency threshold might be about one per cent of this value, i.e. $40\,pM$, which is less than 10^3 atoms per cell, and very difficult to attain experimentally. When bacteria are studied and cell volumes are sometimes only $10\,\mu m^3$ compared to perhaps $10^5\,\mu m^3$ for a plant, it is necessary to decrease the number of atoms per cell by four decades, i.e. to 10–20 atoms per cell, before the same cytoplasmic concentration is reached. Such conditions are likely to be very difficult to attain or quantify analytically. Dinman (1972) independently concluded that for many "organisms" a value of 10^4 atoms per cell was the detectable limit of a growth requirement for any element, though cobalt may be exceptional (Table XIII) and chromium merits more study.

REFERENCES

Abbott, A. J. and Hewitt, E. J. (1966). *Rep. Long Ashton Res. Stn for 1965*, 145–156.
Agarwala, S. C. (1952). A study of the molybdenum nutrition in cauliflower. Ph.D. Thesis, University of Bristol.
Agarwala, S. C. and Hewitt, E. J. (1954). *J. Pomol. hort. Sci.* **29**, 278–290.
Agarwala, S. C. and Hewitt, E. J. (1955). *J. hort. Sci.* **30**, 163–180.
Agarwala, S. C., Bisht, S. S., Sharma, C. P., and Afzal, A. (1976). *Canad. J. Biochem.* **54**, 76–78.
Aggag, M. and Schlegel, H. G. (1974). *Arch. Microbiol.* **100**, 25–39.
Ahmed, S. and Evans, H. J. (1960). *Soil Sci.* **90**, 205–210.
Ahmed, S. and Evans, H. J. (1961). *Proc. natn Acad. Sci. USA* **47**, 24–36.
Allen, M. B. (1952). *Arch. Mikrobiol.* **17**, 34–53.
Allen, M. B. and Arnon, D. I. (1955). *Physiologia Pl.* **8**, 653–660.
Arnon, D. I. (1950). *Lotsya* **3**, 31.

Arnon, D. I. (1958). *In* "Trace Elements", pp. 1–32, (C. A. Lamb, O. G. Bentley, and J. M. Beattie, eds.), Academic Press, New York and London.

Arnon, D. I. and Stout, P. R. (1939). *Pl. Physiol.* **14,** 371–375.

Arnon, D. I. and Wessel, G. (1953). *Nature (Lond.)* **172,** 1039–1040.

Arnon, D. I., Ichioka, P. S., Wessel, G., Fujiwara, A., and Woolley, J. T. (1955). *Physiologia Pl.* **8,** 538–551.

Asada, K., Urano, M. and Takahashi, M-a. (1973). *Eur. J. Biochem.* **36,** 257–266.

Bartha, R. and Ordell, E. J. (1965). *J. Bact.* **89,** 1015–1019.

Beale, S. I. (1978). *Ann. Rev. Pl. Physiol.* **29,** 95–120.

Beale, S. I. and Castelfranco, P. A. (1974). *Pl. Physiol.* **53,** 297–303.

Beam, K. M., Rich, W. E., and Rajagopalan, K. V. (1974). *J. Biol. Chem.* **249,** 7298–7305.

Becking, J. H. (1962). *Pl. Soil* **16,** 171–201.

Benemann, J. R., McKenna, C. E., Lie, R. F., Traylor, T. G., and Kamen, M. D. (1972). *Biochim. Biophys. Acta* **264,** 25–37.

Benkovic, S. J. (1980). *Ann. Rev. Biochem.* **49,** 227–251.

Bertrand, D. (1966). *C. R. Acad. Sci., Paris, Ser. D.* **262,** 2350–2352.

Bertrand, D. (1974). *C. R. Acad. Sci., Paris, Ser. D.* **278,** 2231–2235.

Bertrand, D. and De Wolf, A. (1973). *C. R. Acad. Sci., Paris, Ser. D.* **276,** 1855.

Böger, P. (1969). *Z. Pflanzenphysiol.* **61,** 85–97.

Bollard, E. G. (1956). *In* "Utilization of Nitrogen and its Compounds by Plants", pp. 304–329, (H. K. Porter, ed.) *Symp. 13 Soc. exp. Biol.,* Cambridge University Press.

Bolle-Jones, E. W. and Mallikarjuneswara, V. R. (1957a). *J. Rubber Res. Inst., Malaya* **15,** 128–140.

Bolle-Jones, E. W. and Mallikarjuneswara, V. R. (1957b). *Nature (Lond.)* **179,** 738–739.

Bond, G. and Hewitt, E. J. (1962). *Nature (Lond.)* **195,** 94–95

Bonner, J. (1976). *In* "Plant Biochemistry", pp. 3–14, (J. Bonner, ed)., Academic Press, New York and London.

Bradfield, J. R. G. (1947). *Nature (Lond.)* **159,** 467–468.

Brownell, P. F. (1979). *Adv. Botan. Res.* **7,** 118–224.

Brownell, P. F. and Crossland, C. J. (1972). *Pl. Physiol.* **49,** 794–797.

Brownell, P. F. and Nicholas, D. J. D. (1967). *Pl. Physiol.* **42,** 915–921.

Burk, D. (1934). *Ergebn. Enzymforsch.* **3,** 22–56.

Burns, R. C., Fuchsman, W. H. and Hardy, R. W. F. (1971). *Biochem. Biophys. Res. Commun.* **42,** 353–357.

Chapman, H. D. (ed.) (1966). "Diagnostic Criteria for Plants and Soils". Univ. California, Division of Agricultural Sciences.

Chatt, J. (1980). *In* "Nitrogen Fixation", pp. 1–18, (W. D. P. Stewart and J. R. Gallon, eds), Academic Press, London and New York.

Cheniae, G. M. (1970). *Ann. Rev. Pl. Physiol.* **21,** 467–498.

Cramer, W. A. and Whitmarsh, J. (1977). *Ann. Rev. Pl. Physiol.* **28,** 133–172.

Curdel, A. (1966). *Biochem. Biophys. Res. Commun.* **22,** 357–363.

Curdel, A. and Iwatsubo, M. (1968). *FEBS Letts.* **1,** 133–136.

Da Silva, J. J. R. F. (1978). *In* "New Trends in Bioinorganic Chemistry", pp. 449–484, (R. J. P. Williams and J. J. R. F. Da Silva, eds.), Academic Press, London and New York.

Davies, E. B. and Stockdill, S. M. J. (1956). *Nature (Lond.)* **178,** 866.

Day, R. and Franklyn, J. (1946). *Science* **104,** 363–365.

Delwiche, C. C., Johnson, C. M., and Reisenauer, H. M. (1961). *Pl. Physiol.* **36,** 73–78.

Diekert, G. B., Graf, E. G., and Thauer, R. K. (1979). *Arch. Microbiol.* **122,** 117–120.

Dinman, B. D. (1972). *Science* **175,** 495–497.

Dixon, N. E., Gazzola, C., Blakely, R. L., and Zerner, B. (1975). *J. Am. Chem. Soc.* **97,** 4131–4133.

Dixon, N. E., Gazzola, C., Blakely, R. L., and Zerner, B. (1976). *Science* **191,** 1144–1150.

Doernermann, D. and Senger, H. (1980). *Biochim. Biophys. Acta* **628,** 35–45.

Elliott, B. B. and Mortenson, L. E. (1975). *J. Bacteriol.* **124,** 1295–1301.

Elliott, B. B. and Mortenson, L. E. (1976). *J. Bacteriol.* **127,** 770–779.

Fellner, S. K. (1963). *Biochim. Biophys. Acta* **77,** 155–156.

Fido, R. J., Gundry, C. S., Hewitt, E. J., and Notton, B. A. (1977). *Aust. J. Pl. Physiol.* **4,** 675–689.

Fishbein, W. N., Smith, M. J., Nagarajan, K., and Scurzi, W. (1976). Federation Proc. **35,** No. 1643.

Ford, S. H., Friedmann, H. C. (1979). *Biochim. Biophys. Acta* **569,** 153–158.

Fridovich, I. (1974). *In* "Molecular Mechanisms of Oxygen Activation", pp. 453–477, (O. Hayaishi, ed.), Academic Press, New York and London.

Gassman, M., Pluscec, J. and Bogorad, L. (1968). *Pl. Physiol.* **43,** 1411–1414.

Glusker, J. P. (1968). *J. Molec. Biol.* **38,** 149–162.

Gough, S. P. and Kannangara, C. G. (1976). *Carlsberg Res. Commun.* **41,** 183–190.

Gough, S. P. and Kannangara, C. G. (1977). *Carlsberg Res. Commun.* **42,** 459–464.

Green, D. E. and Silman, I. (1967). *Ann. Rev. Pl. Physiol.* **18,** 147–178.

Gutierrez, M., Gracen, V. E., and Edwards, G. E. (1974). *Planta, Berl.* **119,** 279–300.

Hallsworth, E. G., Wilson, S. B. and Adams, W. A. (1965). *Nature (Lond.)* **205,** 307–308.

Hampp, R., Sankhla, N., and Huber, W. (1975). *Physiologia Pl.* **33,** 53–57.

Harel. E. (1978) *In* "Progress in Phytochemistry", vol 5, pp. 127–180, (L. Reinhold, J. B. Harborne and T. Swain, eds.). Pergamon Press, Oxford and New York.

Harris, C. E., Kobe, R. D., Teller, D. C. and Rutter, W. J. (1969). *Biochemistry,* **8,** 2442–2454.

Hatch, M. D. and Kagawa, T. (1974). *Arch. Biochem. Biophys.* **160,** 346–349.

Hatch, M. D., Kagawa, T. and Graig, S. (1975). *Aust. J. Pl. Physiol.* **111–** 128.

Heimer, Y. M. and Filner, P. (1971). *Biochim. Biophys. Acta* **230,** 362–372.

Hertogh, H. H. De, Mayeux, P. A., and Evans, H. J. (1964a). *J. Bacteriol.* **87,** 746–747.

Hertogh, H. H. De, Mayeux, P. A., and Evans, H. J. (1964b). *J. Biol. Chem.* **239,** 2446–2453.

Hesse, M. (1974). *Planta (Berl.)* **120,** 135–146.

Hewitt, E. J. (1951). *Rep. Long Ashton Res. Stn for 1950,* 64–70.

Hewitt, E. J. (1952). "Sand and Water Culture: Methods Used in the Study of Plant Nutrition", 1st Edition, Commonwealth Agricultural Bureau, Farnham Royal.

Hewitt, E. J. (1956). *Soil Sci.* **81,** 159–171.

Hewitt, E. J. (1958). "Encyclopaedia of Plant Physiology", Vol. 4, pp. 427–481, (W. Ruhland, ed.), Springer, Berlin.

Hewitt, E. J. (1966). "Sand and Water Culture: Methods Used in the Study of Plant Nutrition", 2nd Edition, Commonwealth Agricultural Bureau, Farnham Royal.

Hewitt, E. J. (1979). In "Chemistry and Agriculture". Royal Society of Chemistry, Special Pub. No. 36, pp. 91–127.

Hewitt, E. J. and Bond, G. (1966). J. exp. Bot. 17, 480–491.

Hewitt, E. J. and Gundry, C. S. (1970). J. hort. Sci. 45, 351–358.

Hewitt, E. J. and Notton, B. A. (1980). In "Molybdenum and Molybdenum-Containing Enzymes", pp. 275–325, (M. P. Coughlan, ed.), Pergamon Press, Oxford and New York.

Hewitt, E. J., Notton, B. A., and Rucklidge, G. J. (1977). J. Less Common Met. 54, 537–553.

Hill, S. and Postgate, J. R. (1969). J. gen. Microbiol. 58, 277–285.

Holme-Hansen, O., Gerloff, G. C., and Skoog, F. (1954). Physiologia Pl. 7, 665–675.

Horner, C. K., Burk, D., Allison, F. E., and Sherman, M. S. (1942). J. agric. Res. 65, 173–193.

Huffman, E. W. D. and Allaway, W. H. (1973). Plant Physiol. 52, 72–75.

Hutner, S. H. (1972). Ann. Rev. Microbiol. 26, 313–346.

Hutner, S. H., Provasoli, L., Schatz, A., and Haskins, C. P. (1950). Proc. Am. Phil. Soc. 94, 152–170.

Ichioka, P. S. and Arnon, D. I. (1955). Physiologia Pl. 8, 552–560.

Jendrisak, J. J. and Burgess, R. R. (1975). Biochemistry 14, 4639–4645.

Johnson, J. L. (1980). In "Molybdenum and Molybdenum-Containing Enzymes", pp. 347–383, (M. P. Coughlan, ed.), Pergamon Press, Oxford and New York.

Johnson, J. L. Cohen, H. J., and Rajagopalan, K. V. (1974). J. Biol. Chem. 249, 5046–5055.

Johnson, J. L., Jones, H. P., and Rajagopalan, K. V. (1977). J. Biol. Chem. 252, 4994–5003.

Johnson, J. L., Hainline, B. E., and Rajagopalan, K. V. (1980). J. Biol. Chem. 255, 1783–1786.

Jones, H. P., Johnson, J. L., and Rajagopalan, K. V. (1977). J. Biol. Chem. 252, 4988–4993.

Jones, R. W., Abbott, A. J., Hewitt, E. J., and Watson, E. F. (1978). Planta 141, 183–189.

Joshi, G., Dolan, T., Gee, R., and Saltman, P. (1962). Pl. Physiol. 37, 446–449.

Kandel, M., Gornall, A. G., Cybulsky, D. L., and Kandel, S. I. (1978). J. Biol. Chem. 253, 679–685.

Kessler, D. L. and Rajagopalan, K. V. (1972). J. Biol. Chem. 247, 6566–6573.

Kipe-Nolt, J. A. and Stevens, S. E. Jr. (1980). Pl. Physiol. 65, 126–128.

Kirk, J. T. O. and Tilney-Bassett, R. A. E. (1967). "The Plastids". W. H. Freeman, London and San Francisco.

Kisiel, W. and Graf, G. (1972). Phytochemistry 11, 113–117.

Klein, O. and Senger, H. (1978a). Pl. Physiol. 62, 10–13.

Klein, O. and Senger, H. (1978b). Photochem. Photobiol. 27, 303–308.

Kliewer, M. and Evans, H. J. (1962). Archiv. Biochem. Biophys. 97, 428–429.

Kondo, K., Chiba, H., and Kawai, F. (1952a). Bull. Res. Inst. Fd Sci. Kyoto Univ. No. 8, 17–27.

Kondo, K., Chiba, H., and Kawai, F. (1952b). Bull. Res. Inst. Fd Sci. Kyoto Univ. No. 88, 28–35.

Kneifel, H. and Bayer, E. (1973). *Angew. Chem. Internat. Edn* **12**, 508.

Kratz, W. A. and Meyers, J. (1955). *Am. J. Bot.* **42**, 282–287.

Lagoutte, B. and Duranton, J. (1975). *FEBS Lett.* **51**, 21–24.

Lee, J., Reeves, R. D., Brooks, R. R., and Jaffie, T. (1977). *Phytochemistry* **16**, 1503–1505.

Lee, K-Y., Pan, S-S., Erickson, R., and Nason, A. (1974). *J. Biol. Chem.* **249**, 3941–3952.

Leonhardt, U. and Andreesen, J. R. (1977). *Arch. Microbiol.* **115**, 227–284.

Lester-Smith, E. (1965). "Vitamin B_{12}", 3rd edition, Methuen, London; Wiley, New York.

Ljungdahl, C. G. (1980). *In* "Molybdenum and Molybdenum-Containing Enzymes", pp. 463–486, (M. P. Coughlin, ed.), Pergamon Press, Oxford and New York.

Lohr, J. B. and Friedmann, H. C. (1976). *Biochem. Biophys. Res. Commun.* **69**, 908–913.

Loomis, W. E. and Shull, C. A. (1937). "Methods in Plant Physiology". McGraw-Hill, New York and London.

Loustalot, A. J., Burrows, F. W., Gilbert, S. G., and Nason, A. (1945). *Pl. Physiol.* **20**, 283–288.

Lowe, R. H. and Evans, H. J. (1962). *J. Bacteriol.* **83**, 210–211.

Lowe, R. H., Evans, H. J., and Ahmed, S. (1961). *Biochem. Biophys. Res. Commun.* **3**, 675–678.

Lyon, G. L., Peterson, P. J., and Brooks, R. R. (1969). *Planta (Berl.)* **88**, 282–287.

Mackay, E. M. and Pateman, J. A. (1980). *J. gen Microbiol.* **116**, 249–251.

Maskell, E. J., Evans, H., and Murray, D. B. (1953). *Rep. Cocoa Res.* Imp. Coll. Agric., Trinidad, 1945–1951, 53–64.

Mathys, W. (1977). *Physiologia Pl.* **40**, 130–136.

McKenna, C. E., Benemann, J. R., and Traylor, T. G. (1970). *Biochem. Biophys. Res. Commun.* **41**, 1501–1508.

Meisch, H-U. and Bauer, J. (1978a). *Arch. Microbiol.* **117**, 49–52.

Meisch, H-U. and Benzschawel, H. (1978b). *Arch. Microbiol.* **116**, 91–95.

Meisch, H-U. and Becker, L. J. M. (1981). *Biochim. Biophys Acta* **636**, 119–125.

Meisch, H-U. and Bellmann, I. (1980a). *Z. Pflangenphysiol.* **96**, 143–151.

Meisch, H-U. and Bielig, H-J. (1975). *Arch. Microbiol.* **105**, 77–82.

Meisch, H-U. and Bielig, H-J. (1980b). *Basic Res. Cardiol.* **75**, 413–417.

Meisch, H-U., Becker, L. J. M., and Schwab, D. (1980c). *Protoplasma* **103**, 273–280.

Meisch, H-U., Benzschawel, H., and Bielig, H-J. (1977). *Arch. Microbiol.* **114**, 67–70.

Meisch, H-U., Hoffmann, H., and Reinle, W. (1978c). *Z. fur Naturforsch.* **33c**, 623–628.

Meisch, H-U., Schmitt, J. A., and Reinle, W. (1978d). *Z. fur Naturforsch.* **33c**, 1–6.

Moss, T. H. and Fee, J. A. (1975). *Biochem. Biophys. Res. Commun.* **66**, 799–808.

Nguyen, J. and Feierabend, J. (1978). *Pl. Sci. Letts.* **13**, 125–132.

Nicholas, D. J. D., Kobayashi, M., and Wilson, P. W. (1962). *Proc. natl Acad. Sci. USA*, **48**, 1537–1542.

Nolan, W. G. and Bishop, D. G. (1974). *Proc. Aust. Biochem. Soc.* **7**, 61.

Northrop, D. B. and Wood, H. G. (1969). *Biol. Chem.* **244**, 5801–5807.

Notton, B. A. and Hewitt, E. J. (1971). *Biochem. Biophys. Res. Commun.* **44**, 702–710.

Notton, B. A., Graf, L., Hewitt, E. J., and Povey, R. C. (1974). *Biochim. Biophys. Acta* **364**, 45–58.

O'Brien, R. W., Frost, G. M., and Stern, J. R. (1969). *J. Bacteriol.* **99**, 395–400.

Opik, H. (1968). *In* "Plant Cell Organelles", pp. 47–88, (J. B. Pridham, ed.), Academic Press, London and New York.

Palmer, W. G. (1954). "Experimental Inorganic Chemistry", pp. 386, Cambridge University Press.

Payer, H-D. and Trültsch, U. (1972). *Arch. Mikrobiol.* **84**, 43–53.

Peck, H. D. Jr. and LeGall, J. (1982). *Phil. Trans. R. Soc. Lond.,* B **298**, 443–466.

Pelosi, P., Fiorentini, R., and Galopinni, C. (1976). *Agric. Biol. Chem.* **40**, 1641–1642.

Petranyi, P., Jendrisak, J. J., and Burgess, R. R. (1977). *Biochem. Biophys. Res. Commun.* **74**, 1031–1038.

Pienkos, T. T., Shah, V. K., and Brill, W. J. (1977). *Proc. natl. Acad. Sci. USA* **74**, 5468–5471.

Plesnicar, M. and Bendall, D. S. (1970). *Biochim. Biophys. Acta* **216**, 192–199.

Pocker, Y. and Ng, J. S. Y. (1973). *Biochemistry* **12**, 5127–5134.

Polacco, J. C. (1977a). *Pl. Physiol.* **59**, 827–830.

Polacco, J. C. (1977b). *Pl. Sci. Lett.* **10**, 249–255.

Pope, M. T., Still, E. R., and Williams, R. J. P. (1980). *In* "Molybdenum and Molybdenum-Containing Enzymes", pp. 3–40, (M. P. Coughlan, ed.), Pergamon Press, Oxford and NewYork.

Porra, R. J. and Grimme, L. H. (1974). *Arch. Biochem. Biophys.* **164**, 312–321.

Porra, R. J. and Grimme, L. H. (1978). *Internat. J. Biochem.* **9**, 883–886.

Postgate, J. R. (ed.) (1971). "The Chemistry and Biochemistry of Nitrogen Fixation", Plenum Press, London and New York.

Quin, B. F. and Hoglund, J. H. (1976). *Pl. Soil* **45**, 201–212.

Quinlan-Watson, F. (1953). *Biochem. J.* **53**, 457–460.

Ramaswamy, M. K. and Madhusudanan-Nair, P. (1973). *Biochim. Biophys. Acta,* **293**, 269–277.

Raven, J. A. and Glidewell, S. M. (1978). *Pl. Cell Environ.* **1**, 185–197.

Reisenauer, H. M. (1960). *Nature (Lond.)* **186**, 375–376.

Repaske, R. and Ambrose, C. A. (1970). *Bacteriol. Proc.* p. **66**, as cited by Hutner, 1972.

Rijven, A. H. G. C. (1958). *Aust. J. Biol. Sci. B.* **11**, 142–154.

Roach, W. A. and Barclay, C. (1946). *Nature (Lond.)* **157**, 696–697.

Roitman, I., Travassos, L. R., Azevedo, H. P., and Cury, A. (1969). *Sabowraudia. J. Int. Soc. Human Anim. Biol.* **7**, 15–19.

Rossi, C., Chersi, A., and Coltivo, M. (1969). *In* "CO$_2$ Chemical, Biochemical and Physiological Aspects", pp. 131–138, (R. E. Forster, J. T. Edsall, A. B. Otis and F. J. W. Roughton, eds), National Aeronautics Space Admin. (NASA), Washington.

Rucklidge, G. J., Notton, B. A., and Hewitt, E. J. (1976). *Biochem. Soc. Trans.* **4**, 77–80.

Salvador, G. F. (1978). *Pl. Sci. Letts* **13**, 351–355.

Sandmann, G. and Böger, P. (1980). *Pl. Sci. Letts* **17**, 417–424.

Scawen, M. D., Hewitt, E. J., and James, D. M. (1975). *Phytochemistry* **14**, 1225–1233.

Schönheit, P., Moll, J., and Thauer, R. K. (1979). *Arch. Microbiol.* **123**, 105–107.

Seckbach, J. (1969). *Pl. Physiol.* **44**, 816–820.
Sevilla, F., Lopez-Gorge, J., Gomez, J., and de Rio, L. A. (1980). *Planta* **150**, 153–157.
Shemin, D. (1956). *In* "Essays in Biochemistry", pp. 241–248, (S. Graff, ed.), John Wiley, New York.
Shomer-Ilan, A. and Waisel, Y. (1973). *Physiologia Pl.* **29**, 190–193.
Singh, H. N., Vaishampayan, A., and Sonie, K. C. (1978). *Mutation Res.* **50**, 427–432.
Spector, M. and Winget, G. D. (1980). *Proc. natn. Acad. Sci. USA*, **77**, 957–959.
Stary, J. (1964). "The Solvent Extraction of Metal Chelates", pp. 122–127. Pergamon Press, Oxford and New York.
Stern, J. R. (1967). *Biochemistry* **6**, 3545–3551.
Sweeney, W. V. and Rabinowitz, J. C. (1980). *Anv. Rev. Biochem.* **49**, 139–161.
Tabillion, R. and Kaltwasser, H. (1977). *Arch. Microbiol.* **113**, 145–151.
Takahashi, M. and Asada, K. (1977). *Pl. Cell Physiol.* **18**, 807–814.
Tobin, A. J. (1969). *In* "CO$_2$, Chemical, Biochemical and Physiological Aspects", p. 139, (R. E. Forster, J. T. Edsall, A. B. Otis and F. J. W. Roughton, eds.), Natn Aeronautics and Space Admin. (NASA), Washington.
Tobin, A. J. (1970). *J. Biol. Chem.* **245**, 2656–2666.
Tyagi, V. V. S. (1974). *Annal. Bot.* **38**, 485–491.
Van Baalen, C. and O'Donnell, R. (1978). *J. gen. Microbiol.* **105**, 351–353.
Warburg, O., Krippahl, G., and Buckholz, W. (1955). *Z. Naturforsch.* **10b**, 422.
Ward, A. K. and Wetzel, R. G. (1975). *J. Physiol.* **11**, 357–363.
Weiland, R. T., Noble, R. D., and Crang, R. E. (1975). *Am. J. Bot.* **62**, 501–508.
Weinstein, J. D. and Castelfranco, P. A. (1978). *Arch. Biochem. Biophys.* **186**, 376–382.
Welch, R. M. and Huffman, E. W. D. Jr. (1973). *Pl. Physiol.* **52**, 183–185.
Wider de Xifra, E. A., Battle, A. M. del C., and Tigier, H. A. (1971). *Biochim. Biophys. Acta* **235**, 511–517.
Wilhelm, C. and Wild, A. (1980). *Biochem. Physiol. Pflangen.* **175**, 163–171.
Williams, R. J. P. and Da Silva, J. J. R. F. (1978). *In* "New Trends in Bioinorganic Chemistry", pp. 121–171, (R. J. P. Williams and J. J. R. F. Da Silva, eds.), Academic Press, London and New York.
Wilson, L. A. (1963). Growth of tomato and radish species in relation to zinc supply, light intensity and day length. Ph.D. Thesis, University of Bristol.
Wilson, S. B. and Hallsworth, E. G. (1962). *Pl. Soil* **22**, 260–279.
Wilson, S. B. and Nicholas, D. J. D. (1967). *Phytochemistry* **6**, 1057–1066.
Wood, J. G. and Sibly, P. M. (1952). *Aust. J. Sci. Res, B.* **4**, 500–510.
Woolhouse, H. W. (1966). *New Phytol.* **65**, 22–31.
Zuber, H. (1968). *Hoppe Seiler, Z. Physiol. Chem.* **349**, 1337–1352.

Epilogue

If it be true that good wine needs no bush, 'tis true that a good play needs no epilogue. Yet to good wine they do use good bushes; and good plays prove the better by the help of good epilogues . . .
(Shakespeare, As You Like It)

Epilogue

N. W. PIRIE

Immanuel Kant found most cause for amazement in the stars above him and the moral law within. Biologists get unfailing amazement from more parochial phenomena. Prominent among them is the enterprise shown by organisms in using and synthesizing an immense range of molecules while often showing extreme conservatism in retaining apparently trivial morphological details for millions of years, in contrast with their lack of enterprise in using less than a third of the naturally occurring elements. This conference was concerned with the metals that are used in bulk by all organisms for many purposes, and with those (about one-sixth of the total number of elements) that have more specialized roles for which minute quantities suffice. As an epilogue it may be worth while searching for unifying features which may have led organisms to select this group of trace elements from among the much larger group of elements that seems to be unused.

There is no reason to expect that such an attempt will be successful. Nature is often strikingly illogical. For example: hens produce 20 times as many ova in a year as women and use only one ovary for the job; plants with the CAM system of photosynthesis seem to gain little from not having gases diffusing through stomata in opposite directions at the same time; and the arrangement of retina and blood vessels in the human eye seems to be back-to-front. With increased knowledge, the advantages of such seemingly irrational arrangements may become apparent. Ecologists constantly remind us of the truth of Albrecht von Haller's dictum: "*Natura in reticulum sua genera connexit, non in catenum . . .*". Still-to-be-discovered strands of the net of interrelated metabolic processes may make more sense than the chain of connections that first catches our attention.

The "evolutionary tree" is a convenient metaphor which misleads us into writing of "higher" and "lower" organisms. Darwin justifiably condemned that anthropocentric usage, because surviving organisms near the bottom of the metaphorical tree have faced a hostile environment for longer than those near the top. What little is known about the biochemistry of these survivors suggests that their main metabolic processes depend on much the same group of elements as is used by organisms at the top of the tree. Only the large-scale morphological features of organisms justify the terms "higher" and "lower": in internal structure, the cells of all eukaryotes are of comparable intricacy. Clearly therefore, even the most "primitive" surviving organisms come near the end of a prolonged period of biochemical evolution. Neither they, nor fossils, give any evidence about the elements that were initially involved in biopoesis. This leaves an unfortunate gap in knowledge. It would be very interesting to know whether an extensive range of reactions and elements was at first involved (Pirie, 1957), from which the more efficient parts have gradually been selected. Or whether, as is generally assumed, life initially depended on a small group of processes, similar to those that are now dominant, to which new members and new elements have gradually been accreted.

One principle governing the choice of elements seems to be clear: those with less than half the median atomic number (AN) are preferred. Mo (AN = 42) is the largest that is used consistently. Even allowing for the relative scarcity of the heavier elements, and for the inflation of the heavier half of the atomic sequence by the 30 rare earths, the generalization has validity. However, it states no more than a preference. Pb and Sn may be essential elements, and W (AN = 74) can be used when Mo is not available.

There is nothing surprising in the use of Mn, Fe, and Cu in oxidations and reductions; they are effective catalysts for such processes *in vitro*. It is more surprising that V is rarely used. Although it is relatively scarce, it is more abundant than Cu, Ni, Co, or Cr (Taylor and McLennan, 1981). Co and Ni are essential but are not known to operate by means of a valency change. Each has a role in the structure of such substances as vitamin B_{12} and urease as specialized as the role of I in vertebrates. Equally surprising is the non-use of Ti in spite of its abundance. By contrast, Zn, which is much less abundant than Ti, is present in more than 100 enzymes. Williams (1981) argues that its role in hydrolytic enzymes is comprehensible because it can act as an acid. Biologists would welcome a full and comprehensible explanation of the precise feature(s) of the Zn atom that fit it for this pre-eminent position. The possibility should also be borne in mind that a very early organism happened to use Zn and that, by evolution and endosymbiosis, the use spread.

Organisms show particularly interesting versatility when they exploit for a special purpose an unusual feature of an element which is generally used in other ways. Several carnivorous animals reflect light inside their eyes with Zn-cysteine in the tapetum. Such an arrangement is not known in plants although it could improve conditions for plants adapted to living in dim light. The magnetic property of Fe_3O_4 is used by a bacterium to detect dip in Earth's magnetic field and thus the direction in which to swim to find anaerobic mud. Much publicity has recently been given to the use of the horizontal component of the field by animals. Immobile plants would have little use for that property of Fe_3O_4, but most of them have mechanisms for distinguishing up from down. For organisms as small as bacteria, it may be sensible to use magnetic dip: gravity is more universal. Fe_3O_4, with a density of 5·18, would be better material in a statolith than the starch grains which are often used—or even the $BaSO_4$, with a density of 4·5, which *Chara* has been enterprising enough to exploit (Schröter *et al.*, 1975). Some invertebrates make their hard radular teeth from Fe_3O_4; silica may supply hardness adequate for the needs of plants. Nevertheless, Fe, which is as versatile in biology as in technology, seems to be under-exploited by them.

Plants absorb, to varying extents, elements for which they have little or no use, or which may be toxic. The universal presence of Si in soil gives a special interest to the trivial use made of it. In early fossils the hard parts of plants and animals are silicious; calcium phosphate came later. Si is still essential but its use is regarded as a primitive trait and some plants are so embarrassed by the quantity which they absorb that they sequestrate it as vegetable opal (tabasheer). The situation with Al is similar; enormous concretions of aluminium succinate are deposited in some trees. These elements seem either to be relatively harmless or to be easily sequestrated. An impoverished flora usually occupies sites such as waste dumps from mines exploiting the more toxic elements. However, given time, remarkable adaptation is possible. The ash from a shrub growing in New Caledonia, where presumably the flora was never ravaged by ice ages, contains 44% Ni.

If all the scattered references in the literature are taken at their face value, more than half the elements are accumulated by some plant and, judging from the rate at which more elements have been found to be essential in recent years, some more will prove to be essential. One pitfall must be borne in mind: because there is some elasticity in the composition of enzymes, one element can sometimes replace another of similar size. Therefore, when one of the generally accepted trace elements is scarce in an environment, another which can take its place may seem to be essential. A further possible complication is that, in animals at any rate, one trace element can counteract the toxicity of another if that is present in slight excess. Thus As appears to be essential, but it increases the excretion of Se and so improves the performance of animals fed on seleniferous fodder (references in Diplock, 1981). Although plants do not excrete as animals do, some analogous system could operate. In the very artificial conditions necessitated by the rigid exclusion of the element that is being studied, it is not impossible that the balance between other elements may be upset. If harmful substances were completely absent, as might not be needed. Similarly, as our own diets become increasingly artificial, it may be wise to watch the supply of trace elements. Anxiety is directed in both directions! Mertz (1981) in the USA fears trouble from too many sources of I, Davies (1981) from too few sources of As.

REFERENCES

Davies, N. T. (1981). *Phil. Trans. Roy. Soc.* B. **213,** 171.
Diplock, A. T. (1981), *Phil. Trans. Roy. Soc.* B. **294,** 105.
Mertz, W. (1981). *Science* **213,** 1332.
Pirie, N. W. (1957). *In* "The Origin of Life on Earth", p. 55, Acad. Sci. Moscow. And: *Nature* **180,** 886.
Schröter, K., Lauchli, A., and Sievers, A. (1975). *Planta* **122,** 213.
Taylor, S. R. and McLennan, S. M. (1981). *Phil. Trans. Roy. Soc.* A **301,** 381.
Williams, R. J. P. (1981). *Proc. Roy. Soc.* B. **213,** 361.

Index